Alternative Cars in the 21st Century

A New Personal Transportation Paradigm

Robert Q. Riley

Published by
Society of Automotive Engineers, Inc.
400 Commonwealth Drive
Warrendale, PA 15096-0001
U.S.A.
Phone: (412) 776-4841
Fax: (412) 776-5760

Library of Congress Cataloging-in-Publication Data

Riley, Robert Q., 1940-
Alternative cars in the 21st century : a new personal transportation paradigm / Robert Q. Riley.
p. cm.
Includes bibliographical references and index.
ISBN 1-56091-519-6 : $39.00
1. Automobile industry and trade--Technological innovations. 2. Automobiles--Technological innovations. I. Title. II. Title: Alternative cars in the twenty-first century.
HD9710.A2R55 1994
338.4'7629222--dc20 94-19325
CIP

Color rendering of futuristic cars on front cover by Barbara Munger.

ISBN 1-56091-519-6

SAE Order No. R-139

To Beverley

Said the Eye one day, "I see beyond
these valleys a mountain veiled with blue mist.
Is it not beautiful?"
The Ear listened and after listening intently
awhile, said, "But where is any mountain?
I do not hear it."
Then the Hand spoke and said, "I am trying
in vain to feel it or touch it, and I can find
no mountain."
And the Nose said, "There is no mountain,
I cannot smell it."
Then the Eye turned the other way, and
they all began to talk together about the Eye's
strange delusion. And they said, "Something
must be the matter with the Eye."

Kahlil Gibran, "The Madman"

Acknowledgments

I would like to thank the countless people who have contributed to this work. A special thanks is extended to Don Goodsell, SAE's European Editorial Consultant, who labored through three drafts and provided a nearly continuous flow of insightful literary and technical observations. Dr. Paul Van Valkenberg also provided a number of recommendations that were incorporated into the final version. John Brooks, a brilliant engineer with a lifetime of experience in powerplant design at Macullogh, Chrysler, and Garret, reviewed early drafts and offered invaluable input on powerplant engineering.

I would also like to thank the companies and individuals who provided graphics, as well as technical and supportive information on a variety of subjects. These include Chrysler Corporation, Diahatsu Motor Company, Ford Motor Company, General Motors Corporation, Honda Motor Company, Nissan Motor Co., Renault, Saab-Scania of America, Subaru of America, Fugi Heavy Industries (Subaru, Japan), Volkswagen AG, and others. Both Harley-Davidson and Bob McKee of McKee Engineering Corp. provided information on Trihawk. The market research companies, J.D. Power & Associates and R.L. Polk Company, contributed market demographics. Dan Sperling at the Institute of Transportation Studies at Berkeley, California, Lee Schipper, co-author of Energy Efficiency and Human Activity: Past Trends, Future Prospects, Dr. Paul MacCready of Gossamer Condor and Sunraycer fame, Al Sorbey, William Garrison, and many others freely offered their information and personal insights about the important subject of future transportation alternatives. *Car Styling* magazine in Japan, Seymour/Powell in London, and Syd Mead, Inc., in Los Angeles each contributed graphics. Without the help of Mari Mayo and Kevin Michael Studio in San Marino, California, the photographic reproductions of student renderings from the Art Center College of Design in Pasadena, California, could not have been included.

And there are people who have made significant, though perhaps more indirect contributions in other ways. The uniquely talented members of the Quincy-Lynn Enterprises team each contributed a piece of their personal excellence to the products we created, and also to the author's personal perspective on the subject of alternative cars. Without the personal contributions of Ed Kay, the spirit of enterprise and determination that helped energize Quincy-Lynn Enterprises, as well as this book, may not have existed in sufficient measure.

And the most essential contribution of all came from my wife, friend, and co-traveler, Beverley. It was only through her unending confidence and support that the liberty to make such a long-term personal investment was possible in the first place.

Table of Contents

Introduction

Land transportation has played a significant role in the development of modern economies. The ability to freely and inexpensively move goods and people is a fundamental link in the economic chain. In addition, transportation also contributes directly to a nation's economy. In the U.S., one in seven jobs is related to the automobile industry, and one in five retail dollars is spent on automotive or automotive-related products. In OECD countries, up to 15 percent of disposable income is spent on transportation. Growth in the transportation sector is largely collinear with the growth in GDP. In the process of providing positive benefits, transportation also produces negative by-products. Direct effects, both positive and negative, are linked to the sector's enormous consumption of energy, resources, goods, and labor. Consumption stimulates the economy, but it also depletes resources and its by-products pollute the environment. In the process of facilitating consumption in general, the transportation sector has emerged as an enormous economic block, as well as a major consumer of natural resources and producer of environmental pollution on its own.

In a more industrialized world, the same natural resources are necessarily spread over much greater populations. Resources are therefore more rapidly depleted and the capacity of ecosystems to absorb industrial by-products more quickly reaches saturation. Just as deficit spending can produce temporary affluence at the expense of long-term health, spending the world's capital of natural resources without regard to global limitations can produce short-term benefits in ways that may ultimately be unhealthy for the entire system.

Attempts to quantify the environmental impact of increased consumer populations produce widely divergent results. Results depend on estimates of future populations, projections of economic trends, and assumptions regarding future technology, as well as on assessments of ecosystem tolerance. Although quantification may be elusive, almost no one believes that the present rate of consumption can be sustained if the current rate of industrialization and economic growth is extrapolated to a world of double or triple the existing population. Ecological limitations to increasing consumption undoubtedly exist at some level. Based on projections by the World3 computer simulation, Donella Meadows, co-author of Beyond The Limits, suggests that with environmentally friendly technology the world can support some eight billion inhabitants at roughly the standard of living of Western Europe in the 1990s. Dr. David Pimentel estimates the population sustainable at an equivalent of U.S. lifestyle to be about 2 billion.[1] However, the most optimistic demographers believe that population will continue to grow until it reaches at least 9-10 billion near the middle of the twenty-first century. Those who are less optimistic project a stabilization level of 14-15 billion inhabitants; and most of these new inhabitants will live in undeveloped or developing regions. It is the developing countries that will be responsible for the most rapid increases in consumption in the coming decades, and they are also the countries that can least afford environmentally friendly technologies.

The idea of a world economy that produces, consumes, and pollutes at 20 times the present rate stretches the imagination.* Trends point to just such an event, although the physical limitations of the planet are likely to prevent it from actually materializing. There is a story about two fellow travelers who fell out of an airplane. On the way down, one of them begins to complain about falling. Finally the other fellow looks over and says: "If you think falling is bad, suppose we survive until we reach the ground?" Likewise, suppose the world survives on its present course, relatively intact and prosperous? How might that affect our planet, its resources, and the quality of life for its inhabitants? Fulfilling the transportation needs of this emerging new world represents a significant component in the overall production/consumption/pollution cycle. Transportation already consumes more than 20 percent of the world's total primary energy. According to Ove Sviden, EKI, University of Linkoping, Sweden, if energy consumption continues to grow as it has in the past, by the year 2100 total primary energy required to meet the world's needs will be ten times greater than in 1990, and transportation will be consuming 40 percent of this much larger pool. The energy consumed to move goods and people will then have to be 20 times cleaner, just to match the present burden that transportation's energy consumption places on the environment.[2]

Light-duty vehicles are the single greatest consumer of transportation energy. Reducing private transportation's energy intensity may be the single most important strategy in averting the impending collision between the positive economic benefits of worldwide mobility, the world's limited natural resources, and the ever-diminishing capability of the environment to absorb the by-products of energy production and consumption.

Many have shared their ideas for accommodating the inevitable human population while reducing the demand on private transportation. One of the most popular is the idea of eliminating or reducing the need to travel at all. As Don Goodsell, SAE's European Editorial Consultant points out, if one has to travel more than twenty minutes to work, either one's residence or one's place of employ is in the wrong place. The ascendancy of telecommuting, working at

* A United Nations projection for the year 2100, *Newsweek*, June 1, 1992.

home and communicating by modem, is predicted by a number of prognosticators, including David Rothman, author of Silicon Jungle. In the U.S., 7.6 million office workers already spend at least part of their workday at remote or mobile electronic offices. It is also possible that in many of the world's burgeoning cities private cars may become largely outmoded by expanded transit systems. Still others suggest that self-contained, car-free cities like those envisioned by Paolo Soleri might replace the sprawling conurbations of today. Dr. Paul MacCready envisions the widespread use of electric sub-cars as relieving the demand on conventional, multi-purpose cars. University of California's William Garrison sees a future served by small neighborhood cars for local travel, with commuting workers driving one- and two-passenger commuter cars along separate roadways, perhaps elevated along freeway medians or attached along the sides by outriggers.

Good ideas are abundant and the future will likely incorporate many of them to varying degrees. This book reviews the problems associated with private transportation, along with alternative modes and vehicle designs, as well as alternative energy sources. An obvious bias is expressed for solutions that modify the existing system in ways that create little demand for infrastructure or lifestyle change. The idea of personal transportation is a good one. Society is well adapted to the concept and the infrastructure is in place. Relatively minor modifications to the design and mix of vehicles can significantly reduce energy consumption and pollution. To accomplish this requires that appealing, energy-efficient transportation products, such as downsized urban cars and commuter cars, are developed and appropriately marketed to consumers. Driving and commuting habits can thereby remain essentially unchanged, investment in new facilities will be minimal, and cities will not have to be restructured around a new transportation architecture. Its success relies primarily on product theme and marketing appeals, rather than on advancements in basic technology or large-scale infrastructure overhaul. Disadvantages arise from the idea of leading the market, rather than following it.

Energy-efficient urban cars and commuter cars are of course not the only solution to transportation's energy and pollution difficulties. Technical solutions to the task of supplying clean energy to meet the appetites of high-mass cars may be possible over the long term. But most technical challenges are not resolved by one sweeping answer. Instead, solutions tend to be comprised of a number of synergistic measures that in combination produce the desired result. Vehicle weight reduction is a basic approach that encompasses a variety of technologies. High-tech composites and ultralight structures are sure to be a part of the technology of tomorrow's energy-efficient automobile designs. More directly, integrating low-mass urban cars and commuter cars into the fleet of private automobiles is a relatively low-technology, marketing-oriented approach that can produce a near-term and substantial decrease in transportation's energy consumption, as well as in the resulting pollution. Improvements that flow from vehicle reconfiguration would then be complemented by long-term improvements in basic technology and lighter, more cost-effective materials. A variety of other implements, such as telecommuting, restructured cities, and expanded transit systems also can make important contributions.

This perspective regarding downsized alternative cars is based on personal beliefs that evolved from small-car development exercises conducted over an eight-year period at the development firm, Quincy-Lynn Enterprises, Inc. A number of designs for energy-efficient commuter cars and urban cars were developed (see Appendix). In the process, a vision of alternative automobiles and their place in the market emerged. Measures suggested in the text for integrating low-mass cars are heavily biased toward the human factors, which are often unyielding to purely pragmatic persuasions. Consequently, while providing information on a variety of subjects related to low-mass cars, alternative fuels, and electric vehicles, a new perspective on designing, marketing and integrating low-mass vehicles into the fleet is also encouraged.

Naturally, a substantial motivation must exist for industry to consider embarking on a risky pioneering effort when the traditional approach is obviously working. Unless there is compelling evidence to the contrary, there is little justification for pioneering new products in high-risk ventures, only to compete with existing products that can be marketed with little risk. However, the status quo appears to contain great risk when one examines the subject from a global perspective. A compelling reason for change is implied by the global limitations that are sure to impact the future health of the planet, and therefore the future health of the market. Consequently, the subject is first approached as a global overview. It then progresses to a discussion that combines the pragmatic justifications for smaller cars with the marketing implications that follow.

The technical aspects of markedly downsized vehicles and the degree to which downsizing impacts energy consumption are also reviewed. Low-mass cars can have the greatest effect on reducing transportation energy intensity in the urban environment. Alternative fuels also are reviewed because new motor fuels are sure to be widely adopted for future vehicles. Regardless of one's assessment of the abundance of petroleum, deposits are finite and motor fuel must ultimately come from a source other than petroleum. Even if economically recoverable petroleum reserves are three time greater than today's known reserves, we will probably have to abandon oil as a primary energy source for transportation by the year 2020. It is also possible that alternative fuels may ultimately be utilized more as fuels for electrical generation, rather than fuels consumed by individual vehicles. Electric cars would then complete the loop. A separate chapter therefore reviews EVs and their potential to reduce overall energy consumption, as well as atmospheric pollution.

The subject of three-wheel cars is included because the configuration may ultimately be adapted to special-purpose vehicles such as urban cars. Urban car designs can benefit from the three-wheel configuration. However, the incidence of rollover accidents may be greater with three-wheel urban cars, regardless of the margin of safety against rollover. And more generally, significant concerns regarding vehicle safety arise along with the idea of integrating low-mass cars, of either three or four wheels, into traffic with conventional high-mass vehicles. A chapter on low-mass vehicle safety is therefore included. Although it is technically feasible to protect occupants during a small-car/large-car involvement, the consequences of the unfavorable transfer of energy to the smaller vehicle are inescapable. Urban cars might be restricted in speed and operating environment to exert a downward pressure on accident and injury statistics. But absolute safety must ultimately be provided by an increased emphasis on technical solutions for low-mass vehicles. This approach will likely result in higher costs for occupant protection and crash avoidance systems.

It is tempting to relate the market potential of special-purpose urban cars and commuter cars to microcar market failures in the past. Extremely small, fuel-efficient cars have often produced a poor sales history, both in North America and in Europe. Chapter Eight therefore reviews the success and failures of low-mass cars that were produced in Europe after WWII. The approach is first to place market failures in their historical perspective, and then to elaborate on the impact of product theme and marketing appeals, as reviewed earlier in the text.

And finally there is this question of practical and dispassionate choices about lifestyles. There are too few areas in which we must spend money and invest personal energy, but receive nothing but utility in return. Automobiles are remarkable transportation tools, but they are also one of the few tools of necessity that can be enjoyed for the end as well as for the means. True, they provide a measure of mobility and freedom that is unavailable in subways and on buses. But they also provide a *sense* of freedom and mobility that cannot be weighed in terms of expedience. And there is also the pleasure of owning and admiring an exquisite work of art, which certainly applies to a well-designed and finely made machine. A beautiful automobile gives us more than mere utility. Therefore, the energy-efficient automobiles discussed on the following pages also embrace the prospect of inventing a new pleasure, rather than destroying an old one.

1. David Pimentel, *et al.*, "Natural Resources and an Optimum Human Population," *Population and the Environment: A Journal of Interdisciplinary Studies*, Vol. 15, No. 5, May 1994.

2. Ove Sviden, EKI, University of Linkoping, "Sustainable Mobility: A Systems Approach to Determining the Role of Electric Vehicles," published in OECD Document, *The Urban Electric Vehicle: Policy Options, Technology Trends, and Market Prospects*, ISBN 92-64-13752-1.

Chapter One

Private Cars Under Siege

as the Cause of Pollution, Traffic Congestion and Resource Depletion

I do not believe that we are to be flung back into the abysmal darkness by those fiercesome discoveries which human genius has made. Let us make sure that they are our servants, not our masters.

Winston Churchill

Worldwide, the benefits and liabilities of private automobiles are being more closely scrutinized. Cars have been targeted because of three offensive traits: their appetite for fuel, the traffic congestion they cause, and the pollution they produce. For the first time, it seems clear that there are limits to the number of automobiles that cities can assimilate. Following nearly half a century of pro-automobile transportation planning, a renaissance of urban and intercity transit may be poised to emerge in Europe. Planners are turning more to trains, streetcars, buses and even bicycles, and away from private cars. Even on a grass-roots level, a consensus against unlimited growth in the automobile population is gaining momentum. Cities are rebelling, and the private automobile is already an unwelcomed guest in a growing number of densely packed metropolitan areas. The U.S. is not far behind Europe in terms of traffic congestion and commuter burnout. Consequently, pressure for a similar re-think is building in traffic-polluted metropolitan areas like L.A., Chicago and Washington D.C. Because of poor air quality in Los Angeles, the State of California has already embarked on a program designed to force a technology shift in transportation, and several other U.S. cities are following suit.

After 20 years of intensive efforts to economize or find new solutions, viable alternatives to gasoline-fueled private cars remain elusive and transportation fuel consumption is increasing across the board. Energy consumption is up in most industrialized countries, and remarkably, transportation energy intensity is also generally greater. On the basis of energy per passenger kilometer, many industrialized nations consume more energy moving people a given distance today than before the '73-'74 oil crisis. Moreover, the automobile is invariably implicated in political difficulties involving petroleum. Regardless of rhetoric to the contrary, the massive military assault of Operation Desert Storm was undertaken primarily to insure the price and flow of oil into the world's mechanized veins. As for the automobile itself, what began as a love affair is rapidly turning into a calculated business deal throughout the industrialized world.

Problems associated with personal transportation are likely to continue to compound worldwide as population grows, industrialization spreads, resources diminish, and the environment becomes more overloaded. Global trends toward industrialization and higher personal incomes translate into more cars and greater energy consumption. When a society industrializes, the burro and the rickshaw are quickly discarded in favor of their mechanized brothers. People invariably favor private cars where the local economy can support them. Consequently, the automobile's impact on the environment and on the world's resource base will become more sweeping and global, and less a distinguishing attribute of a few wealthy nations. Trends have already, and are likely to continue in the future to offset traditionally modest and incremental improvements in vehicle fuel economy. The automobile's problems are unlikely to yield to token design changes. Instead, the machine itself may need a holistic renewal, a wholesale overhaul, in order to remain the central component of our modern society's transportation system.

Converging Dynamics of Population Growth and Industrialization

In 1968, Paul Ehrlich's first book, The Population Bomb, arrived on the newsstands and within a few months everyone was talking about a new end-of-the-world. This version would happen not by fire, but by an ever-increasing population of gluttonous human beings who would literally consume the entire planet. Ehrlich pointed to an imaginary world of the distant future in which exponential population growth extends beyond the possible into a nightmare of some 60 million billion people who would stand at 100 persons per square yard on the earth's surface.[1] Of course, Ehrlich's subject was not a hypothetical world of the distant future, but a more real world of ecological collapse that he foresaw just around the corner. At the basis of his warning was a very real runaway growth in the world's population. Today, after having increased in numbers by nearly 60 percent in the 25 years since The Population Bomb, the world is now inhabited by some 5-1/2 billion people. Sometime in the twenty-first century, the population will reach approximately twice that number before it stabilizes (Figure 1.1).

If fears of an environmental or population catastrophe prove exaggerated, the results tend to cast a darker cloud over the transportation sector. Assuming that the world community survives reasonably intact for another half century, the combination of exponential population growth and expanding industrialization is expected to create an enormous increase in worldwide demand for resources, along with associated pressures on the environment. Either component (population or industrialization) by itself contains alarming numbers for the future of transportation. Combined, they forecast an overwhelming growth rate in the world's automobile population, in traffic congestion and pollution, and in the worldwide appetite for motor fuel.

By itself, rapid population growth does not necessarily mean an equal increase in consumption. It can just as easily produce a corresponding decline in living conditions instead. In many regions of the world where population has been growing, economic conditions have been stagnant or declining. Populations in undeveloped regions are substantially outside the loop where they neither participate in the world economy, nor do they consume substantial resources.

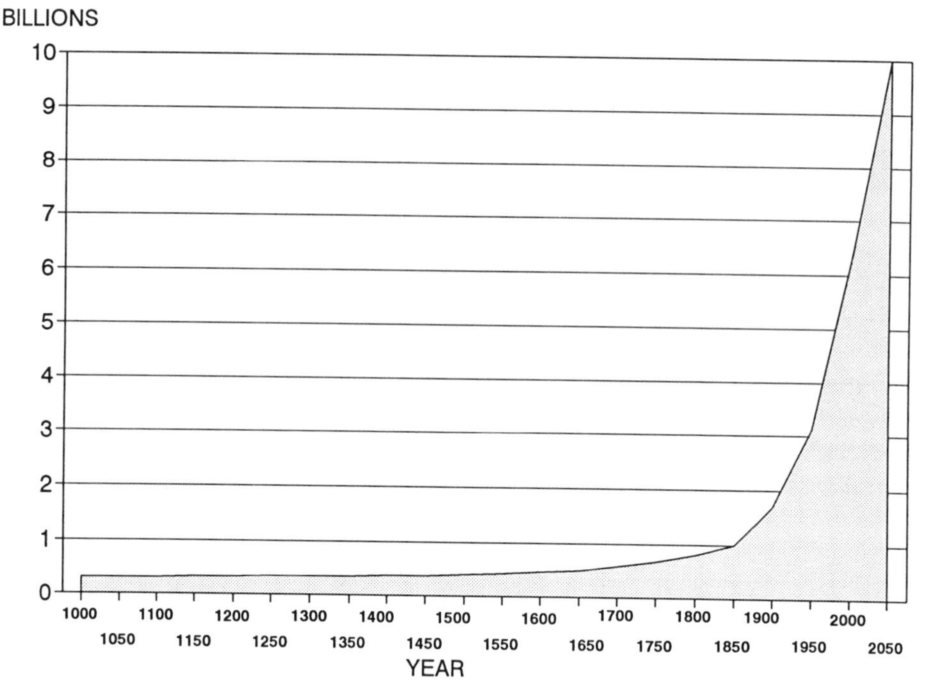

Fig. 1.1. Population Growth — 1000 AD to 2050 AD. (Source: Ref. [1])

This changes quickly when they begin to industrialize. Industrialization promotes economic growth and consumption follows suit, increasing in step with the growth in Gross Domestic Product (GDP). As a rule, nations do not choose to remain undeveloped. Adopting technology generally improves a population's living conditions and helps put distance between the society and potential poverty. Technology provides some fragment of control over the ruthlessness of nature.

Over the past quarter century the vast populations of Asia have been making the transition to industrialized economies. During the '80s these developing countries maintained a steady increase in per capita income while the Western economies remained relatively flat. Although worldwide recessions may cause deviations, the upward economic trend in the developing world is expected to continue. Economists do not predict an apocalyptic collapse in the world's economy. Over the long term, economic tribulation in the developing world is not likely to preserve the planet to a rich few. Growth has been strongest, and will likely continue, in China, Hong Kong, Indonesia, Malaysia, Singapore, Taiwan, Korea and Thailand.

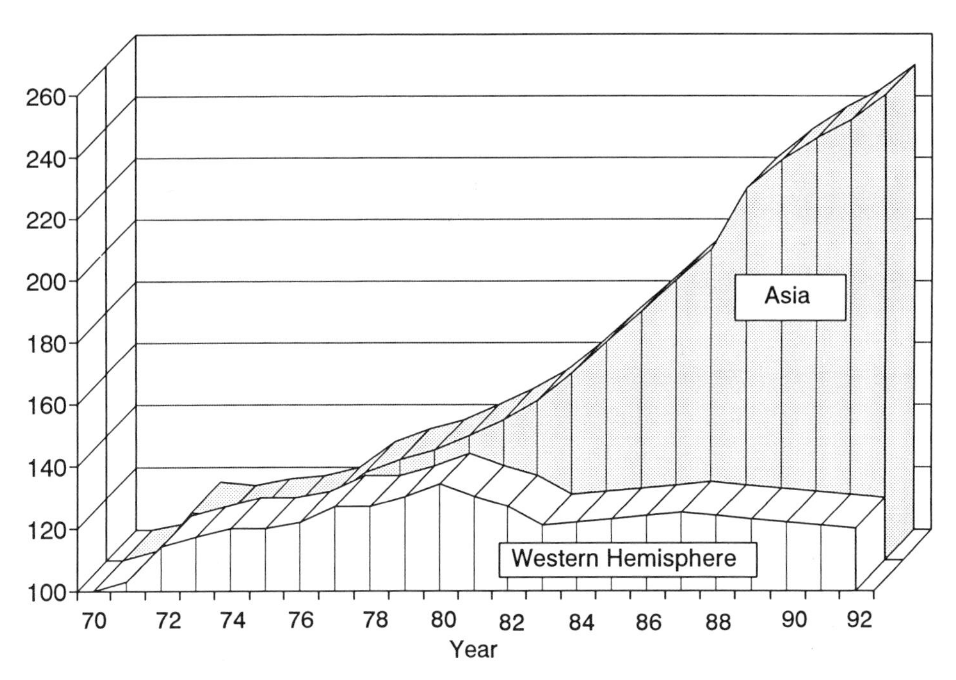

Fig. 1.2. Growth in Real Gross Domestic Product. (Source: U.S. DOE, Energy Information Administration)

Today, only 28 percent of the world's inhabitants live in developed regions. The other 72 percent reside in undeveloped and developing regions. It is this latter group of nations that have the potential, to varying degrees, for rapid increases in consumption. In a transitional economy, the growth in consumption far exceeds the simple growth of the population. Nations of people who were previously non-consumers transform into consumers almost overnight, far outpacing their ability to multiply in numbers. Growth in consumption increases in step with the growth in GDP. The move toward market economies in Eastern Europe and the countries of the former U.S.S.R. may ultimately underwrite their tremendous human resource base and stimulate robust economic development in that relatively dormant region. Politically, it is undoubtedly in the world's best interest to encourage economic renewal in Russia. Environmentalists, however, shudder at the thought of the rapid increase in consumption that will likely accompany an economic awakening in densely populated and previously stifled regions. China may contain some of the world's most abundant deposits of petroleum. Opening Chinese petroleum resources to Western development will surely propel that nation into a new age of economic development. Even partial successes will add ever increasing numbers to new armies of competitors for the earth's resources.

If the rest of the world ate as Americans do, it would require double the food produced in the record harvest years of 1985 and 1986. If the rest of the world drove automobiles as Americans do, all proven petroleum reserves would be exhausted within a decade. If a Vespa motorscooter were given to everyone in China, the world would be plunged into a new energy crisis overnight. Even at today's 5-1/2 billion inhabitants, many argue that the world cannot support worldwide consumption at a level equal to that of Americans in the '90s. One can only wonder at the potential lifestyle in a world of some 10-14 billion inhabitants forecast for the mid to late twenty-first century. Worldwide consumption equal to today's level in the U.S. may never be more than a frightening idea. The superlatives do, however, provide a sense of the pressures that may be placed on society's infrastructures and resources, as well as on the ecological stability of the planet. Moreover, it is not relevant that one can argue with a particular forecast, or that select data may not be precise. The trend and the absolute enormity of the results are undeniable. Anticipation of rapidly increasing pressures on all aspects of the transportation infrastructure is already driving an increasingly intense search for a new paradigm regarding energy and transportation.

An Expanding World with Contracting Petroleum Resources

A continuous supply of economical energy is fundamental to industrialized economies. Throughout the twentieth century, the developed countries in the West have been blessed with cheap and abundant supplies of energy. The U.S. has enjoyed the benefits of the lowest energy prices in the world. Changes in the price or threats to the supply of energy send shock waves rolling through the world's financial systems. Energy is an intrinsic part of every aspect of our lives, from heating, cooling and lighting our homes, to providing power to manufacture goods, and fuel to propel our cars. Automobiles are especially large users of energy. Moreover, they are almost totally dependent on oil as a source of motor fuel. And fuel made from today's inexpensive petroleum has proven difficult to replace. In the future, a universally growing demand for energy is likely to continue to drive petroleum prices upward. In a 1993 report, the U.S. Energy Information Administration (EIA) provided the following estimates of crude oil prices through 2010.[2]

TABLE 1.1. OIL PRICES PROJECTED THROUGH 2010
In 1991 Dollars

Year	1990	1995	2000	2005	2010
1990 $/barrel	22.54	19.90	22.90	26.10	29.30

Source: U.S. Energy Information Administration

The economy, as well as the consumption of energy, are expected to grow worldwide, but much more rapidly in developing countries than in the developed regions of the world. Developed countries will continue to move heavy industrial operations into underdeveloped and developing regions, which will stimulate economic growth and increase energy consumption in those areas of the world. Technological progress will continue to improve energy efficiency, primarily in developed countries. As a result, energy consumption is expected to grow slightly less rapidly than the world economy and the rate of industrialization. However, total world energy consumption will markedly increase through at least the first half of the twenty-first century as improving economic conditions

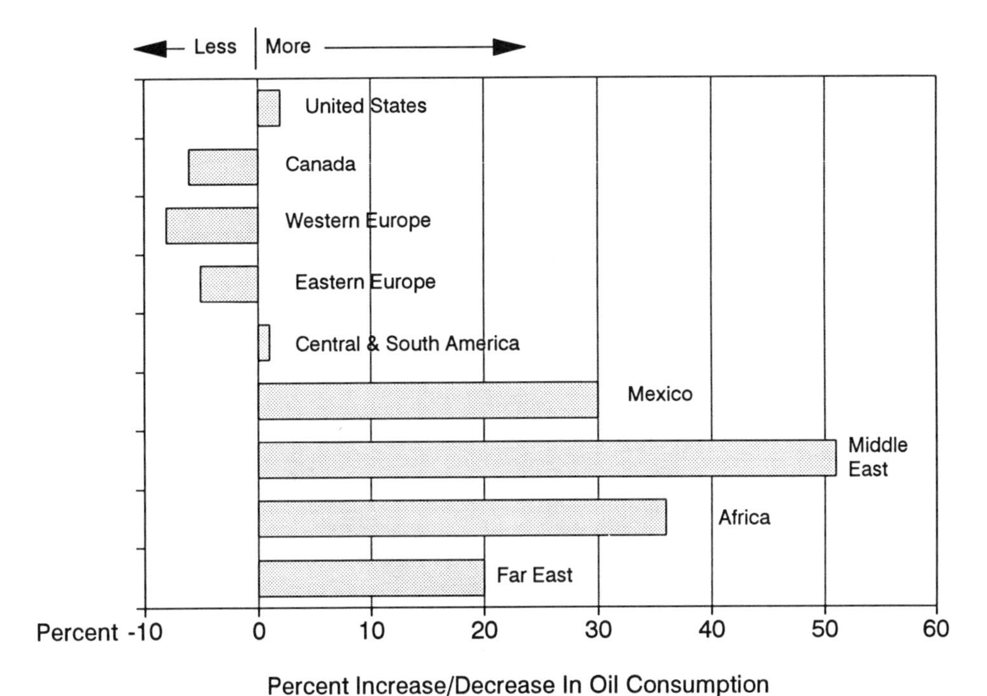

Fig 1.3. Oil Consumption Trends by Region, 1980-1989. (Source: U.S. Energy Information Administration)

in developing countries create vast new populations of consumers of motor vehicles and electrical power and all the other trappings of an industrialized society. Oil will remain the most important single source of energy through at least the first half of the next century. Transportation will continue to be largely dependent on petroleum fuels.

In 1989, the world consumed approximately 66 million barrels of oil per day. During the same year, the U.S. alone consumed more than 17 million barrels per day, or approximately 25 percent of the world's production of oil. The U.S. appetite for oil is unparalleled in the community of developed nations. In the U.S. in 1989, petroleum supplied 42 percent of the nation's total energy needs, which is approximately twice as much as either coal or natural gas and four times as much as nuclear, hydroelectric and all other energy sources combined. Nearly 11 liters (3 gal.) of oil are consumed in the U.S. each day for each person. Average consumption in the rest of the world amounts to less than 1.9 liters (0.5 gal.) per day. Barring some natural or economic disaster, world oil consumption will continue to increase in step with the expanding economies of developing nations. Without a significant change in transportation's energy source and rate of consumption, oil will have to be recovered and refined at a significantly greater pace if we are to meet the demands of an expanding population's energy and transportation needs. Based on today's rate of production and known reserves, most of the remaining reserves of easily recoverable oil may be within the control of the OPEC nations within 10-15 years.

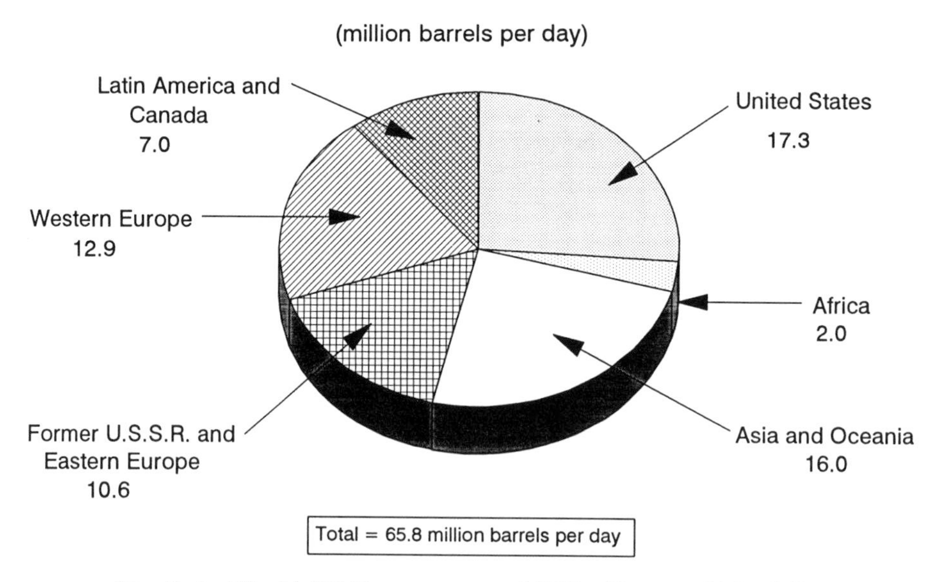

Fig. 1.4. World Oil Consumption 1989. (Source: U.S. Office of Technology Assessment)

Comparing current regional production to proven reserves provides a graphic illustration of what is in store (Figure 1.5). Despite the 17 percent contribution that North America makes to the world's oil supply, that continent possesses only four percent of the proven global oil reserves. At the present rate of production, those reserves could be essentially depleted in approximately 10-15 years. Western Europe accounts for six percent of the world's oil production but has less than two percent of the reserves. Their reserves could be depleted in as little as 13 years. North America and Western Europe consumed nearly half of the world's oil output in 1989 and within a decade it is unlikely that either region will have significant oil reserves of their own. On the other hand, oil producers of the Persian Gulf account for 4.5 percent of the world's oil consumption but possess 65 percent of proven oil reserves. OPEC as a whole, which includes seven oil producers outside the Gulf area, controls approximately 75 percent of the world's oil reserves. At the present rate of production their proven reserves will last another 100 years.

It is often argued that proven reserves have always been limited, and that as reserves are consumed new deposits are discovered. According to the argument, new deposits will always be found because they have always been found in the past. Unfortunately, a history of discoveries does not guarantee a similar future. As for domestic supplies, the prospects of finding new oil fields in the U.S. are slim. The lower 48 states are the most explored areas of the world. The U.S.

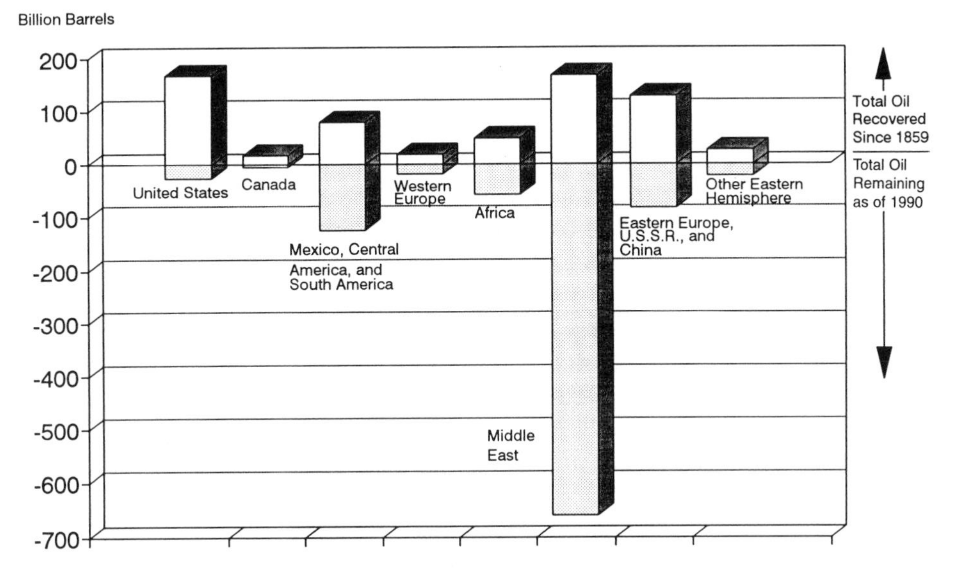

Fig. 1.5. Worldwide Petroleum Reserves. (Source: U.S. Office of Technology Assessment)

Department of Energy (DOE) reported that as of 1986 approximately 80 percent of all wells ever drilled worldwide (2.9 million wells) were drilled in the U.S. Moreover, experts estimate that 80 percent of all oil and gas ever to be found in the U.S. are in fields that have already been discovered. Already, the United States has entered a period of declining domestic oil production and increasing dependency on foreign oil imports. U.S. drilling activity has slowed almost to a halt and the cost of extracting oil from the existing U.S. wells is steadily rising. The industry drilled 17,500 exploratory wells in 1981 and 7150 in 1986. In 1989 approximately 5220 exploratory wells were drilled.[3] Exploratory activity in the U.S. is expected to continue to plummet.

Poor economics is the cause of the decline in U.S. oil exploration. Imports provide the least expensive source of oil. Drilling in the U.S. is a very capital-intensive operation. If oil is not found, the investment is a total loss. If oil is found, U.S. wells tend to be poor producers compared to wells in other regions. For example, the average well in the U.S. produces about 10 barrels of oil per day. In Saudi Arabia a well can produce as much as 2000 barrels per day. U.S. oil is the world's most expensive to produce. Consequently, investment capital for new U.S. wells has virtually disappeared.

Political and economic vulnerability to the highly volatile OPEC nations is expected to significantly increase over the next 25 years unless new motor fuels or new petroleum deposits are found. The U.S. Office of Technology Assessment (OTA), in a 1991 report, indicated that by the year 2020 the U.S. will be consuming nearly 25 million barrels of oil per day, and that as much as 75 percent of U.S. supplies will be imported.[3]

OTA points out that the U.S. can reduce vulnerability to supply disruptions and economical and political pressures through a number of alternatives. Free commerce can be promoted by encouraging the diversification of oil supplies. For example, although Russia has significant oil reserves, the potential of the region has not been fully developed. Providing technology and capital for exploring and developing Russian oil resources would not only diversify the world's resource base, but it would also provide the income so badly needed to restore stability and maintain peaceful progress toward market economies. A similar opportunity for joint venture exists in Venezuela where some of their huge oil reserves (the heavy oils in the Orinoco Basin, for example) can benefit from U.S. technology. China may offer another alternative source of inexpensive oil.

Shifting to resources such as coal and natural gas will also exert a downward pressure on the price of oil and provide additional energy security to industrial-

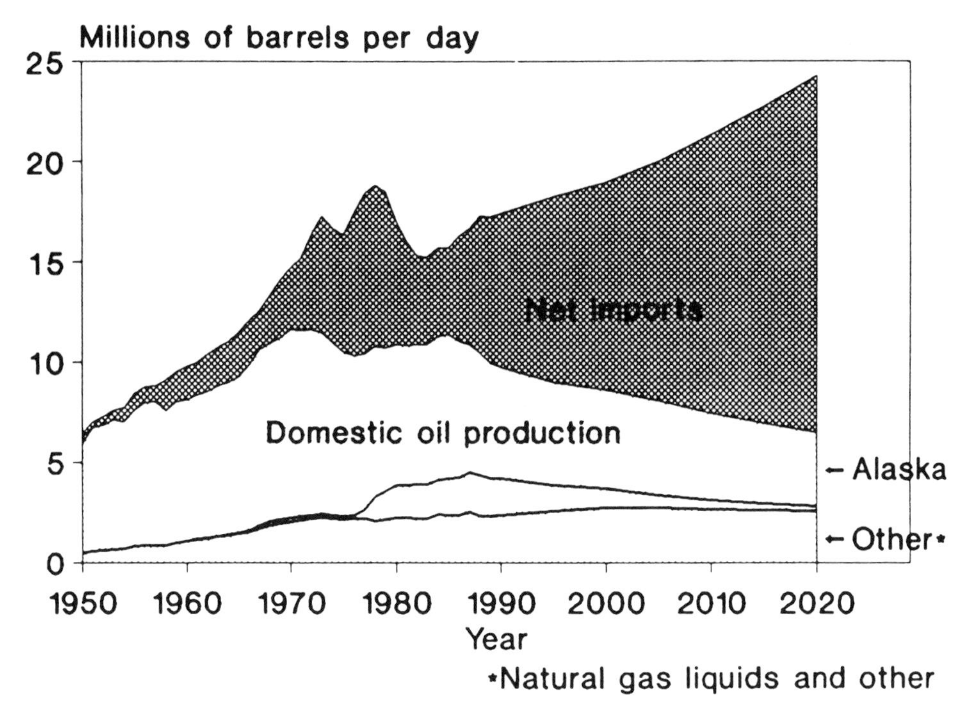

Fig. 1.6. U.S. Oil Import Trends. (Source: U.S. Office of Technology Assessment)

ized nations. However, a longer-term solution must be comprised of more fundamental changes in the way in which energy is produced and consumed. OTA recommends increased efforts to develop and integrate domestically available alternative fuels, and to accelerate efforts to improve vehicle fuel efficiency. According to OTA, alternative fuels and improved vehicle fuel efficiency are two essential ingredients of any long-term resolution to the impending crisis over dwindling petroleum supplies.[3] Although many options exist for fixed-site applications, an alternative fuel or energy source for transportation presents a unique set of difficulties that are not easily resolved.

The Impact of Transportation Energy Use

By 1990, the world was consuming energy at the rate of 361.16 ExaJoules (342.39 quads) annually. Although petroleum supplies only 40 percent of the world's energy, the transportation sector is almost totally dependent on oil.

TABLE 1.2. 1990 UNITED STATES AND WORLD ENERGY USE BY SOURCE
In ExaJoules (Quads)

Energy Source	World		United States		U.S. Percent of World Use
Petroleum	142.85	(135.39)	35.40	(33.55)	25
Natural Gas	76.40	(72.41)	20.36	(19.29)	27
Coal	97.03	(91.97)	20.15	(19.10)	21
Hydroelectric	23.55	(22.32)	3.11	(2.95)	13
Nuclear Electric	21.41	(20.30)	6.50	(6.54)	30
Total	361.16	(342.39)	85.52	(81.43)	24

Source: DOE International Energy Annual (1991)

In the U.S., the transportation sector consumes approximately 23.21 ExaJoules (22 quads) annually, or about 27 percent of the total domestic energy consumption. However, transportation is responsible for 65 percent of all petroleum consumed in the U.S., and personal transportation vehicles consume over half of this oil (Figure 1.7).[4] Gasoline fuel for cars, light trucks, motorcycles and all other forms of personal transportation account for nearly 35 percent of the nation's total consumption of petroleum.

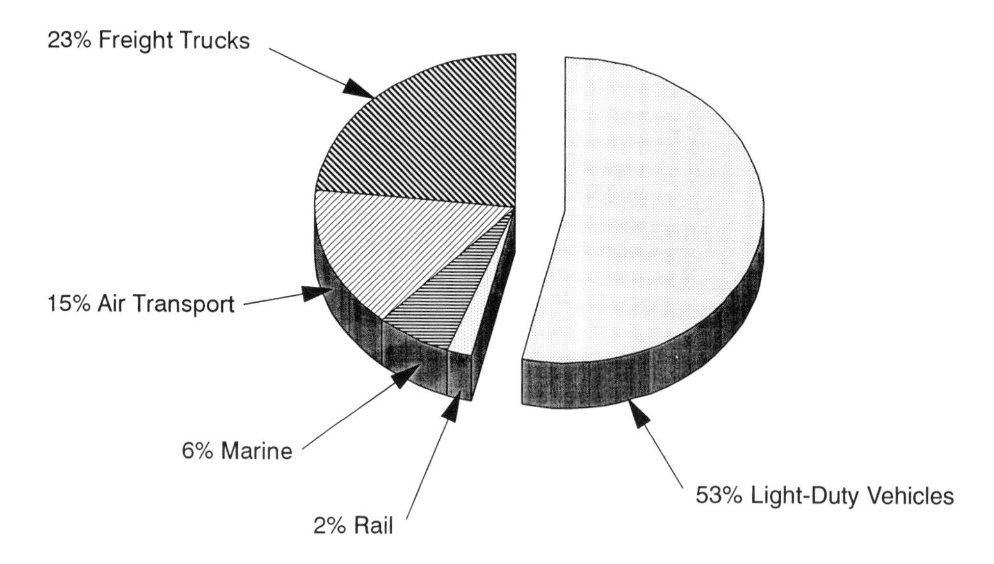

Fig. 1.7. U.S. Transportation Energy Consumption, 1990.

In the developed world, growth and consumption are relatively stable. The number of cars and the consumption of gasoline in the U.S. is expected to grow at a stable but modest rate. The automobile population should grow at about twice the rate of the human population (which now averages 0.6 percent annually in the developed world), and vehicle kilometers traveled is growing at about 4 percent per year in the U.S. Fuel consumption goes up according to the increase in vehicle kilometers traveled, minus improvements in fuel efficiency. Taking into account estimated efficiency gains, along with petroleum consumption trends in other sectors, U.S. oil consumption is expected to grow at slightly less than one percent per year, reaching 20.3 million barrels per day by the year 2010.[4]

Trends in developing regions are toward much greater consumption. In rapidly industrializing nations, the human population is increasing at approximately 2.5 percent per year, but the automobile population is increasing much more rapidly (Figure 1.8). By the year 2010, India is expected to have 36 times more cars than in 1990, and China's automobile population is expected to grow to 91 times the 1990 level. All of Eastern Europe and the countries of the former U.S.S.R. will likely double their automobile population by the year 2010. At the current rate, the automobile population of Mexico will grow 250 percent and the rest of the developing world's automobile population will grow, on average, nearly 300 percent by the year 2010. In comparison, the number of cars in the

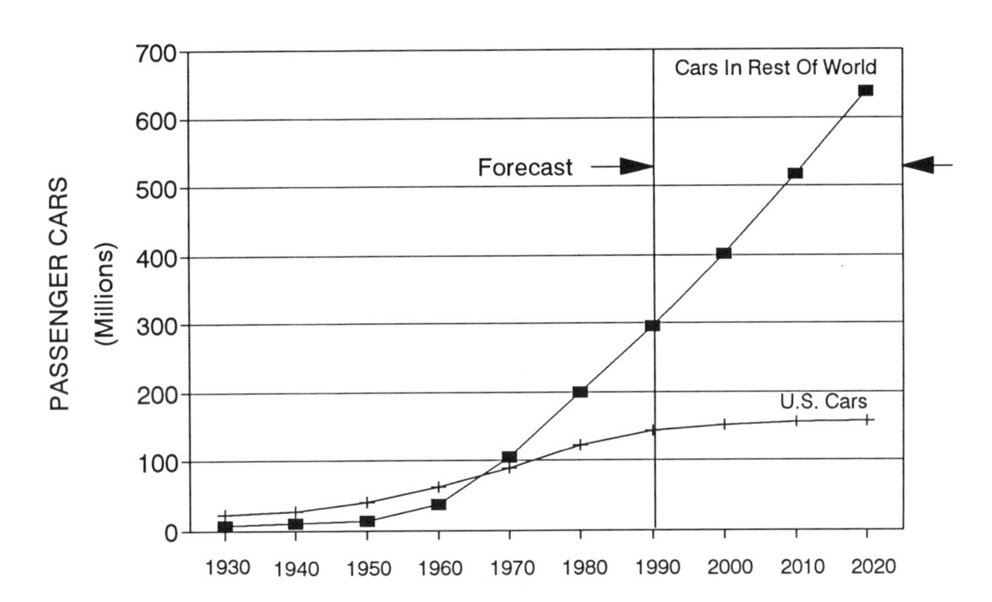

Fig. 1.8. Passenger Cars in U.S. and World.

U.S., Canada, Western Europe and Japan will grow approximately 12 to 15 percent over the same period.[6]

Estimates from various sources tend to agree. The United Nations Fund for Population Activities (UNFPA) recently pointed out that since 1950 the human population has doubled and the world's automobile population has increased sevenfold. They estimate that the world will have twice as many cars, or about 800 million, in 20 years.[7] The World Bank estimates that global motor vehicle population will approach one billion by the year 2030.[5] Fig. 1.8, which is based on American Automobile Manufacturers Association (AAMA) figures, projects 794,300,000 cars by the year 2020. Rapid growth in the world's automobile population is certain and its affect on petroleum consumption is inevitable. If past trends are projected into the future, transportation's share of the world's total primary energy consumption would reach 40 percent by the year 2100. According to Ove Sviden, EKI, University of Linkoping in Sweden, even if economically recoverable petroleum resources are three times larger than today's known reserves, the transportation sector must move away from petroleum motor fuel by the year 2020.[8]

Transportation Energy Use in Developed Countries

Forecasts on a macro level depend on the sum of many long-term and difficult-to-project regional dynamics. One can only look at broad trends and draw general conclusions. Predictions are generally more reliable in developed economies where population growth is lower, transportation infrastructures are in place, and energy use trends are mature. In industrialized nations at the beginning of the oil-embargo-inspired movement toward economizing on energy, planners were optimistic that by applying known economic motivators, energy use could be significantly influenced. Technology would presumably improve energy efficiency, business would yield to economic persuasion and commuting habits would improve. Changes have indeed taken place. According to Lee Schipper, co-author of Energy Efficiency and Human Activity: Past Trends, Future Prospects, in eight OECD countries surveyed, energy use during the 1970-1988 period was generally as follows: Energy consumed for manufacturing has declined, energy consumed for household and services is about the same, and transportation energy use is up.[9] In the final analysis, however, little or no energy was saved during the two decades cited. Except for modest improvements in the U.S., transportation has been the undoing of plans for lower energy appetites in industrialized nations. Mobility in the industrialized world accounts for 18-23 percent of a country's total energy use, and in general, transportation consumes more energy today than before the '73-'74 oil embargo. In the eight OECD countries surveyed above, transportation's total consumption is up, and many countries are consuming more energy on a per capita basis as well.[10]

Automobiles are the predominant mode of transportation throughout the developed world. Private cars are the most convenient mode of travel. In many cities they may be the only reliable mode. Additionally, owning a car is a sign of affluence, worldwide. It naturally follows that automobile ownership tends to grow in step with growth in GDP. Countries with the highest GDP growth rate also have the highest growth rate in the population of private cars (see Table 1.3). When people can afford cars, they buy them. A saturation level of car ownership in developed countries is assumed to exist near the one-car-per-licensed-driver level. The U.S., with slightly more than one car for every licensed driver, has an automobile population growth rate of 0.78 of GDP. In Japan, West Germany and Sweden the ratio of automobile growth to GDP is 1.9, 1.4 and 1.4, respectively. The high growth rate in Japan correlates with the comparatively lower ratio of cars to population and the nation's high economic growth.

A high percentage of passenger kilometers traveled by private car is characteristic of industrialized countries. Schipper found that in the eight OECD countries

TABLE 1.3. PRIVATE CAR OWNERSHIP/USE AND GDP GROWTH, 1970-1987

	Annual Growth in p-km %*	Car Ownership Per 1000 Inhabitants			GPD Annual Growth %	Automobile Fuel Growth % (incl diesel)
		1970	1987	%/yr incl		
Japan	4.3	91	247	5.9	4.2	3.49
U.S.	1.9	420	601	2.1	2.7	0.91
West Germany	2.4	230	457	4.0	2.2	2.18
Sweden	2.1	286	438	2.5	1.8	2.68
Norway	4.6	193	392	4.2	4.0	2.89
France	3.3	240	391	2.9	2.6	3.05
U.K.	3.2	205	307	2.4	2.1	2.59
Italy (1973 -1987)	3.4	190	387	3.2	3.0	3.81

Source: Transportation Science, February 1992
*p-km = passenger-kilometers traveled

TABLE 1.4. TRAVEL BY MODE AND COUNTRY

Country	Cars 1970 - 1987	Buses 1970 - 1987	Rail 1970 - 1987	Air 1970 - 1987	Total 1970 - 1987
Japan					
% Share p-km	42.5 - 55.9	14.7 - 9.3	41.4 - 31.3	1.3 - 3.5	100.0 - 100.0
Eng. Intensity (MJ/p-km)	1.73 - 2.09	0.48 - 0.55	0.20 - 0.20	4.29 - 2.63	0.95 - 1.37
United States					
% Share p-km	90.7 - 85.9	3.7 - 3.5	1.0 - 0.7	4.6 - 9.9	100.0 - 100.0
Eng. Intensity (MJ/p-km)	2.94 - 2.68	0.84 - 0.87	1.73 - 2.10	5.96 - 2.79	2.99 - 2.62
West Germany					
% Share p-km	77.6 - 83.2	10.8 - 8.4	10.9 - 7.6	0.7 - 0.8	100.0 - 100.0
Eng. Intensity (MJ/p-km)	1.91 - 2.31	0.51 - 0.77	0.64 - 0.60	4.36 - 2.64	1.64 - 2.05
Sweden					
% Share p-km	82.4 - 80.0	8.2 - 9.1	8.5 - 7.9	0.9 - 3.0	100.0 - 100.0
Eng. Intensity (MJ/p-km)	1.95 - 2.02	0.92 - 1.03	0.50 - 0.52	6.11 - 3.13	1.79 - 1.84
Norway					
% Share p-km	74.3 - 81.8	15.2 - 7.7	7.9 - 5.3	2.6 - 5.2	100.0 - 100.0
Eng. Intensity (MJ/p-km)	1.72 - 1.69	0.68 - 1.29	1.00 - 0.80	9.19 - 5.97	1.70 - 1.83
France					
% Share p-km	79.9 - 81.9	7.4 - 6.5	12.4 - 10.6	0.3 - 1.1	100.0 - 100.0
Eng. Intensity (MJ/p-km)	1.49 - 1.53	0.52 - 0.70	0.46 - 0.35	4.53 - 2.13	1.30 - 1.36
UK					
% Share p-km	76.2 - 85.6	13.9 - 7.0	9.3 - 6.7	0.5 - 0.7	100.0 - 100.0
Eng. Intensity (MJ/p-km)	1.85 - 1.90	0.73 - 1.13	0.98 - 0.84	4.62 - 3.86	1.63 - 1.79
Italy					
% Share p-km	75.1 - 77.8	13.8 - 13.6	10.6 - 7.7	0.5 - 0.9	100.0 - 100.0
Eng. Intensity (MJ/p-km)	1.37 - 1.47	0.46 - 0.77	0.47 - 0.45	6.13 - 3.98	1.18 - 1.32

Source: Transportation Science, February 1992

surveyed, more than 75 percent of all passenger kilometers traveled are by automobile, except for Japan which reported 55.9 percent passenger kilometers traveled by car (see Table 1.4).[10] In all countries, the absolute passenger kilometers traveled is also growing. Next to air travel, automobiles are the most energy-intensive form of transportation. It naturally follows that affluent, industrialized countries are likely to experience a consistently high energy consumption for personal transportation.

Statistics suggest that switching to more fuel-efficient transportation modes will improve overall energy efficiency. However, it is important to understand that modal energy intensity is most significantly affected by vehicle utilization. For example, the improvement in air transport energy intensity between 1970 and 1987 resulted partly from improved technology, but primarily from flights that were operated at greater load factors. Declining profits and increased competition forced airlines to become more efficient carriers. Scheduling and vehicle size have therefore been modified so that aircraft flying today have fewer empty seats. This approach is potentially just as effective with land transportation. Modal switching can reduce passenger-kilometer energy consumption by leaving lightly loaded automobiles at home and filling the currently unoccupied seats aboard buses and subways. Automobile energy intensity will improve also if cars are better utilized (unused payload capacity is reduced).

Automobiles in industrialized nations typically operate with a load factor of 1.6 to 1.8 occupants. Doubling the load factor will have the effect of reducing the energy per passenger kilometer by 50 percent. Essentially, this is the thrust of car pooling. The same fuel consumption is thereby distributed over more passengers. (Increasing the passenger count on-board light-duty vehicles has little effect on fuel consumption.) However, car pooling has the negative effect of also reducing the driver's independence. Drivers may no longer come and go as they please. Instead, they are restricted to the schedules of their fellow commuters. Additionally, more than 70 percent of suburban office workers use their cars to run errands on the way to and from work, and 80 percent of office workers use them at some point during the workday.[11] As a result, the reasons for the automobile's popularity, the convenience and independence it affords, are largely lost by car pooling.

For transportation engineers and planners, energy efficiency is not the only consideration in modal preference. Emissions may be more easily controlled in vehicles designed for mass transit, and traffic congestion is also somewhat relieved. Although commuters may resist, many argue that emphasizing mass transit systems is ultimately the best way to help defuse the modern-day crisis in urban traffic.

The Blight of Urban Traffic

In America they call it "gridlock," in Germany it is *Verkehrsinfarkt* (traffic infarction). Regardless of the label, there is nothing quite as effective at making automobiles unattractive as being trapped in bumper-to-bumper traffic. Waiting at the entrance to the Lincoln Tunnel or creeping along at 5:00 pm on the Ventura Freeway, one quickly begins to question whether technology might be better utilized. Romance and the automobile seem suddenly incompatible. The racehorse sleekness of shining steel melts into an oddly ridiculous sight when it becomes a burden that keeps one locked in place, rather than an instrument of mobility. There is also the practical matter of squandered resources and wasted time.

Idling along in traffic congestion is wasting the world's fuel and creating tons of extra pollution. The U.S. Department of Transportation estimates that fuel wasted in traffic congestion accounted for more than 11.4 billion liters (3 billion gal.) of fuel, or four percent of the nation's total consumption of gasoline in 1984. They estimated that by the year 2005, approximately 26.5 billion liters (7 billion gal.) per year will be wasted on traffic congestion alone.[12] Cars stopped in traffic have a fuel efficiency of zero. When they begin to creep along, efficiency climbs slightly above zero, but still remains dismal. One minute at idle can consume enough fuel to propel a full-size family sedan a kilometer or more.

Time is becoming the most valuable commodity of modern man. Nowhere is time more ruefully wasted than in traffic jams. As a percentage of the traditional eight-hour work day, 1-1/2 hours in traffic (45 minutes each way, for example) equates to nearly 20 percent of the day. Traffic in most cities is moving more slowly and trip times are increasing. In London, commuting speeds have already been reduced to 13 km/h (8 mph). Commuting traffic in the L.A. basin now moves an average of 53 km/h (33 mph). Although commuting speed in Los Angeles is slightly up in recent years, many believe that speed will ultimately drop to 24 km/h (15 mph), perhaps some time around the year 2000 or soon thereafter. Twenty-four km/h (15 mph) traffic would theoretically more than double the time in transit. The breakdown in personal transportation is showing up also on corporate spreadsheets. Already, the average U.S. business pays approximately $1035 annually, per employee, to compensate for lost productivity of workers stuck in traffic jams.[13] The Commission on California State Government Organization and Economy warned that growing congestion has placed California on the brink of a "transportation crisis which will affect the economy and prosperity of the state."[12]

Sitting in traffic is sure to continue as one of the most annoying and wasteful ways to spend an otherwise pleasant morning or afternoon. But if you must, Ken Orski, President of Urban Mobility Corporation, probably has the best attitude. In his closing remarks before the 22nd Anniversary Luncheon of the Traffic Improvement Association of Oakland County, California, he had this advice about how to survive traffic jams: "Get yourself a comfortable car with plenty of creature comforts—a tape deck, a telephone, a fax machine, and a glove compartment-sized microwave oven (yes they really exist)—and at least suffer in comfort!" According to Orski, commuters should learn to cope with conditions because they are not likely to get much better.

The sight of private automobiles stuck in traffic, each with their single occupant, is a common occurrence throughout the developed world. The striking architecture of Los Angeles with its grand network of serpentine freeways is a supreme example of a metropolis built around the automobile. The system has been the envy of many of the world's larger metropolitan areas. Europe in general, however, lacked the wide open spaces typical of the U.S. Their traffic systems were built in traditionally more confined and densely populated environments. Many were converted from streets that were already in place before cars existed. As a result, European city traffic systems have deteriorated more rapidly and more profoundly.

Traffic congestion is especially resistant to solutions. In its most basic sense, traffic is expressed in terms of flow, or vehicles per hour per lane. In free-flowing traffic, doubling the number of vehicles on the road will produce twice the flow, provided that headway is proportionally reduced and speed remains unchanged. But in general, the more tightly vehicles are packed, the slower the traffic moves. Consequently, adding vehicles increases flow until the effects of reduced speed become greater than those of additional vehicles, at which point flow begins to drop. Adding more vehicles then accelerates the drop in flow. At some density level, traffic will actually stop and flow becomes zero. In high-density signalized city traffic, signal timing and sequencing come into play. Signals are sequenced so that drivers do not have to stop as long as they maintain a synchronization speed. Newer computerized systems can count cars and adjust signal timing to account for traffic variations. Human controllers in Los Angeles and New York can interrupt the computer system and manually defuse stacked-up traffic. Gridlock occurs when there is no unoccupied space available for queuing. For example, when a light turns green, there must be unoccupied space beyond the intersection for cars that were stopped at the signal. If the space is filled with cars, traffic cannot move. This will occur at some density level in every signalized system of roadways.

In an automobile-oriented transportation system, the number of vehicles in transit is essentially relative to the number of people and varies according to the time at which they all pour into their automobiles. Therefore, a high-density population that relies on automobiles for transportation will invariably produce traffic congestion, as well as the accompanying air pollution, noise and accidents. Regardless of improvements in traffic management and facilities, there is an absolute limit to the number of cars that can fit into cities. Theoretically, one can expand the size of streets and add more parking facilities, but at some point, there may be no downtown left. In the U.S., nearly half of all urban space is already dedicated to parking, servicing, and providing roadways for cars. Worldwide, the average city devotes about one-third of its land mass to the automobile.

The problems of the cities will increasingly be the problems of the automobile. Before the turn of the century, the world will have more city dwellers than country dwellers. If conurbations with over one million inhabitants are included, metropolitan areas will include approximately 60 percent of the urban population, or more than 1.5 billion people worldwide. According to United Nations projections, in 2025 there will be 93 metropolitan areas with populations greater than five million. In 1984, there were only 34 such cities. In 2025, Los Angeles will have 15 million, Mexico City will have 30 million, Tokyo will have 20 million and London will have 10 million. Already, in developed countries far more people live in urban areas (see Table 1.5).

TABLE 1.5. PERCENT OF POPULATION LIVING IN URBAN AREAS (1990)

United Kingdom	91.5	Sweden	83.4
Denmark	88.9	Bahrain	81.7
Netherlands	88.4	Germany	81.0
Qatar	88.0	Luxembourg	81.0
Venezuela	86.6	Lebanon	80.1
Australia	85.5	United Arab Emirates	77.8
Malta	85.3	Japan	76.5
Argentina	84.6	Canada	75.9
Uruguay	84.6	Spain	75.8
New Zealand	83.7	United States	73.9
Chile	83.6		

Conventional Ideas for Taming the Automobile

Many ideas for improving the automobile's fuel efficiency or reducing traffic, pollution, and space requirements are based on the idea of restricting or eliminating the car itself. To analysts and commuters alike, a switch to public transportation is often seen as the ultimate answer to city traffic congestion. The idea of at least partially giving up cars is receiving support even from within the automotive industry. Pehr Gyllenhammer, General Manager of Sweden's Volvo Corporation, is unequivocal: "Die Innenstadt (the downtown area of the city) will have to be closed to individual automobile traffic. For the car to survive as the most versatile means of transportation, it will have to be taken out of situations in which it cannot prove its advantage."[14] Fiat's Gianni Angelli states: "Something has to be done to reorganize traffic into and around the central cities, to keep the number of cars within reason."

Undoubtedly, switching from cars to mass transit will reduce traffic congestion in many large cities. But when people have a choice, they overwhelmingly choose the independence of the automobile. As a result, ideas for penalizing or restricting cars in the cities and concentrating more heavily on public transportation are becoming more popular, even with the population. Germans, who are second only to the U.S. in per capita car ownership and perhaps first in their love of fine machines, are becoming increasingly disenchanted with cars in the inner cities. In a 1989 poll in Munich, 93 percent of the population supported giving public transportation priority, rather than spending money on improving systems to support the automobile. Politicians predicted a different result. In a parallel poll, 62 percent of the local politicians said that people would vote in favor of cars.[14] In Munich the situation is critical. Downtown Munich has to cope with more than a million cars each day and traffic is virtually at a standstill during morning and afternoon rush hours. Conditions are equally bad in Frankfurt where 200,000 people pour into the city each day—most of them by car. As a result of burgeoning commuter traffic, many German cities are threatened with a total collapse of automobile traffic systems. According to a recent poll of inhabitants of the Old Federal States, 85 percent of voters favor significant restrictions on city traffic and 53 percent would like to see the automobile completely banned from downtown areas.[15]

Similar trends are developing throughout all of Europe. In 1985, voters in Milan, Italy, overwhelmingly approved a plan to close the city to all automobiles during rush hour. Sienna and Bologna now prohibit all motor vehicles during certain business hours, except for taxis, buses, delivery trucks, and cars owned by local residents. Athens uses odd-and-even rules keyed to license numbers to prohibit cars from entering the central city. Other options also are being tried. In Paris, every business with more than 20 employees pays a public transport levy. In Lorrach, Germany, workers who drive their cars are charged punitive parking fees. In many cities the bicycle is enjoying a renaissance. In Asia, bicycles are the predominant means of local transportation. And Europeans, who never entirely abandoned the bicycle, are turning on to cycling more than ever. The Swiss Pharmaceutical company, Ciba-Geigy, gives a new bicycle to employees who give up their parking spaces. In Denmark it is common for commuters to own two bicycles, one for commuting from home to the train station, and a second at the other end for riding from the station to work.

In the UK, after decades of pro-automobile planning, it is the railways that may ultimately win favor with commuters. Driving in downtown London is nearly impossible and the prospect of finding a parking place is equally as dismal. An acquaintance who lives near London visited the U.S. several years ago and actually drove an automobile into New York city: a travesty unparalleled in American driving experiences. When asked about the encounter, he replied that "compared to London it was actually quite pleasant." Intercity travel in the UK is also best by rail. On the 88-kilometer (55-mile) trip from Faversham to London, for example, a round trip during off-peak hours costs £7.90 ($11 in 1993), and trains run four to the hour throughout the day. For the price of a ticket, one gets unlimited use of the railways in the city for the day. A similar trip by car would cost about £17.50 ($25) for fuel and £14 ($20) for parking, in addition to the aggravation of driving in London's congested traffic.

Totally car-free cities are probably not the wave of the future. However, a shift in values for those who learn to rely less on private cars may be inevitable. Cities that turn increasingly toward transit systems may become a sort of training ground for commuters—churning out graduates who no longer place the automobile at the same high level of importance. I quote an American journalist, Peter Tautfest, who lives and works in Hannover, Germany:[14]

> "I remember going out on rainy nights and driving to the movies, the theater, or a restaurant only to find that on returning I would spend up to 45 minutes circling the neighborhood—finally parking my car a half hour's walk from my apartment. I learned, and started taking taxis. After doing this the third time, I became aware of the absurdity of owning a car that I couldn't use. So when our trusty Volkswagen broke down, I took it to the junkyard instead of the mechanic and thought I'd try living without a car.
>
> "Now I ride my bike to work, whizzing past long lines of cars and beating my car-driving colleagues. On rainy days I take the bus,

> comfortably reading the newspaper while my colleagues are stuck in the proverbial gridlock listening to the latest traffic update."

Selling an automobile to Mr. Tautfest will be a much different task now than before he abandoned his car. Marketing automobiles within a transit-adapted culture to future transit-riding commuters may also be much different than it is today. The idea of future inner cities free of private cars, with people brought to the city by transit systems and served while there by trams and people movers, is certainly attractive and valid. It is easy to envision a modern, artful network of sweeping tramways and mobile walkways connecting buildings and offices. Parks might occupy the central city and sculptures and fountains would enhance the open, cosmopolitan ambiance. Automobiles, for those who feel they need them, might be reserved for the occasional motor trip or outing to the country. More practical inhabitants, would of course travel intercity by rail or air. In a modern city so conceived, the privately owned family sedan may no longer be the centerpiece of the society's transportation system. Instead, community cars could be available on an as-needed basis, much like the Mobility Enterprise program developed at Purdue University in the '80s.[16]

Such an image is probably more universally European than American. In speaking of the city in the U.S. it is easy to become confused. U.S. cities are more often like an amoeba with ill-defined edges and no discernible head. For years I lived and worked in San Fernando Valley, which is classified as a suburb of Los Angeles. However, San Fernando Valley is home to more people than live in the entire state of Arizona. The "city" to residents of San Fernando Valley is just as likely to mean places like Van Nuys or Burbank, which are central business districts of areas that began as independent suburbs. Downtown Los Angeles is far away and largely irrelevant to the lifestyle of many inhabitants of the "Valley." Today, the busiest stretch of L.A. freeway is the Ventura Freeway in Encino, which is located in south/central San Fernando Valley, far from downtown Los Angeles.

What happened in Los Angeles is classical of developments in many U.S. cities. In effect, cities have turned inside out as employers followed workers into the suburbs. The result is sprawling residential districts spotted with business and industry and with little or no traditional hub. Traffic has not been alleviated in the process. Instead, it has engulfed the suburbs as well. The idea of a car-free downtown area therefore has little meaning. In the case of Los Angeles, banning cars from the central city would have virtually no effect on traffic conditions in the Los Angeles basin. Moreover, public transportation would be hard-pressed to accommodate commuting needs county-wide. Different solutions are therefore required in U.S. conurbations, and some form of personal transportation device appears to be the only answer.

Ken Orski suggests that we can no longer think in terms of "solving" traffic congestion. The problem is one of management, which in a medical dictionary is defined as "keeping an incurable disease under control." His suggested management techniques fall under three categories, which include incrementally expanding road capacity, reducing the growth in demand for transportation, and controlling the pace of development. According to Orski, additional roadway capacity should be built, but at nowhere near the pace of the '60s and '70s. Costs are too high today and new roadways usually mean tearing out already-developed areas. Controlling demand, according to Orski, would involve efforts to increase car pooling, encourage flexible work schedules, turning more to telecommuting (working at home and communicating by modem), and building more transit systems. In order to control growth, communities would be limited to the growth of the supporting facilities. Building permits are not granted without adequate sewage and utilities, and they should not be granted if the transportation system cannot support additional development. Developments would therefore be staged along with the supportive transportation infrastructure.[11]

Controlling the Automobile with Higher Fuel Costs

Ideas for controlling the automobile are often framed in terms of controlling the costs associated with driving. A popular argument against the benefits of more fuel-efficient cars has to do with the price of fuel. As a percentage of the total costs of owning and operating an automobile, fuel costs are relatively minor, especially in the U.S. According to the argument, improving fuel economy from 13 km/L (30 mpg) to 26 km/L (60 mpg), for example, does not result in a large savings in the overall cost of operating an automobile. Therefore, there is no overriding benefit to consumers to economize on fuel by driving less, or to drive more fuel-efficient cars. Unfortunately, the impact of transportation costs on individual budgets is largely hidden. If the real costs of driving private cars were transferred directly to drivers, economics would become a much more significant factor.

Costs associated with environmental damage are typically subsidized by government and industry, and thereby spread over society as a whole or transferred to other sectors. Examining a number of hidden costs, the Ecological Research Institute in Heidelberg, Germany, has calculated that in order for cars to pay for all their related expenses, gasoline would have to sell for $12 per gallon

($3.17/L). In a similar exercise, Carlo Rubia, winner of the 1984 Nobel Prize in physics, estimated that for automobiles to pay their own way, the worldwide price of gasoline should be on the order of $16 per gallon ($4.22/L).[14] If the hidden costs were no longer subsidized by society they would have to rise to the surface where they would affect transportation choices. At $12-$16 per gallon ($3.17-$4.22/L), for example, consumers could no longer remain unaware and attitudes toward automobile fuel economy, as well as private cars in general, would quickly change.

On a more moderate level, higher fuel prices in the U.S. may be inevitable. The U.S. has for years enjoyed artificially low fuel prices. Lee Schipper suggests that the electric car equates to a conventional car at $6 per gallon of fuel ($1.58/L). That is, if gasoline were $6 per gallon, consumers would more likely see electric cars as much more attractive alternatives. The reduced range of EVs (or the reduced payload capacity of significantly smaller IC cars) might then appear less important in comparison with the benefits of reduced energy appetites. To continue with artificially low fuel prices, then lament the nation's high consumption of fuel, is a policy at odds with itself. Schipper notes a connectivity between fuel costs, vehicle fuel economy, and passenger kilometers traveled in industrialized countries. Although Europeans drive only 60-80 percent as much as Americans, and European cars typically achieve higher fuel economy, Europeans and Americans spend roughly the same portion of their incomes on motor fuel.[17] This suggests that there may be a cost tolerance threshold. When that threshold is breached by higher fuel prices, reduced consumption will likely result.

TABLE 1.6. PER-GALLON GASOLINE PRICES FOR SELECTED COUNTRIES (1991)
(Per liter price in parentheses)*

Japan	$3.90 ($1.02)
Sweden	$4.45 ($1.17)
Canada	$2.06 ($0.54)
France	$3.86 ($1.02)
UK	$2.55 ($0.67)
U.S.	$1.43 ($0.38)
Italy	$5.10 ($1.35)
W. Germany	$2.87 ($0.76)

Source: Transportation Energy Data Book, Edition 13
* Liter prices calculated by the author.

Overloading the Environment with Waste Products

Greater consumption means that the by-products of consumption also are produced at a greater rate. In general, consumption's by-products turn out to be harmful to the environment. Pollution is loosely linked to the growth of GDP and therefore goes up with improved standards of living. If living standards grow at just two thirds the rate of the past 25 years, consumption will quadruple in 70 years. By the year 2100, the world economy could be producing 20 times as much as it is today.[18] Most analysts doubt that the world can sustain that level of production/consumption, or that the environment can assimilate the resulting pollution. In terms of total environmental impact, reducing energy consumption may be the single most important strategy.

Manufacturing automobiles produces environmental pollution on an industrial level. Once the car is in service, it then becomes a continuous source of environmental pollution throughout its service life. Petroleum fuels are one of the major contributors to atmospheric pollution around the world. In a typical U.S. city, motor vehicle emissions account for 30-50 percent of hydrocarbon (HC), 80-90 percent of carbon monoxide (CO), and 40-60 percent of nitrogen oxide (NO_x) emissions. Table 1.7 shows transportation's average contribution to total emissions in the U.S.[19] In OECD countries as a whole, about 39 percent of HC emissions, 66 percent of CO emissions and 47 percent of NO_x emissions come from motor vehicles.[5]

Additionally, several studies have shown that emissions from automobiles increase significantly at the low speeds typical of urban traffic. The U.S. Environmental Protection Agency (EPA) estimates that IC engine emissions at a vehicle speed of 8 km/h (5 mph) are two to three times greater than at the Federal Urban Driving Schedule (FUDS) average speed of 31.5 km/h (19.6

TABLE 1.7. PERCENTAGE OF TRANSPORTATION'S CONTRIBUTION TO TOTAL U.S. EMISSIONS

Year	Suspended Particulate	Sulfur Oxide	Carbon Monoxide	Nitrogen Oxides	Volatile Organic Compound	Lead
1980	15.3	3.8	70.5	46.9	39.6	84.1
1985	19.2	4.3	47.9	44.7	37.6	74.2
1989	20.8	4.7	65.7	39.7	34.6	30.6

Source: U.S. Environmental Protection Agency

mph). The California Air Resources Board estimates that at 8 km/h (5 mph) HC emissions are 23 times greater, CO emissions are 13 times greater, and NO_x emissions are about 4 times greater than at 32 km/h (20 mph).[20] Different studies indicate different magnitudes of increased harmful emissions; but universally, they show much higher emissions at city-traffic speeds.

Global warming is an environmental wildcard. The decade of the '80s was the hottest on record. However, an accurate record of temperatures has been kept only since 1880. Consequently, "the hottest decade on record" may be within the standard deviation of natural variability. At this point, no one can predict with certainty just how much effect the industrial production of greenhouse gases will have on global climate. However, laboratory tests and computer models indicate that the significant rise in atmospheric carbon dioxide (CO_2), methane, and other greenhouse gases will ultimately raise global temperatures and, for better or for worse, change the world's climate. Figure 1.9 shows the contribution to greenhouse gases in the U.S. by sector.

Automobiles are especially efficient at producing greenhouse gases. Cars and light trucks are responsible for approximately 20 percent of all atmospheric CO_2

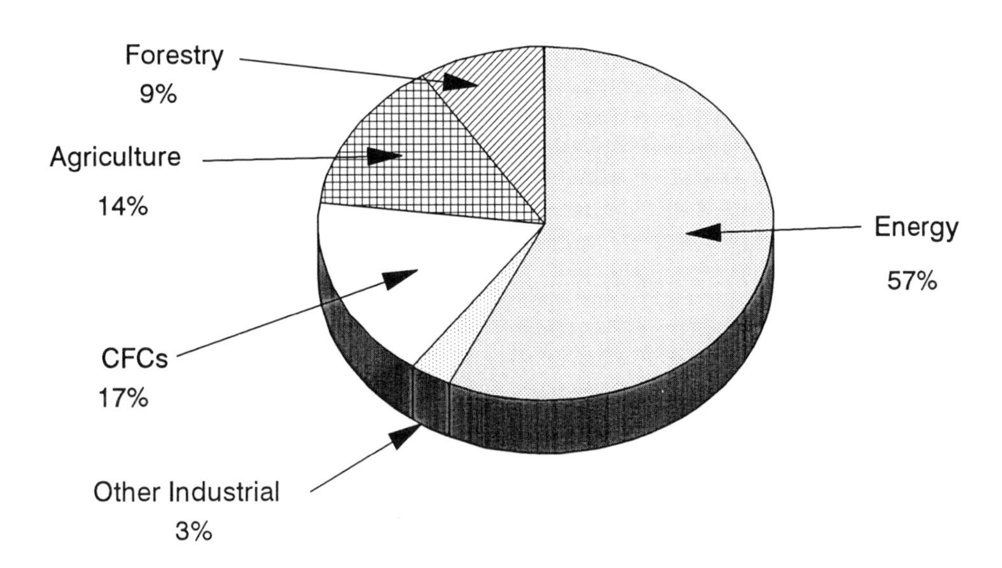

Fig. 1.9. Contributors to Global Warming. (Source: U.S. Environmental Protection Agency, 1989)

in the U.S., or about five percent of the total CO_2 produced from all human activities worldwide. The United Nations Fund for Population Activities warns that, because of rapidly increasing automobile populations, developing countries will be emitting 15.06 billion t (16.6 billion U.S. tons) of CO_2 per year by 2025. At that point, developing nations will be emitting four times as much CO_2 as the developed countries. Motor vehicle carbon emissions are essentially proportional to total fuel consumed.

The planet's ozone layer is deteriorating more rapidly than expected. In recent years, a hole has opened in the southern hemisphere every summer, and there are signs of similar events brewing in the northern hemisphere. By the year 2000, all temperate regions of the world can expect a 10 percent depletion in the ozone layer above them. The reduced ozone layer is expected to produce a 26 percent increase in the most common forms of skin cancer. As a result, efforts to ban the use of chlorofluorocarbons (CFCs) have been accelerated worldwide. Before recent legislation, approximately 25 percent of the CFCs used in the U.S. had been for vehicle air conditioners.

Automobiles are not alone when it comes to negative effects associated with producing positive benefits. There are many facets of industrialization that end up creating liabilities that may not have been anticipated. Even in the fine art of learning, producing paper for a society's printed matter has turned more than one pristine river into an ecological disaster. Advancements in medicine backfire by producing antibiotic-resistant strains of disease-causing micro-organisms, and food producing technologies rapidly exhaust the ability of land to grow crops. Benefits always have a cost. However, many of these environmental costs are hidden from the average consumer. Over the short term, it may be easy to ignore the polluted river or the no-longer-productive cropland. When it comes to transportation, the environmental costs of mobility are especially visible. Daily, the consumer breathes the air, sees the smog, and endures the traffic congestion. As individuals, consumers dispense fuel into cars then knowingly spew the by-products of combustion into the atmosphere. It is therefore more difficult to remain unaware of the consequences. As a result, automobile-related issues have become especially emotional and highly politicized.

Alternative Forms of Personal Transportation May Be Inevitable

As Orski suggested about traffic congestion, thinking in terms of a solution to the difficulties associated with the increasing demand for transportation may not be realistic. Management, or "keeping an incurable disease under control," is probably the most feasible approach. As part of the management strategy, smaller, more fuel-efficient cars would make it possible to maintain the present level of mobility while reducing the costs to individuals, nations, and the planet. Several studies have indicated that smaller vehicles can help relieve congestion, as well as a variety of other traffic-related ills. Studies by Garrison and Pitstick, using GM's Lean Machine as a basis, indicated that roadway capacity is significantly improved when very small vehicles are substituted for large ones.[21] Other associated benefits include improved use of land and parking facilities, and reduced pollution and fuel wasted at idle. Smaller cars have a direct effect on lowering energy requirements over the urban driving cycle (see Chapter Three). Also, smaller cars are significantly less dangerous to pedestrians (see Chapter Seven).

Transportation plays a predominant role in the worldwide consumption of oil and production of harmful emissions. In developing regions, it is expected that oil consumption will increase at an accelerated rate over the coming decades. In the U.S., where one-third of the world's automobiles exist, a movement toward vehicle types with significantly improved fuel economy as well as electric-powered cars, can markedly reduced that nation's demand for oil. Enormous benefits can be achieved with existing technologies and with little sacrifice in lifestyles if the mix of personal transportation vehicles were expanded to include alternative vehicle types with modest appetites for energy.

In the final analysis, imposing unreachable fuel economy standards on large, under-utilized cars may not be the wisest use of legislative power. Thoughtfully developed policies designed to encourage the development of cars that fit commuting needs should be linked to incentives that also encourage a shift in consumer values. New attitudes toward transportation in general, as well as a switch to more fuel-efficient vehicles, are feasible steps that can be taken to moderate the cycle of ever-increasing consumption and pollution. Ideas such as the electronic office and reconfiguring cities for better use of public transportation also can help. Consider the bleak hypothesis offered by William Lee, CEO of Duke Power in North Carolina: "Suppose that the world population stabilizes at the low end of the United Nations estimate of nine billion. Next, suppose that a majority of the world's residents reach a standard of living equal to half that enjoyed by Americans. Finally, suppose that all energy uses, from factories to sports cars, grow twice as efficient. Under these optimistic assumptions, the world will need to generate three times as much power."[22]

References

1. Paul Ehrlich, The Population Bomb, Random House, 1968.

2. Energy Information Administration, "1993 Annual Energy Outlook With Projections to 2010," DOE/EIA-0383(93).

3. U.S. Congress, Office of Technology Assessment, "U.S. Oil Import Vulnerability: The Technical Replacement Capability," OTA-E-503, U.S. Government Printing Office, Washington, D.C., October 1991.

4. Energy Information Administration, "Petroleum: An Energy Profile," DOE/EIA-0545(91).

5. Asif Faiz, *et al.*, "Automotive Air Pollution: Issues and Options for Developing Countries," Infrastructure and Urban Development Department, The World Bank, August 1990, (WPS 492).

6. John L. Mason, Energy and Transportation, SAE SP-869, Society of Automotive Engineers, Warrendale, PA, 1991.

7. "Car Population Explosion," *The Futurist*, January-February, 1991.

8. Ove Sviden, "Sustaining Mobility: A Systems Approach to Determining the Role of Electric Vehicles," paper presented at the OECD/IEA Conference, Stockholm, Sweden, 25th-27th May, 1992, published in OECD Document, *The Urban Electric Vehicle: Policy Options, Technology Trends, and Market Prospects*, ISBN 92-64-13752-1.

9. Lee Schipper, Steve Meyers, *et al.*, Energy Efficiency and Human Activity: Past Trends, Future Prospects, Cambridge University Press, 1992.

10. Lee Schipper, *et al.*, "Energy Use In Passenger Transport in OECD Countries: Changes Since 1970," *Transportation Science*, Vol. 19, No. 1, February 1992.

11. C. Kenneth Orski, "A Common Sense Look at the Problem of Traffic Congestion," *Vital Speeches of the Day*, Jan. 1, 1990.

12. Michael Renner, "Transportation Tomorrow: Rethinking the Role of the Automobile," *The Futurist*, March-April, 1989.

13. *Bicycling Magazine*, March 1993, p. 28.

14. Peter Tautfest, "Clearing Up The Euro-Jam," *World Monitor*, March 1991.

15. Matthias Huthmacher, "Slowly Moving Forward," *Scala*, Aug./Sept., 1991, Frankfurt, Germany.

16. F.T. Sparrow, Robert K. Whitford, "Automotive Transportation Productivity: Feasibility and Safety Concepts of the Urban Automobile," Final Report to the Lilly Endowment, Inc., Purdue University, December 1984. (Mobility Enterprise was an experimental program designed to explore the feasibility of a fleet of small urban cars supplemented by larger community cars that could be used by participants on an as-needed basis. Several interim reports were produced. The final report is cited here.)

17. Lee Schipper, *et al.*, "Fuel Prices, Automobile Fuel Economy, and Fuel Use For Land Travel: Preliminary Findings from an International Comparison," unpublished draft #LBL 32699, for International Energy Studies, Energy Analysis Program, Lawrence Berkeley Laboratory, Berkeley, California.

18. Sharon Begley, "Is It Apocalypse Now?," *Newsweek*, June 1, 1992, p. 37 (United Nations estimates cited).

19. United States Environmental Protection Agency, Office of Air Quality Planning and Standards, "National Air Quality and Emissions Trends Report, 1989," Research Triangle Park, NC.

20. Quanlu Wang and Danilo L. Santini, "Magnitude and Value of Electric Vehicle Emissions Reductions for Six Driving Cycles in Four U.S. Cities with Varying Air Quality Problems," University of Transportation Studies, 11-22-92.

21. William L Garrison and Mark E. Pitstick, "Lean Vehicles—Strategies for Introduction Emphasizing Adjustment to Parking and Road Facilities," SAE Paper No. 901485, Society of Automotive Engineers, Warrendale, PA, 1990.

22. Gregg Easterbrook, "A House of Cards," *Newsweek*, June 1, 1992.

Chapter Two

Personal Transportation Vehicles for the 21st Century

Courtesy: General Motors Corp.

In truth, the ideas and images in men's minds are the invisible powers that constantly govern them.

John Locke

Few would argue with the idea that the world would be better off if energy were conserved and harmful emissions were reduced. It is also clear that smaller, lighter, more fuel-efficient commuter cars and electric urban cars would help. But exactly what kind of cars do these labels describe and how do we get consumers to purchase them? Electric cars and fuel-efficient runabouts are old ideas that have been consistently rejected by consumers. Significantly different vehicle types must be matched by new consumer attitudes or we will be pushing sand up a hill to no avail. One cannot promote new cures without cultivating social and political values that make the cures palatable. And on a more fundamental level, where are the boundaries to industry's role? Does industry champion social and political causes, or is it the job of automobile manufacturers to make cars for profit?

Traditionally, neither political nor social issues are appropriate arenas for industry. But strategically, it is better to lead rather than follow when change appears inevitable. And from a narrow marketing perspective, consumers are becoming increasingly more sophisticated and worldly, and new attitudes offer new marketing tools as well as new challenges. In today's environment, business is no longer just selling products. Through their products they are expressing company values and asking consumers to subscribe to those values. Consumers are also looking for values to which they can subscribe. When a company connects with attractive products that address the broad concerns of today's consumers, both prestige and sales are likely to soar. In 1990, Global Business Network conducted a six-week-long computer forum between a distinguished group of ecologists, business planners, physicists, and community leaders. The subject was the "responsible corporation" of the '90s and how new consumer attitudes might affect Nissan's strategic planning. The results, published as *The Nissan Report*, suggests that a new relationship is emerging between business and consumers.

> "A new consumer is rewriting the rules of the marketplace. The greedy 'me generation' consumer of the eighties is disappearing; in her place is emerging a more community-focused and responsibility-minded consumer. This new consumer cares not only about a product and what it can do for her (or say about her), she cares about its place in society as well.... To these new consumers, we are not just selling cars (any more than other companies are simply selling soap or hamburgers or long-distance service), we are speaking to their values, their beliefs, their ways of being."[1]

Projects such as GM's Impact, Chrysler's TEVan, and Ford's Ecostar speak directly to this new dynamic. Their value as consumer products may ultimately be overshadowed by their more broad benefit in portraying their creators as environmentally conscious leaders. Environmental and energy issues are closely associated with the automobile industry. Innovation and leadership in these areas will therefore become increasingly more important as we move into the twenty-first century. In this context, it makes little sense to debate industry's role or the appropriateness of innovation and leadership in broad issues. Whether one chooses the label of "leading rather than following change," or that of "implementing new marketing tools," the benefits of recognizing new socio-marketing dynamics and moving into them on one's own terms are enormous.

Selling vastly different vehicle types will require a period of pioneering, which can be foreshortened by strong and visible support from government (public acknowledgement, appropriate fuel prices, traffic-restricted downtown areas, tax incentives for energy-efficient cars, special lanes for narrow-lane vehicles, etc.). In this regard, government can be an active partner, and is already being solicited in many North American and European cities to provide special freeway lanes, parking facilities and tax incentives for electric cars. Support from government will increase consumer confidence and in many ways assist, and even partially underwrite, pioneering efforts. By leading, industry can engage government on global interests in ways that may be difficult to resist. Conversely, following rather than leading government initiative would be the least desirable alternative.

Expanding the transportation options to include significantly smaller, energy-efficient vehicles provides an automobile-oriented solution to an automobile-oriented problem. In addition, it is an alternative that emphasizes vehicle packaging and marketing techniques, rather than advanced technology. Developing significantly smaller vehicle types and effective marketing appeals requires neither new technologies nor new marketing principles. Both can be accomplished within the state of the art.

Plans to integrate unconventional vehicle types should minimize the need for innovation in basic technology and maximize innovation in vehicle configuration. Innovations that require minimal or no infrastructure change will be most easily accommodated, and the polarization of industry, governments, and consumers thus created will then compel the infrastructure to follow. As more widely divergent vehicle types proliferate, direct comparisons with standard automobiles will grow less valid. The technical benefits of smaller, mission-specific cars flow from the conceptual underpinnings of the size of cars in relation to the missions they perform.

Smaller Cars Reduce Hardware Overhead and Improve Load Factors

Automobile transportation systems worldwide have two fundamental characteristics that invite energy savings. The first is that automobiles are often too large and powerful for their mission. The second actually compounds the first to the degree to which they co-exist. Automobiles are operated on average at about 30 percent of their payload capacity (1.6-1.8 occupants) wherever they exist in the industrialized world. Consistently poor load factors result in high energy intensity for automobiles worldwide. As it turns out, most of the energy used by the automobile is consumed to transport itself. The energy used to move the occupant is almost insignificant by comparison. Undoubtedly, an alien life form arriving on Earth could observe our automobile population and conclude that people are devices created by cars so that cars can navigate through traffic without crashing. The idea that it's the other way around, that it's the people and not the cars that are the object of the energy, might seem grossly out of step with what is actually taking place.

The fundamental mismatch between vehicle mass/size in relation to that of its payload is often unspoken; perhaps because it is so universally accepted and difficult to quantify. Much depends on the synergy of the system, and even more on the operating conditions at the time the energy trail is audited. On the most simple level, when a 1600 kg (3500 lb) machine transports an 80 kg (175 lb) occupant on a local trip to the market, the available-energy pie is divided so that approximately 95 percent gets the car to the market and the remaining five percent gets the occupant there. More specifically, about 82 percent of the latent energy in gasoline is wasted when it is converted into mechanical power, which just pollutes the air and gets no one to the market. Of the 18 percent left, about a third goes to overcoming air resistance and the other two-thirds is consumed by inertia and rolling resistance, of which the occupant accounts for a small portion. In this scenario the occupant gets 0.006 of the fuel's energy, the car gets 0.174, and 0.820 is wasted. Since the automobile is responsible for 99.4 percent of the total energy consumed, and it tenaciously resists improvements in energy efficiency, minimizing the car itself is the most straightforward way to reduce its portion of the energy budget.

In the U.S., the car itself is being slowly minimized as curb weight is reduced. Average new car curb weight has dropped by approximately 500 kg (1100 lb) over the past two decades. Today, roughly 25 percent less mass is providing equal transportation to Americans who own newer cars. Extrapolated to the entire U.S. automobile fleet, reduced vehicle weight calculates out to approximately 75 billion kg (165 billion lb) less automobile on the road, or 75 billion kg

(165 billion lb) less mass to move along the nation's roadways. For those who like superlatives, that is roughly equal to the combined weight of nearly 1 billion adults, or about 18 percent of the world's human population. One can hardly argue with the economics of leaving that much weight sitting at the curbside. Smaller vehicles directly reduce the workload, and equally smaller engines and drivetrains work hand-in-hand to reduce the overall energy appetite.

The other major component is that of load factor. Changes in payload have little effect on fuel consumption once the automobile has been defined. Consequently, when vehicles operate more closely to full capacity, fuel consumed in relation to passenger kilometers traveled is significantly reduced. Unfortunately, vehicles are consistently under-utilized throughout the industrialized world. As noted in Chapter One, automobiles in industrialized countries operate at roughly the same load factor: about 1.6 to 1.8 occupants. The trend in industrialized countries is toward even lower load factors. A breakdown of U.S. vehicle occupancy rates by trip category is shown in Table 2.1. The 1990 Nationwide Personal Transportation Study (NPTS) revealed that the magnitude of automobile under-utilization is growing. The average occupancy rate of U.S. passenger cars was 1.6 persons in 1990, down from 1.9 in 1977.[2] That represents a 16 percent drop in passenger car utilization. When vehicle occupancy drops, energy intensity goes up by a corresponding amount. The vehicle will consume roughly the same energy, regardless of whether seats are occupied or vacant. If vehicle trip profiles do not allow for adequate load factors, then vehicles should be redesigned to fit travel preferences.

TABLE 2.1. AVERAGE VEHICLE OCCUPANCY BY TRIP PURPOSE

	1977		1983		1990	
Trip Purpose	Aver. Occup.	% v-km	Aver. Occup.	% v-km	Aver. Occup.	% v-km
Work Related	1.3	39.3	1.3	34.3	1.1	35.6
Shopping	2.1	11.1	1.8	13.4	1.7	11.9
Family & Personal	2.0	13.8	1.8	17.0	1.7	21.4
School & Church	2.0	5.2	2.1	4.1	1.7	4.5
Social & Recreational	2.4	27.3	2.1	30.0	2.1	25.8
Other	2.0	3.3	1.9	1.2	1.5	0.8
Average Occupancy/Total	1.9	100.0	1.7	100.0	1.6	100.0

Source: U.S. Office of Highway Information Management

Note: v-km (vehicle-kilometers) is reported in the U.S. as vehicle miles traveled.

One cannot simply throw out seats and affect energy intensity. The entire vehicle must be downsized accordingly. Size and mass can be reduced by the greatest amount when seating capacity is restricted to two. If seating capacity is increased to three, for example, vehicle size becomes roughly equivalent to that of a four-seater. Two-occupant and single-occupant layouts therefore offer the greatest opportunity for reducing vehicle size and mass. Although two-passenger and single-passenger cars are generally assumed to have marginal utility, this is not supported by vehicle use patterns.

NPTS distribution of vehicle trips by number of occupants reveals that a two-place vehicle will accommodate 87.2 percent of all trips, which accounts for 83 percent of all passenger car vehicle-kilometers traveled in the U.S. (Figure 2.1). Trip diaries maintained during the Mobility Enterprise studies at Purdue University in the early '80s revealed that 91.1 percent of the participants' trips were made with two or fewer occupants.[3] In Los Angeles, approximately 80 percent of all commuting trips by private car take place with a single occupant aboard. According to GM's Al Sobey, focus groups conducted in Los Angeles revealed that commuters would actually use (not could use, but would use) a single-place vehicle for 20-30 percent of their commuting trips.[4] When vehicle seating capacity is increased above two, benefits in terms of increased trip accommodation become increasingly marginal. For example, a three-place vehicle will accommodate only 7.4 percent additional trips, which will result in a vehicle that is adequate for 94.6 percent of all trips. Expanding vehicle capacity from three to four occupants results in an additional 2.4 percent trip accommodation, which increases the total accommodation to 98 percent of trips. To accommodate the final two percent requires vehicles of varying capacities, up to, and even greater than nine occupants.[5]

Driving over-powered cars and chronic under-utilization of multipurpose vehicles are the two most wasteful habits affecting the world's consumption of transportation energy, of which about 95 percent comes from petroleum. Using vehicles that fit driving patterns will greatly improve vehicle utilization and significantly reduce the energy intensity of personal transportation.

Figure 2.2 compares the energy intensity of a hypothetical commuter car with that of a full-size car in various trip modes. The two-place commuter car is assumed to have fuel economy of 42 km/L (100 mpg). For a vehicle in the 350-450 kg range (770-990 lb), this is well within the capability of existing technology. The comparison vehicle is a standard 1990 Chevrolet Caprice; EPA rated at 26 mpg (11.03 km/L) highway and 17 mpg (7.22 km/L) city. Energy consumption does slightly increase as payload is added; however, its effect on fuel consumption is minimal. Fuel consumption is therefore assumed to remain unchanged, regardless of the load. Essentially, fuel consumption is

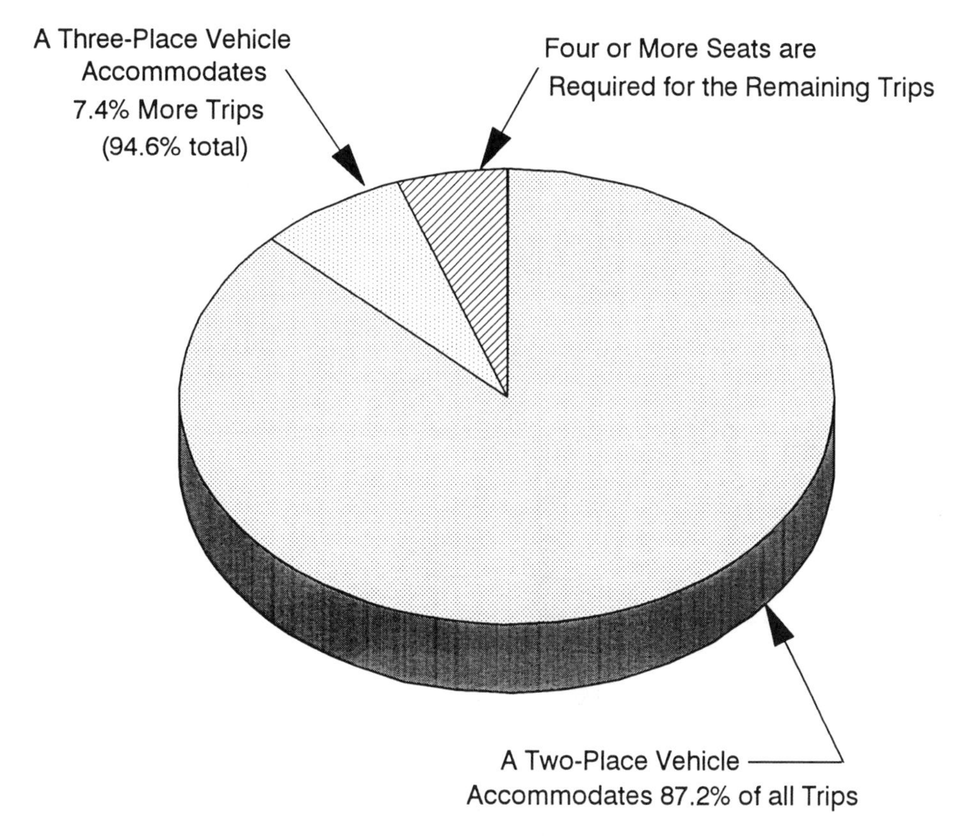

Fig. 2.1. Vehicle Capacity to Accommodate Trips. (Source: U.S. DOT, Federal Highway Administration)

determined in advance by overall system design, then occupants come along almost for free.

The commuter car excels at reduced load factors and in city driving. As passenger count decreases, the commuter car's advantage increases. In the fully laden highway mode, however, the commuter car's energy advantage over the full-size car drops to about 35 percent, on a per passenger kilometer basis. Nevertheless, its reduced fuel consumption still produces significant savings. Factored into U.S. national energy consumption, fuel savings even at the reduced highway advantage would still displace approximately one quarter of all foreign oil imported in 1991. Undoubtedly, the extra energy budget of the full-size car underwrites its greater mass, as well as its presumably greater use of power-assist features and space conditioning. This may not be a poor tradeoff for intercity trips with the entire family, but intercity trips are not appropriate missions for a commuter car. In the suburbs or city, during single-occupant

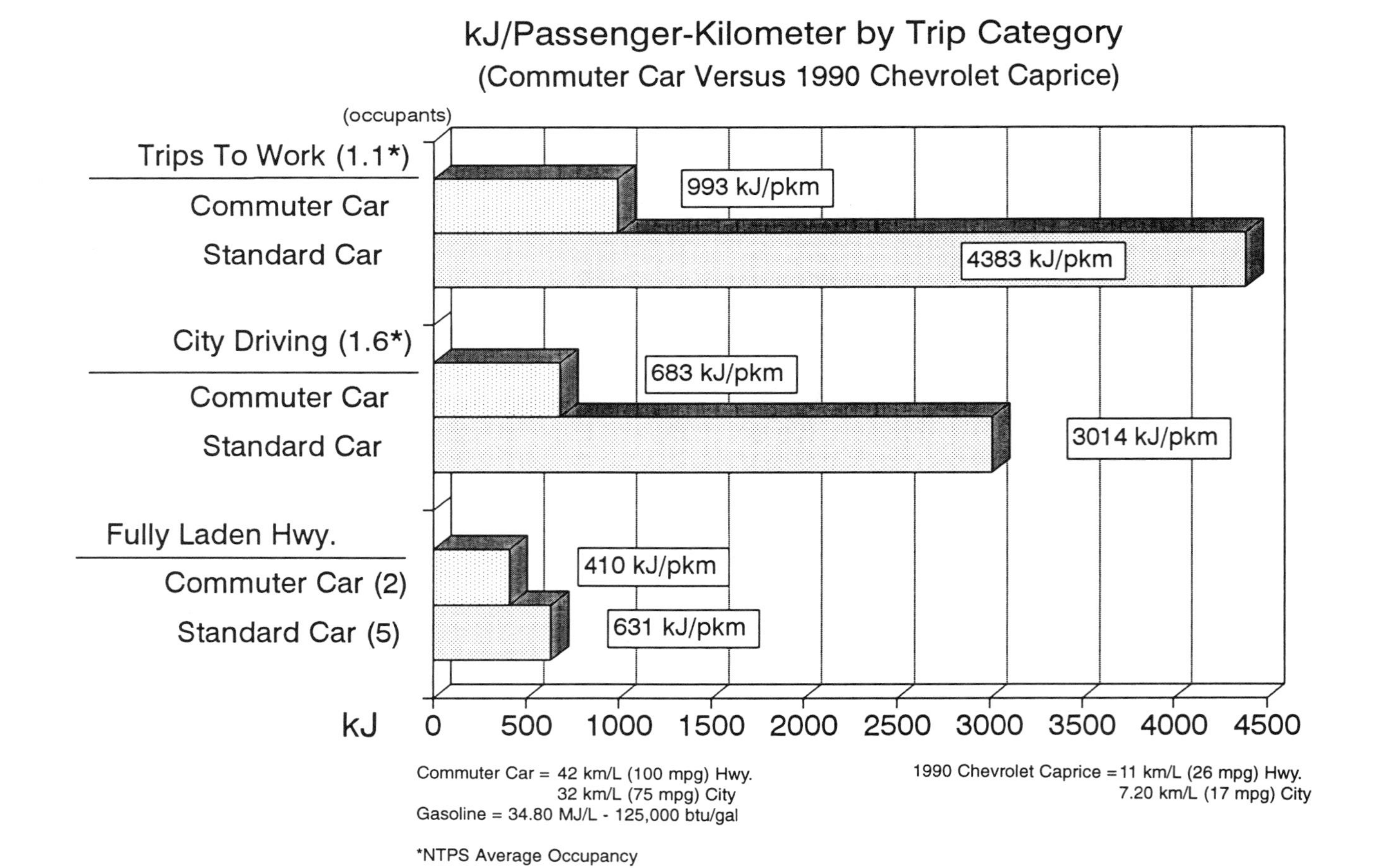

Fig. 2.2. Commuter/Standard Car Energy Comparison.

shopping trips, errands or commutes to work, the under-utilized full-size car is vastly outmatched by the smaller and lighter special-purpose car.

The cost of mobility, in terms of traffic congestion, pollution and resource depletion, can be significantly lowered by reducing the hardware overhead of the private automobile. On the most fundamental level, lower system mass translates into reduced transportation energy intensity. The idea of improving load factors by car pooling is valid but difficult to implement. Travel statistics merely reflect the needs and wants of travelers in an essentially free-choice environment, and choices run strongly against car pooling. Attempts to modify travel habits and preferences have been largely unsuccessful. Government policies do, however, influence local model preferences where consumers are rewarded or penalized according to their choice of vehicles. In Japan, for example, Kei-car (midget car) sales increased from 4.1 percent market share in 1988 to 17.3 percent market share in 1991 when tax reforms and new models encouraged consumers to buy smaller cars.[6]

Commercial transportation has also successfully improved load factors, especially in aviation. Policies have not been successful, however, in changing the load factors of private automobiles. As noted earlier, low load factors are not unique to any particular region. Vehicle occupancy is relatively consistent wherever automobiles exist. And almost universally, travel statistics in developed nations show a trend toward lower, rather than higher load factors. One can therefore assume that automobile use patterns are more fundamental, rather than idiosyncratic. The private automobile, because it is a private and personal means of transportation, encourages low occupancy. Low occupancy rates are consistent with the definition of the product.

The Effects of Small Cars on Traffic Congestion

Improving and expanding transit systems is often mentioned as an option for relieving urban traffic congestion. However, transit systems are not a panacea for urban traffic ills, especially in sprawling U.S. conurbations. As an alternative to mass transit, private cars offer a degree of independence and privacy that cannot be matched, regardless of how well the transit system may be designed. An alternative solution might therefore support the existing system of independent cars, yet reduce the magnitude of their associated problems. Smaller cars can improve traffic conditions in inner cities. But to make a significant difference, cars must be significantly smaller. Moreover, cars designed for the urban environment, that is, urban cars, may be so different from standard automobiles that a new category of personal transportation vehicles will naturally evolve.

Several studies have indicated that smaller vehicles relieve traffic congestion. Increased parking capacity, reduced fuel consumption and air pollution are other benefits of smaller cars. Also, smaller cars are less dangerous to pedestrians (see Chapter Seven), which is especially important in urban traffic where pedestrians and vehicles closely intermingle. Studies by Garrison and Pitstick using the GM Lean Machine as a basis indicated that roadway capacity is significantly improved when very small vehicles are substituted for large ones.[7] They were quick to point out that the actual improvement in traffic flow is greater than the raw geometric effect of smaller vehicles. This is due primarily to the decreased headway for drivers of smaller cars, which was documented by other researchers, including Wasielewski in an earlier report for General Motors.[8] Garrison and Pitstick also cite another study in England where cars the size of GM's Lean Machine provided capacity increases of 10-15 percent in mixed traffic (small and large vehicles together), and 100 percent in segregated traffic (special lanes for small vehicles). In an earlier study in the U.S., McClenahan and Simkowitz also noted that in free-flowing freeway traffic, shorter cars have a small effect on reducing traffic congestion.[9]

In signalized traffic, the effects are much greater. Simple theoretical models developed at the University of Pennsylvania showed that at a single signal, cars 50 percent shorter improved traffic flow by approximately 10-15 percent. However, McClenahan and Simkowitz discovered that benefits compounded in urban traffic because of the synergistic effects of signalized intersections in series. Using a computer model, half-length cars increased traffic flow up to 70 percent, depending on the ratio of small to large cars (see Table 2.2).[9]

TABLE 2.2. THE EFFECTS OF SHORT CARS ON TRAFFIC FLOW AT VARIOUS TRAFFIC DENSITIES

Number of Cars at Light	Percentage of 10 ft (3.05 m) Cars. Remaining cars are 20 ft (6.1 m)									
	0%		10%		33%		50%		100%	
	Flow (a)	Vel (b)	Flow	Vel	Flow	Vel	Flow	Vel	Flow	Vel
5	1154	13.2	—	—	—	—	—	—	—	—
10	955	6.3	—	—	—	—	—	—	—	—
15	732	3.5	745	3.4	855	4.0	1005	4.5	1240	5.5
20	745	3.6	—	—	—	—	—	—	—	—

(a) Flow converted to vehicles per hour of green light.
(b) Velocity in ft/s (1 ft/s = 0.305 m/s).
Source: Transportation Science

Land use is another important factor that normally goes unrecognized. As individuals, we would surely resist the idea of giving up half our valuable real estate. As a society, that is exactly what automobiles do to our cities. An enormous amount of real estate is dedicated to the automobile infrastructure. The average city, worldwide, devotes about one-third of its land mass to the automobile. In the U.S., nearly half of all urban space is dedicated to parking, servicing, and providing roadways for automobiles. Garrison and Pitstick presented ideas for increasing roadway and parking lot utilization by restripping existing facilities to take advantage of vehicles the size of GM's Lean Machine.[7] Smaller cars can free up expensive land for other productive uses. With no change in land apportionment, smaller cars can relieve congestion and increase capacity at no additional cost.

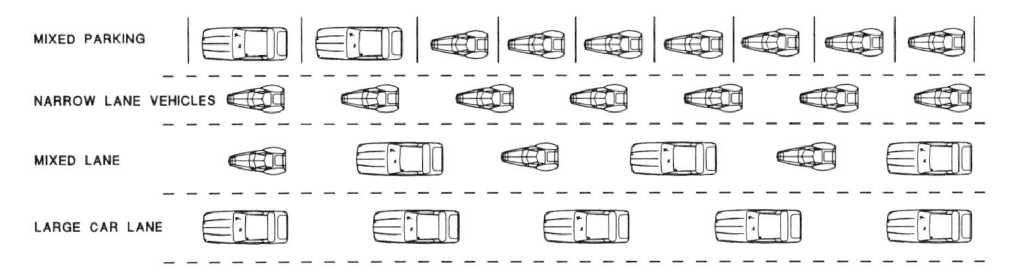

Fig. 2.3. The Effect of Short Cars on Free-Flowing Traffic.

The Effect of Low-Mass, Fuel-Efficient Vehicles on Emissions

Given equal manufacturing materials and processes, low-mass cars require less industrial input and therefore produce less of the by-products of industrial processes. Reducing mass by two-thirds (to 450 kg, for example) may not actually reduce manufacturing input by a full two-thirds. However, there is a strong correlation between product mass and the resources required to manufacture, finish and deliver essentially identical products.

Given equal technology, low-mass, fuel-efficient cars produce fewer harmful exhaust emissions. Again, all harmful exhaust emissions do not necessarily decline in step with improved fuel economy. Carbon emissions are essentially proportional to fuel consumption, and there is a strong correlation between fuel consumption and other harmful exhaust emissions, provided vehicles of equal technology are compared. Operating cycle also has a large effect on emissions. The EPA Urban Driving Cycle includes approximately 18 percent time at idle,

39.7 percent time accelerating, 22.7 percent time decelerating at close throttle, 12 percent time decelerating under power, and only 7.6 percent time at a steady-state speed.[10] Periods of acceleration and deceleration produce extremely high exhaust emissions, and time at idle produces emissions with no benefits in distance traveled. Table 2.3 shows the effects of driving mode on exhaust emissions. Most urban driving time is spent in operating modes that produce extremely high emissions. Smaller cars with smaller engines directly reduce emissions during all driving modes, but most importantly during acceleration and deceleration where emissions are highest and reduced vehicle mass and engine size have the most direct effect on both.

TABLE 2.3. EXHAUST EMISSIONS IN VARIOUS DRIVING MODES U.S. in parts per million

Driving Mode	Hydrocarbons (HC)	Carbon Monoxide (CO)	Carbon Dioxide (CO_2)	Oxides of Nitrogen (NO_X)
Idling	1.34	16.19	68.35	0.11
Acceleration*				
0-15 mph (0-24 km/h)	536.00	2997.00	10,928.00	62.00
0-30 mph (0-48 km/h)	757.00	3773.00	19,118.00	212.00
Cruising*				
15 mph (24 km/h)	5.11	67.36	374.23	0.75
30 mph (48 km/h)	2.99	30.02	323.03	2.00
45 mph (72 km/h)	2.90	27.79	355.55	4.21
60 mph (96.5 km/h)	2.85	28.50	401.60	6.35
Deceleration*				
15-0 mph (24-0 km/h)	344.00	1902.00	5241.00	21.00
30-0 mph (48-0 km/h)	353.00	1390.00	6111.00	41.00

Source: "Automotive Air Pollution: Issues and Options for Developing Countries," The World Bank.
* Metric speeds in parentheses inserted by the author.

Regardless of vehicle use patterns and the ability of small cars to reduce traffic congestion and harmful emissions, if consumers perceive that there is essentially no difference between a large, multi-purpose vehicle and a small mission-specific car, purchasing decisions will likely be made according to old habits. In order for purchasing to match driving patterns, industry and consumers alike must experience a paradigm shift regarding the defining values of the automobile. They must see the automobile in a new perspective. Such a shift in consumer attitudes can be created by vehicle design and marketing approach. Without a new perspective on transportation design, consumers are likely to continue according to past behavior; to purchase oversized cars and then to under-utilize them.

Ultralight Automobiles—A New Category of Vehicles

Future designs may include new categories with cars that are vastly different from today's multi-purpose automobile. Names like "shopper," "neighborhood car," "narrow-lane vehicle," "commuter car" and "urban car" already have become part of the vernacular of a growing new wave of alternative car proponents. Definitions are abundant, and often the nomenclature means whatever it means to the person who is using it. When one speaks of a commuter car or a narrow-lane vehicle, what kind of vehicle does that describe? Another question has to do with the traditional family sedan. At some point a standard car may no longer be *the* standard car. Ideas vary and definitions tend to overlap. In general, vehicle types might be categorized as follows:

1. Passenger Car: A multipurpose automobile having a curb weight greater than 450 kg (990 lb).

2. Commuter Car: A vehicle that is designed primarily for commuting and having a curb weight not greater than 450 kg (990 lb). Must be freeway-capable. Ideally, special Motor Vehicle Safety Standards would apply (see Chapter Seven).

   Specifications:

Maximum Speed:	Up to 120 km/h (75 mph)
Seating Capacity:	Three or fewer
Power:	IC Engine
Fuel Economy:	30-45 km/L (70-105 mpg)

3. Narrow-Lane Vehicle: A commuter car sub-category. A vehicle of significantly reduced tread similar to the GM Lean Machine. May have a seating capacity of one or two occupants.

4. Urban Car: A vehicle that is designed primarily for inner city driving on surface streets and having a curb weight of not greater than 450 kg (990 lb). Also called a city car.

Specifications:

Maximum Speed:	Up to 90 km/h (56 mph)
Seating Capacity:	Two
Power:	Electric or IC Engine
Range:	>80 km (50 miles)

5. Sub-Car — Extremely lightweight and low-powered vehicles that may not be properly classified as automobiles. Examples would include power-assisted bicycles/ tricycles, motorcycles, golf car derivatives and small three- and four-wheel neighborhood vehicles and transit agency station cars.

Passenger Car

No recent design could better define the family sedan of the future than GM's Ultralite. Ultralite is an excellent example of a number of trends in vehicle construction, packaging and personality. Modular packaging, wide use of plastic composites, a compact, fuel-efficient two-stroke engine, adjustable ride height and aerodynamic styling are all part of the formula for the car's year-2000-and-beyond performance profile. The 635 kg (1400 lb) four-place sedan achieves 42 km/L (100 mpg) fuel economy at a steady 80 km/h (50 mph). EPA driving cycles produce 19 km/L (45 mpg) city, and 34 km/L (81 mpg) highway fuel economy. In spite of its diminutive energy appetite, Ultralite can still accelerate from 0 to 97 km/h (0-60 mph) in 7.8 seconds and reach a top speed of 217 km/h (135 mph).

Ultralite (Figure 2.4) seats four adults in a package that is no longer than a Mazda MX-3. Gaping gull-wing doors provide wide-open access to the car's interior, and may be the wave of the future. Ultralite's dynamic shape, small frontal area, and rearward tapering body help produce the vehicle's low 0.192 drag coefficient. Passenger payload can account for 40 percent of gross vehicle weight, which introduces the idea of adjustable ride height to account for payload variables. Once such a system is in place, however, ancillary benefits follow. Ultralite can hunker down on the highway, or adjust front-to-rear ride height relationship as needed to trim the car for minimum aerodynamic drag at higher speeds. Modular construction is exemplified by the self-contained, removable rear power pod. A different drive package can be quickly installed or the existing one can be removed for replacement or servicing. In the future, more vehicles, each with its own personality, will be built on a single platform. Model runs may be lower and economies will therefore come from utilizing a common platform over many different vehicles. Ultralite carries the concept a step further with a power pod that can be changed in the field for service, to switch fuel types, or conceivably to switch from IC engine to electric power for different markets.

Fig. 2.4. The smooth aerodynamic shape and cab-forward layout of GM's Ultralite illustrate a likely trend in future vehicle packaging and styling. (Courtesy: General Motors Corp.)

Ultralite's construction is also on the cutting edge of technology. Made entirely from carbon fiber composites, the body weighs only 191 kg (420 lb). The composite is a hand-laid epoxy/carbon fiber fabric layup that sandwiches a core of urethane foam between outer and inner skins of FRP. The composite sandwich produces an extremely rigid and lightweight body. Automotive application of the FRP/urethane foam composite was pioneered at Quincy-Lynn in the mid-'70s. In those days, we used a much less expensive polyester/fiberglass layup over the urethane core. With either skin material, finishing the surface is a labor-intensive operation which for now makes the system impractical for production. The body must be entirely covered with filler, then hand-finished before it can be painted. Ultralite's epoxy/carbon fiber skin, combined with an array of other high-tech materials, also introduces the problem of high materials costs. The cost of materials alone in the General Motors prototype amounted to $13,000.

Fig. 2.5. Ultralite Body Under Construction: Epoxy/carbon fiber layup over urethane foam results in an extremely light and strong body. (Courtesy: General Motors Corp.)

Commuter Car

As the name implies, a commuter car's purpose is to commute. It is designed for those repetitive one- and two-occupant trips that are typical of the commuting travel profile in most of the world's industrialized nations. The commuter car is built for freeway speeds as well as for urban driving, and is therefore especially suited for the sprawling conurbations typical in the U.S. Because of its performance requirements, the commuter is the most powerful of all the alternative cars, and probably will have the highest curb weight as well. Curb weight may be on the order of 450 kg, and power will most likely be in 18-25 kW range (24-34 hp). An economy car theme is generally presumed, but appeals based on the idea of cost-saving, minimal transportation could spell disaster in the marketplace. A commuter car will likely have more appeal with consumers if it is given a stylish and high-tech personality. Convenience and safety features such as electronic navigation and collision detection and avoidance systems may be standard equipment and thereby promote the car's upscale orientation, as well as contribute to its safety. Molded, flowing interiors, head-up display, and nimble, sure-footed handling would add to the vehicle's high-end theme.

Honda's EP-X concept car is a good example of a commuter car theme, even though the car, as it was conceived, was a bit overweight and overpowered at 620 kg (1300 lb) and 52 kW (70 hp). According to Honda, the EP-X is not just another concept vehicle, but instead represents ideas for future production cars. Occupants are located centrally and in tandem to minimize air resistance and to provide maximum side-impact protection. The interior is fully padded and smoothly contoured to eliminate corners and projections. Front and rear airbags

Fig. 2.6. The Honda EP-X concept car was designed to be highly fuel efficient and fun to drive. The 52 kW (70 hp), 1.0-liter VTEC-E engine utilizes variable lift and variable timing valves. (Courtesy: Honda Motor Company)

protect the occupants. In Honda's words, the car offers "a totally new driving experience in an exciting new package." The appeal is appropriately directed toward the driving experience, not the family spreadsheet. Utility and conservation are not especially marketable attributes. Instead, they serve more as justifications for owning "a totally new driving experience."

Commuter cars by definition are significantly smaller and lighter than passenger cars. Therefore, Motor Vehicle Safety Standards (MVSS) designed for full-size passenger cars may not adequately account for the vehicle's special safety requirements. Because of the performance capability and the freeway operating environment, a special category with Standards specifically developed for commuter vehicles may be necessary. New vehicle categories and improved Safety Standards are more fully explored in Chapter Seven.

Fig. 2.7. The smooth flowing lines of the EP-X interior were designed for comfort and safety. Both driver and passenger are protected by airbags. Note the complete absence of edges or sharp projections. (Courtesy: Honda Motor Company)

Fig. 2.8. Conventional doors are combined with a canopy that tilts to either side for easy access to the tandem seating. The arrangement allows both the driver and the passenger to enter from either side of the vehicle. Exceptionally thick doors provide side-impact protection. (Courtesy: Honda Motor Company)

Narrow-Lane Vehicle

A narrow-lane vehicle fills essentially the same niche as the commuter car, but it is an even greater departure from traditional design. The concept emerged from experiments at General Motors in the early '80s with a vehicle called the Lean Machine. Today, vehicles like the Lean Machine are still being studied as a basis for future products, and transportation officials in California have periodically reviewed such vehicles for their potential to improve roadway capacity and air quality in the L.A. basin.

Lean Machine's passenger pod pivots side-to-side around a longitudinal beam that extends forward from the separate power pod behind. The passenger pod thereby leans into turns like a motorcycle while the power pod corners flat like a conventional car. This ingenious layout produces a very narrow three-wheel vehicle with many of the handling characteristics of a motorcycle and much of the stability of a conventional car. The single-place version shown in Figure 2.9 measures 2616 mm (103 in) long and 1194 mm (47 in) tall. A separate power pod houses the entire drivetrain in a low-profile package within a tread of only

Fig. 2.9. GM's Lean Machine packs awesome performance into a 159 kg (350 lb) package. Motorcycle-like turns produce 1.2g lateral acceleration at 50-degree lean. (Courtesy: General Motors Corp.)

711 mm (28 in). The wheelbase is 1829 mm (72 in) and curb weight is about 159 kg (350 lb). The 11.2 kW (15 hp) engine would push the prototype to about 129 km/h (80 mph) and deliver fuel economy of about 51 km/L (120 mpg) at a steady 64 km/h (40 mph). Engineers at GM computer-modeled an upgraded version with reduced aerodynamic drag and a 28 kW (38 hp) engine. They came up with steady-state fuel economy of more than 85 km/L (200 mpg) and 0-96.5 km/h (0-60 mph) acceleration in 6.8 seconds.

Urban Car

Urban cars, also called city cars, include a category of in-town cars that normally would not have the capability for freeway travel. Quincy-Lynn's Urbacar was based on a similar premise when it was designed in the mid-'70s. Today's urban car would be better represented by Daihatsu's "Move" (Figures 2.10 and 2.11).

Fig. 2.10. The elegantly simple shape of Daihatsu's "Move" Concept Car camouflages the high-tech urban car orientation. The car's electric tractive system, space conditioning, navigational aids, and steer-by-wire controls could provide the 21st-century urban dweller with silent, comfortable and environmentally friendly personal transportation. (Courtesy: Daihatsu Motor Co.)

Because an urban car is specifically tailored to the city environment, space utilization and respect for air quality are primary concerns for the designer. Range requirements are minimal, so ideally, the car will be electric powered. The urban car should be small and space efficient, perhaps with a parking assist feature that would allow one end to retract or to swing the vehicle laterally into a curbside parking space. Recharging might be available at the parking meter and fees for parking and electrical power could be paid electronically through systems already envisioned as part of IVHS in the U.S. and PROMETHEUS in Europe. Navigational aids would also be oriented to the city environment, perhaps showing the most direct route to selected destinations and the location of vacant parking stalls in the vicinity. Infrastructure changes would follow the vehicle into existence. Small, electric powered urban cars could be integrated into the existing environment, then the infrastructure would slowly change to take advantage of new technology.

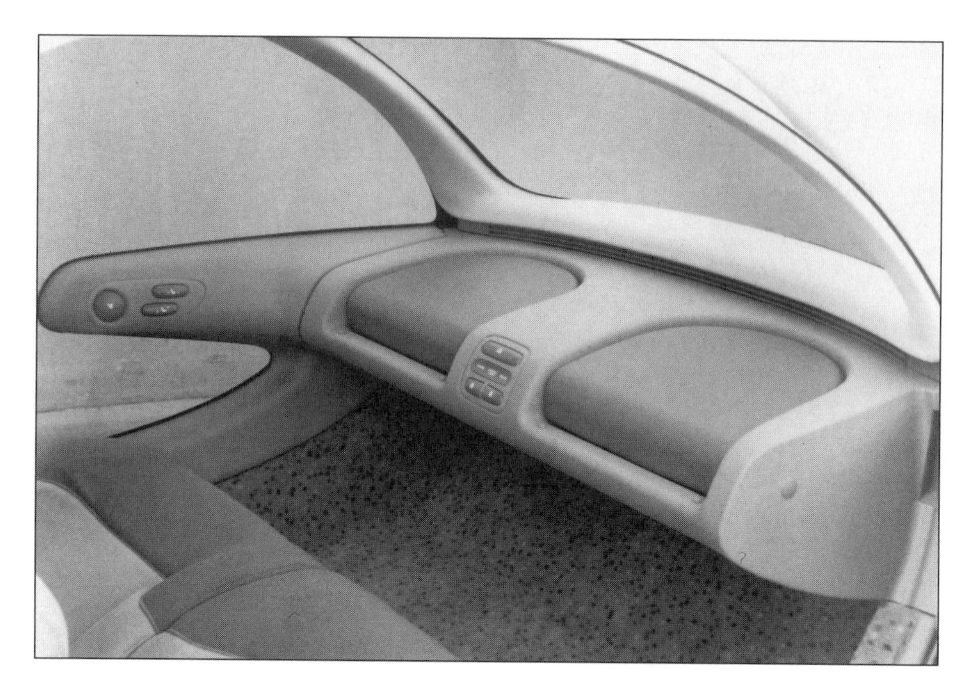

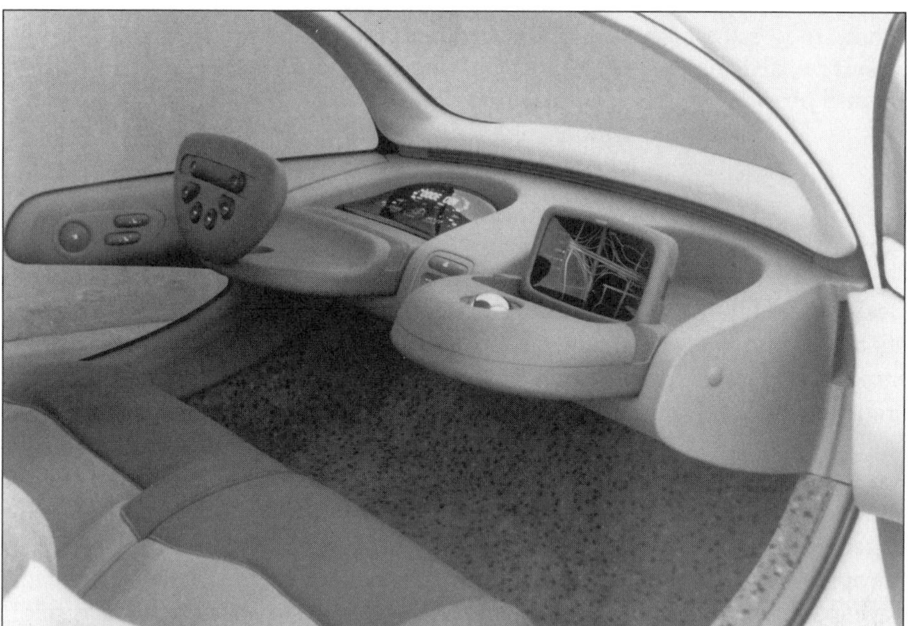

Fig. 2.11. Interior of "Move": Hideaway controls, instruments, and navigational screen open to reveal high-tech electronics and steer-by-wire controls. Note the unobstructed floor. (Courtesy: Daihatsu Motor Co.)

Because urban cars and city cars have limited maximum speed and do not operate on the freeway, a special category with appropriate performance limitations and Safety Standards may be required.

Sub-Car

The sub-car classification applies to a wide range of road vehicles that may not be properly designated as cars. This category would include extremely lightweight two-, three- and four-wheel vehicles designed primarily for local and neighborhood use. Bicycles and mopeds establish the lower extreme and golf car derivatives establish the upper extreme. However, the upper extreme is not

Fig. 2.12. Nexus Human-Assist Vehicle: The 19 kg (42 lb) Nexus incorporates a carbon-fiber frame that entirely encloses the engine and drivetrain. Down-tube serves as the fuel tank with refueling access under the seat. A ceramic-coated, 2.2 kW (3 hp), two-stroke engine runs at very high temperatures for improved efficiency. It propels Nexus to 64 km/h (40 mph) and achieves up to 68 km/L (160 mpg). (Courtesy: Seymour/Powell, London, England)

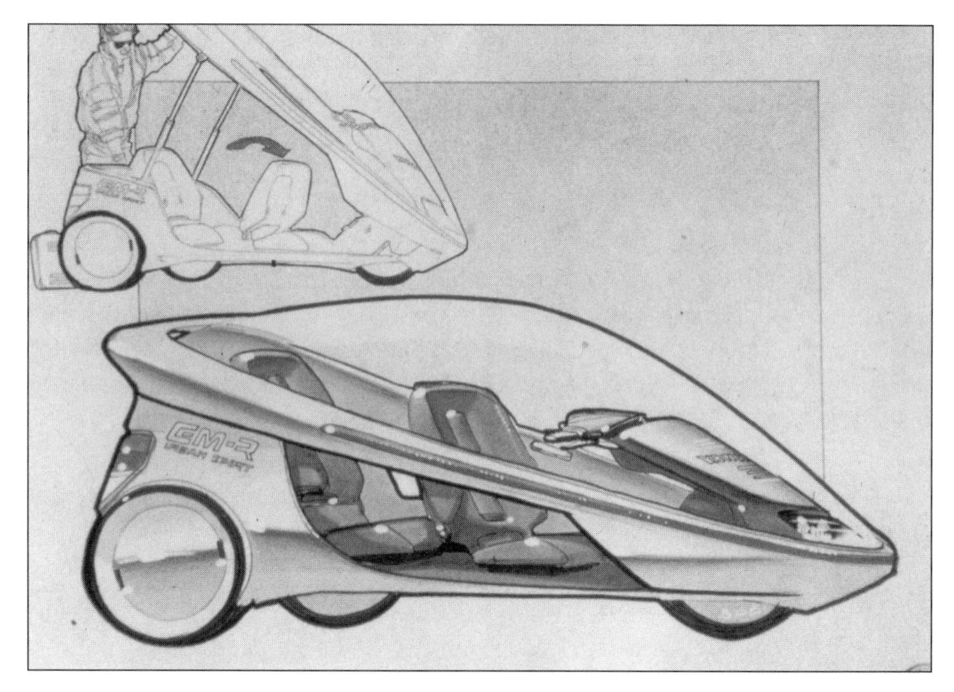

Fig. 2.13. Urban Sport neighborhood car is another concept developed at Seymour/Powell. Because of its small size and limited performance, the vehicle probably would not qualify as an Urban Car. Vehicles of this type might therefore fall into the sub-car category, perhaps as a neighborhood car or local shopper. (Courtesy: Seymour/Powell, London, England)

clearly defined and vehicle types may overlap into other classifications. Although mention of bicycles and golf cars may seem to imply mundane design, this need not be the case. Nexus (Figure 2.12) is a striking example of the bicycle taken to its ultimate powered form. Inspired by the nearly standstill London traffic congestion, designers at Seymour/Powell conceived Nexus as the ultimate efficient and compact personal transportation device.

In most countries, sub-cars are already part of the transportation system, especially in Asia where bicycles are the predominant means of personal transportation. In the U.S., the idea of widespread use of sub-cars brings up significant and far-ranging concerns about liability and safety which must be resolved. Dr. Paul MacCready, designer of Gossamer Albatross, Gossamer Condor, and Sunraycer (and who also assisted with GM's Impact) is leading an ongoing workshop under the code name "Jumpstart." The purpose of Jumpstart is to define sub-car vehicle types, outline potential liability and safety

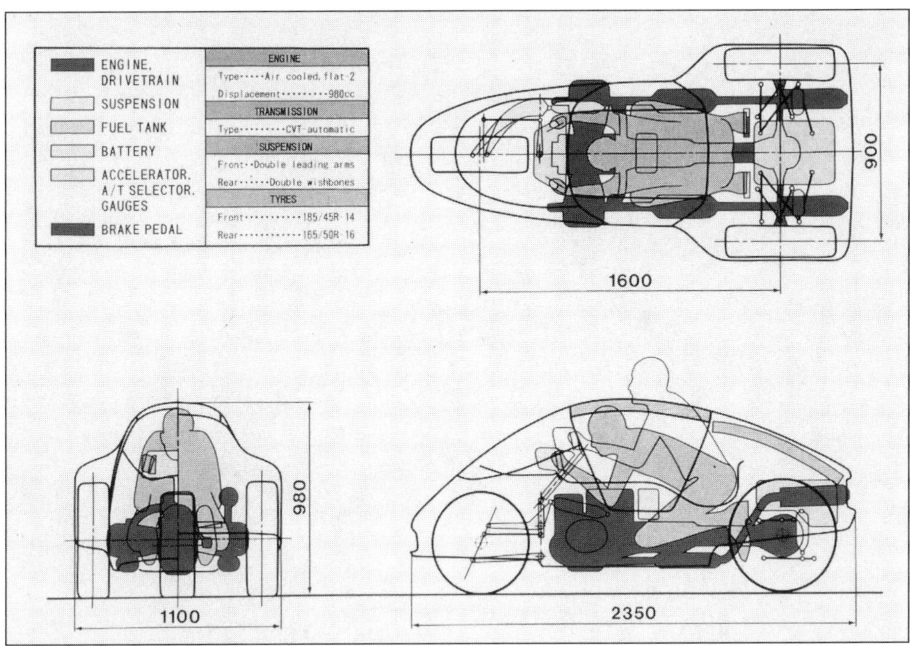

Fig. 2.14. Motor Suite Three-Wheel Sports Cycle is an example of a performance-oriented sub-car. It was designed by Chiaki Maruyama for a contest conducted by Car Styling *magazine, Tokyo, Japan. This three-wheel neighborhood sports-vehicle would be classified as a motorcycle in the U.S., and could therefore intermingle with automobile traffic. (Courtesy:* Car Styling*)*

problems, and develop solutions. If vehicles of this type are encouraged for neighborhood and local shopping trips, they could ultimately displace a significant amount of petroleum. In many U.S. communities, sub-cars (golf cars) are already allowed on certain public roadways where speed is restricted to 40 km/h (25 mph). Mopeds, motorcycles and bicycles are accepted in traffic nearly everywhere, except for minimum power restrictions that apply to freeways. If this category were to become widely accepted as an alternative to larger cars, we can expect to see more vehicles of varying types intermingling with automobile traffic. Jumpstart begins with the premise that such vehicle types can become a safe component of the transportation system. The purpose of the workshop is to define how this might be done. However, the issues are complicated and difficult.

Fig. 2.15. This four-wheel, open-bodied vehicle by Mark Runyan might also qualify as a sub-car. The fact that two- and three-wheel motorcycles are already accepted in traffic with conventional automobiles, while a presumably more stable four-wheel motorcycle derivative is not, seems to ask for a paradigm shift regarding vehicle classifications and safety. (Rendering by Mark Runyan, Courtesy: The Art Center College of Design, Pasadena, CA. Photographic reproduction by Kevin Michael Studio, San Marino, CA)

New Designs and the Marketplace

The idea that smaller, mission-specific vehicles can reduce traffic congestion and conserve natural resources is one thing; selling them is quite another. Consensus within the industry is that the market for special-purpose vehicles and electric cars is extremely limited. Consumers are not enthusiastic about the idea of paying more money for a presumably worse product, just to save a few liters of gasoline. Market estimates that compare vehicle limitations to consumer preferences invariably produce a thumbs down response for a limited-utility, limited-performance car. But such studies are typically based on abstract product ideas and biased by negative hypothetical attributes. They cannot show the market potential of real products that have yet to be developed. Creative product design and innovative appeals are essential if alternative cars are to become popular with consumers. Consumers that reject the idea of limited utility in the abstract, can quickly flip to the other side when a real product offers a new bundle of appeals that might not have been envisioned before. Product clinics with GM's Lean Machine produced about twice the level of favorable responses normally predicted by studies based on abstract product ideas. It would be legitimate to question whether Lean Machine maximized consumer response, or whether a product with more appeal might produce four, six, or ten times the expected favorable response.

The initial market potential for new energy-efficient vehicle types is influenced by many interrelated factors, including those of consumer attitudes and preferences, and patterns of vehicle use. In the final analysis, however, an alternative car's potential for success will be largely determined by the vehicle theme (positioning) and how well the theme is executed. The number of potential sales that are converted to actual sales will then depend on the marketing plan. To the degree that the marketing plan is successful, the new product will exert its own influence on the market as it gains penetration and results will then become synergistic. Ultimately, the new product may influence competitors' offerings and consumer attitudes, and thereby change the market itself. The product's theme and its execution determine the potential of the design for capitalizing on, then influencing, consumer attitudes and motivations.

Vehicle Theme

The vehicle theme establishes the personality of the product. It is the broad product attribute that attracts a particular segment of consumers in the first place. The theme is strongly identified with the purchaser's self image. One segment of consumers may identify with the "independent man-of-the-earth" image and some individuals might carry that out by wearing cowboy boots and driving a four-wheel-drive pickup. A different self image might lead one to the showroom with the Corvettes on display. The basic attraction to vehicle types has to do with the compatibility, and even magnetism, between one's self image and the vehicle theme. That the vehicle is the best choice in terms of utility is often rationalized so that purchases can be made in accordance with self image. Meanings inherent in a product's design go far beyond the product's function.

Peter Dormer, in his fine book The Meanings of Modern Design, discusses the camera as an object of value that transcends its mere function. Dormer describes the camera as having a precision craftsmanship of a quality that can come only from the automated processes of modern production methods: a technology of creating products that goes far beyond the flawed labors of human hands to create a pristine beauty that is uniquely its own. The feel of the camera in the hand, the sound of the shutter as it measures time in thousandths of a second, the immaculately detailed case, and the crystalline purity of the lens are all imbued with the same flawless precision and add to the pleasure of owning and handling the product. According to Dormer, the technical capabilities of the tools of photography far exceeds utility to become an end in itself. The actual photographs become almost a secondary benefit. Photography's accouterments embody a theme that gives the consumer a pleasurable feeling whenever he or she partakes in it. Cars speak to consumers in the same subliminal language.

New energy-efficient vehicle types must have an intrinsic appeal of their own. Conservation and sacrifice are not marketable themes. Energy-efficient vehicles cannot be cheap imitations of "real" cars. The design should exemplify an upscale chic, perhaps like the image of casual designer clothes that everyone knows are not really casual at all. The modern road bicycle is certainly environmentally friendly and energy efficient, but it speaks to the owner's appreciation of quality and fine workmanship in its appeals. A feature by feature tour of a carbon-fiber-framed high-performance bicycle is likely to awaken new eyes of appreciation for beautiful, simple design and attention to detail. Alternative vehicles should also have a defining aesthetic appeal. Designs might convey a sense of solid, high-tech competency that would elevate the product's perceived value. The vehicle should not step down to accommodate scarcity and conservation. Instead, it must move up to conquer waste and pollution through superior technology. A theme of compact precision and technical transcendence thereby replaces the often-envisioned theme of economy and basic utility. These messages must be obvious when consumers experience the design. The vehicle's "green" orientation might then become a type of signature that increasingly symbolizes the modern, world-conscious man and woman of the twenty-first century. The theme and its execution imbue the vehicle with qualities that consumers want to own. A theme that embodies smallness and energy efficiency is not collinear with a spiritless design. New vehicles must be fun to own, touch, care for, and operate. Aesthetic appeal and pride of ownership are independent of vehicle size and energy consumption.

The vehicle theme thereby establishes a connection with the consumer. It conveys through the language of form and texture an image of what the vehicle is and what it does, and it tells the consumer how the product will be experienced. The vehicle theme lets the consumer know whether to expect an exciting driving experience, or just an uneventful trip to work. It also subtly suggests a price category and thereby makes a statement about its affordability and the prestige of ownership. If the vehicle is less expensive than it appears, the purchaser gets prestige at a bargain price. Creating an enticing vehicle theme, one that positions the personal commuter or urban car apart from the conventional automobile and imbues it with a distinctive image of its own, is perhaps the greatest single challenge to the designer. A motor home is not just a large automobile with beds, nor is a motorcycle a small automobile without a body and with a couple of wheels. These vehicles make an emphatic statement of their own. That statement, carried by the appearance, feel, performance and even the sound of the vehicle, establishes the perspective from which the product is seen. It breathes a personality into the machine.

The personality of the personal commuter or the urban car should define a category that removes it from direct comparison with the standard automobile. The defining values of an automobile already exist in the minds of consumers and the label will transfer those values to the new product. An expectation of price will then have been established by comparison to the price/features profile of the standard automobile. New vehicle types therefore should stand apart and create a new category of products with a price/value profile of their own. In theme, they should not separate buyers from existing cars, but instead become a new addition to the household, a new object of ownership and a new experience on the roadways. New vehicle categories should have new descriptive names. Personal watercraft, for example, are indeed miniature boats, but to label them mini-boats would certainly detract from their appeal, and in the process create a size/price comparison with full-size boats. As for executing the theme, a variety of pertinent vehicle attributes provide a natural opportunity to speak to the consumer in the language of design while resolving a number of engineering constraints.

Seating Layout

Seating capacity will have the greatest effect on the ultimate limits to vehicle downsizing. The human body is a relatively stable package and not much can be done to change it. The number of passengers and the seating arrangement establishes much of the overall size and shape of the vehicle. Choices regarding capacity and layout deal with practical issues, but they can also let consumers know that the new design represents a new approach to personal transportation.

A three-place vehicle might be designed around a 2+1 theme, where behind-the-seat storage space would also serve as an emergency third seat. Providing for three occupants will accommodate 7.4 percent more vehicle trips, in comparison to a two-place design. However, several challenges are introduced with a three-place seating arrangement. The obvious difficulty is that of providing adequate occupant space without unduly increasing the size of the vehicle. If the third occupant were placed in a conventional position behind the two front

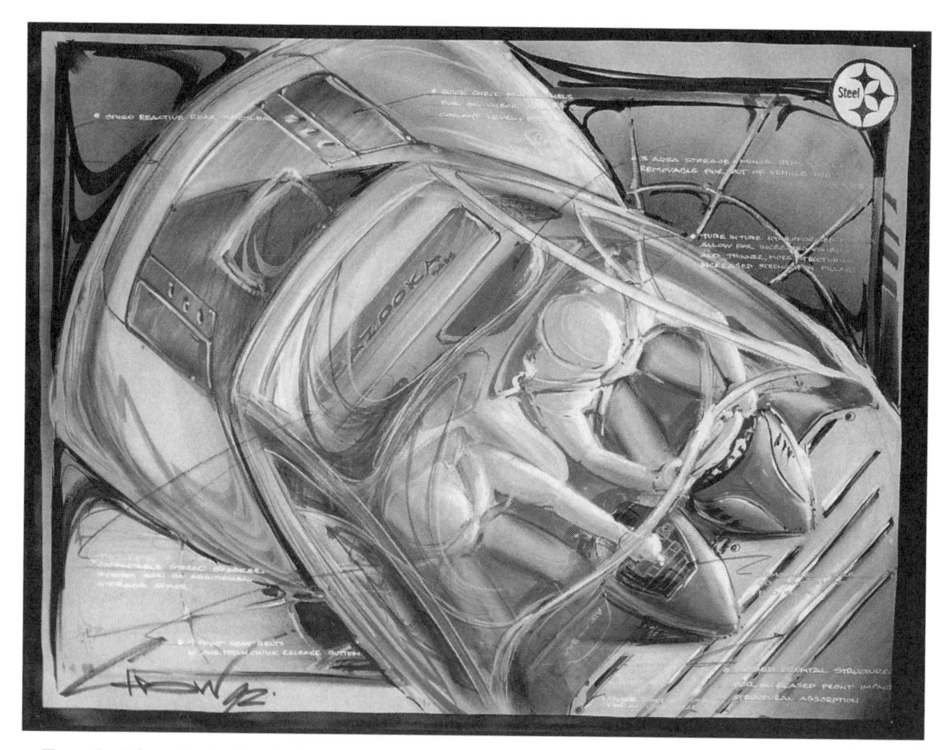

Fig. 2.16. High-Tech Personality Centers on the Occupant. (Rendering by Jimmy Chow, Courtesy: The Art Center College of Design, Pasadena, CA. Photographic reproduction by Kevin Michael Studio, San Marino, CA)

occupants, the passenger compartment might easily double in size. Facing the third occupant rearward can improve space utilization, but it still may not entirely resolve the problem. Room for a fourth seat is likely to be a natural by-product.

Another important consideration of a three-place layout is the 50 percent increase in payload mass over a two-place layout. With a 450 kg (990 lb) vehicle, for example, a payload on the order of 240 kg (three 80-kg occupants) could detrimentally affect ride height and even change the vehicle center of gravity. Wide load variations tend to create difficulties in controlling ride, handling and performance characteristics throughout the vehicle's load range. Adjustable ride height is often considered as the only alternative to an otherwise overly stiff suspension. However, compromises can produce acceptable results. Quincy-Lynn's Urbacar, at 295 kg (650 lb), had an acceptably stable ride height and a relatively smooth ride with either one or two occupants aboard. Urbacar's payload and vehicle mass relationships are roughly equivalent to the hypothetical 450 kg (990 lb), three-place vehicle described above.

A two-place vehicle simplifies the design task and still results in a seating capacity that will accommodate approximately 90 percent of all vehicle trips (in the U.S.). Occupants may be located side-by-side or in tandem. Several interesting designs have been developed with tandem seating. Ford's Cockpit and Honda's EP-X

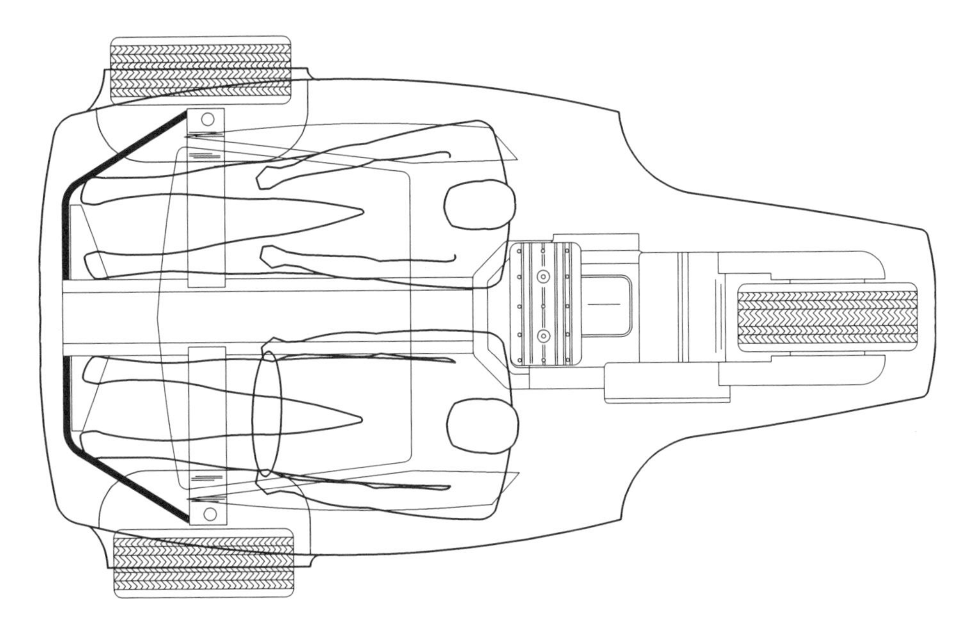

Fig. 2.17. Sitting two abreast limits longitudinal center of gravity displacement between one- and two-up loads.

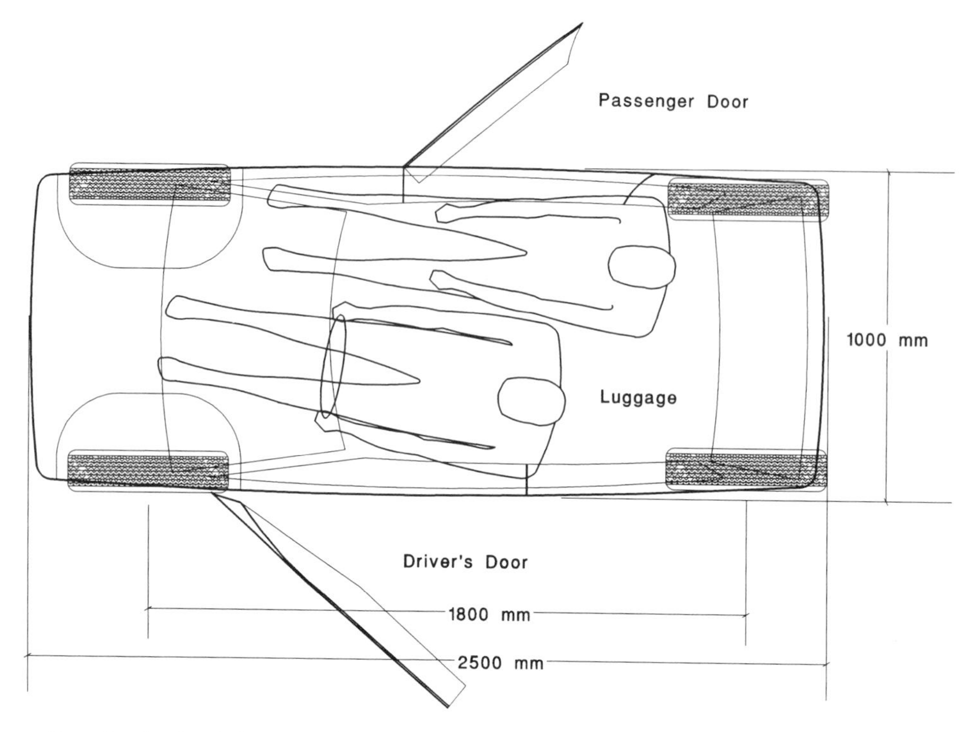

Fig. 2.18. 1+1 Seating Arrangement: By displacing the second occupant slightly toward the rear, a relatively narrow vehicle provides adequate shoulder room.

concept cars are excellent examples. The primary advantage of tandem seating is that it reduces frontal area by about 40 percent. The disadvantage comes from the greater longitudinal displacement of the center of gravity between one- and two-passenger loads. A longitudinally stable center of gravity is an important consideration for a three-wheel design, but not particularly important with a four-wheel platform.

The "1+1" concept is also a promising option (see Figures 2.18 and 2.19). Such a vehicle is essentially a one-place machine with a minimal second seat. GK Industrial Design Associates in Japan produced a 1+1 microcar design as early as 1975. Several mockups were built and a final prototype was designed under the code-name of GK-0. From the Japanese perspective, space is a valuable resource. This idea ultimately leads to the concept of "small, therefore good," which is a theme that runs throughout the range of Japanese products. The GK-0 utilized a slightly staggered seating arrangement that produced a sense of side-by-side seating without the crowded feeling. The resulting vehicle package was exceptionally space-efficient.

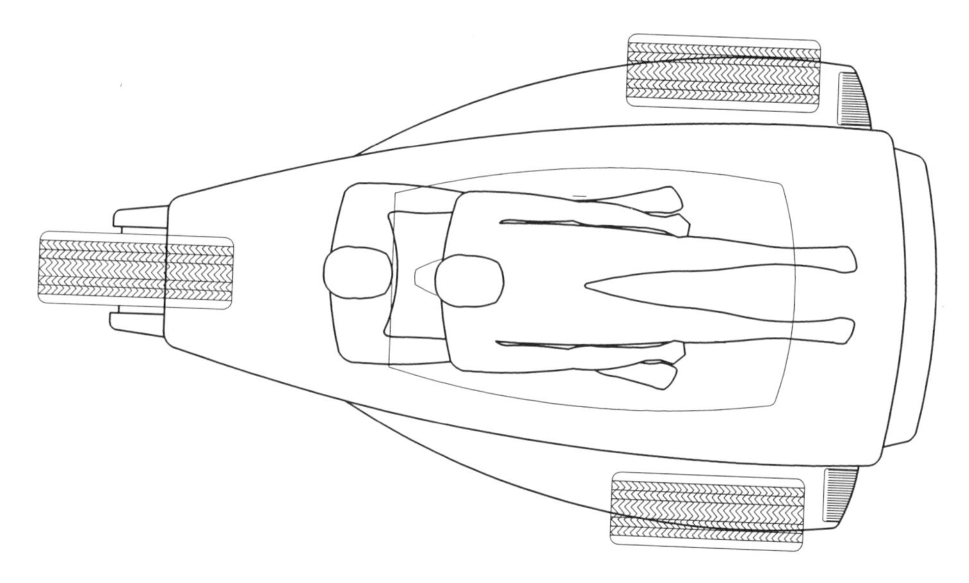

Fig. 2.19. A Condensed Tandem Arrangement: Space is efficiently utilized in a 1+1 layout by allowing the rear passenger's legs to extend along the outside of the driver.

Rumble seats and longitudinal bench seats that allow occupants to squeeze together and make room for an extra passenger are other options. But safety issues will undoubtedly arise with any design that incorporates unconventional seating. Occupant crash protection must be provided equally for all potential passengers. Protecting an occupant in a minimal jump-seat or a rumble seat will likely present difficult engineering challenges. In this regard, an abbreviated tandem arrangement offers interesting possibilities. The rear passenger in a 1+1 tandem layout could be protected by an energy-absorbing device built into the back of the driver's seat.

Seat construction and even the concept of movable seats are items that deserve re-evaluation. Quincy-Lynn's Tri-Magnum, Trimuter and Centurion (see Appendix) were designed with seats formed into the body in a fixed position. Upholstery was then attached directly to the body over high-density foam. This resulted in an interior that presented an attractive molded appearance. Seating receptacles were form-fitting and comfortable even though they had no traditional suspension system. When the seats are fixed, steering, brake and throttle controls are usually mounted to a movable carrier to provide adjustment.

Ingress and Egress

Ingress and egress also present interesting opportunities for distinctive design. The gull-wing door and the clamshell canopy are attractive and practical alternatives to conventional doors. When a vehicle opens like a jet fighter, an otherwise bland design can become exciting. Many of the Quincy-Lynn vehicles utilized either the gull-wing door or the clamshell canopy. The two exceptions in which conventional doors were utilized were partly for a new experience, but primarily because the unconventional systems were simply not practical for the proposed body design.*

On a more practical level, a very small and low-to-the-ground vehicle presents difficulties for the conventional side-entry door. Maneuvering through a conventional door in an exceptionally low vehicle is not the most graceful way to enter. Consequently, alternative methods should be evaluated. Gull-wing doors provide greatly improved access and they usually can be fully opened even in confined parking spaces. Also, lightweight doors that end several inches above the bottom of the body can be removable to create a convertible effect. Removable gull-wing doors were a feature of Quincy-Lynn's Urbacar, as well as the VW Scooter. Lacking a practical justification, gull-wing doors can still stand on their own as a very aesthetically appealing feature. The Mercedes 300SL, the Bricklin and the Delorean are excellent examples of exotic vehicles that were made more intriguing by the appeal of gull-wing doors. On GM's Ultralite, they are both practical and visually striking (Figure 2.20).

When it comes to providing easy access to a very low vehicle, the unconventional clamshell canopy is one of the most attractive, and in many respects the most practical approach. Bob McKee used the clamshell several years ago on his beautifully executed prototype electric car, the Sundancer. GM has incorporated it into several concept cars, and more recently Ford's Cockpit was built with a clamshell canopy (Figure 2.21). It was also a popular feature with a number of Quincy-Lynn cars. With this arrangement, the entire top of the vehicle hinges out of the way to expose a gaping access to the interior. One simply steps over the side of the body and slides into place.

Perhaps the most important decision with the clamshell design is determining the most appropriate cut-line: the point at which the canopy separates from the body. Extending the canopy deeply into the body sides has the benefit of

---

* Both Town Car and Centurion have conventional doors. However, there were many discussions about alternative door designs for both cars and conventional doors were finally adopted with much reluctance (see Appendix).

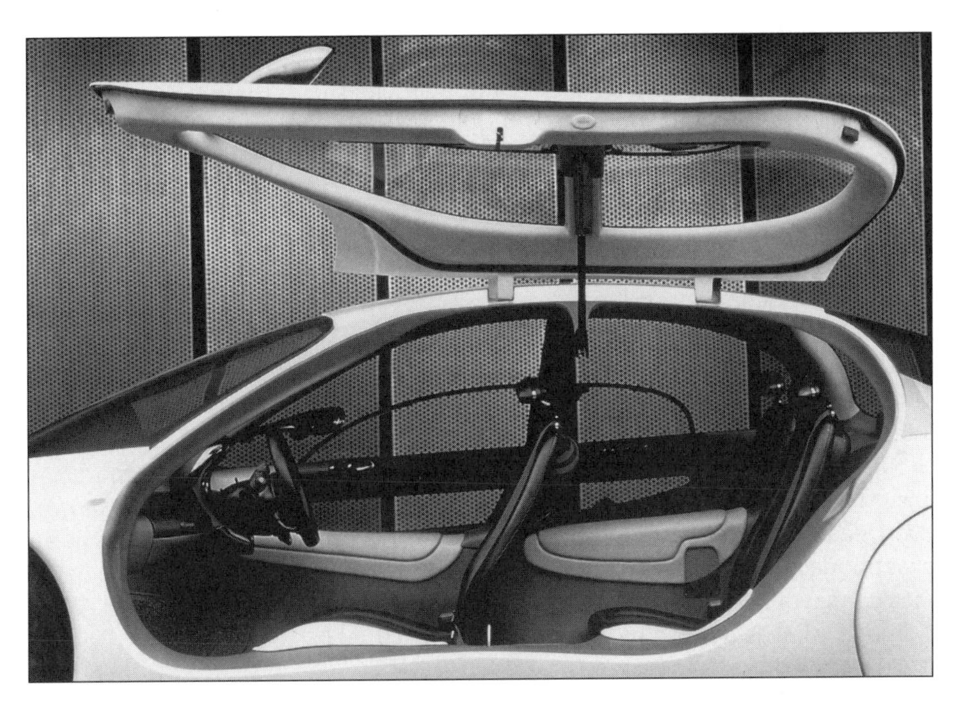

Fig. 2.20. GM's concept car Ultralite utilizes large gull-wing doors to provide a wide, unobstructed entry into the car's interior. (Courtesy: General Motors Corp.)

Fig. 2.21. The tilt-up clamshell canopy provides wide-open access to the interior. Occupants are seated in tandem. (Courtesy: Ford Motor Co.)

reducing the step-over height. In other respects, however, a deeply cut canopy can produce negative results. A canopy cut deep into the body sides compromises structural integrity, creates sealing problems around the perimeter and produces a heavier canopy. Also, the resulting canopy may end up with unnecessarily deep sides that extend into the free access space and actually interfere with an otherwise unobstructed path of entry. Ideally, the canopy should separate at the beltline, or further down along an appropriate character line. The perceived difficulty in stepping over a high body side is often largely overrated. Tri-Magnum's canopy was cut at the beltline which resulted in an exceptionally high 711 mm (28 in) step-over. The high sill gave the car a strong race car flavor that actually enhanced the design's sporty feel. The height of the Trimuter sill was a more moderate 483 mm (19 in) (see Figure 2.22).

Gull-wing and clamshell systems do have disadvantages. With both systems the weight of the door must be counterbalanced in order to open easily. Also, space for retractable windows usually is not available. Windows therefore must be removable, hinged or sliding—all less than ideal solutions. Probably the most

Fig. 2.22. Quincy-Lynn's Trimuter (top) and Tri-Magnum (bottom) incorporated the clamshell canopy for both practical and styling purposes. Occupants are seated side-by-side in both three-wheelers.

difficult problem has to do with safety during a rollover. A door that opens upward will not open when the vehicle is resting on its roof.

Gull-wings and clamshells are not the only options. Subaru's EM-1 and Jo-Car (Figure 2.23) dispense with doors entirely and in their place have simple side-beams designed to keep occupants in and other vehicles out.

Fig. 2.23. Subaru's Jo-Car (top) and EM-1 (bottom) dispense with doors. (Courtesy: Subaru of America)

Modular Design

Modular design is a concept that has been permeating the universe of consumer and industrial products for many years. Today, products are increasingly designed around a basic reconfigurable platform in order to save costs and broaden the product line. Styling often emphasizes, rather than disguises a product's modularity. In this way, the designer uses the language of shape and form to speak to the consumer about the product's capabilities. Modularity also suggests a type of efficient engineering that can imply a respect for the environment. Concept drawings of the Esoro E301 electric urban car developed by the Swiss firm Esoro AG in Zurich shows a basic platform that can be reconfigured to produce as many as five different vehicles, three of which are shown in Figure 2.24.

Ford's Zig and Zag are two vehicles with completely different orientations (Figure 2.25). However, both vehicles are built on the same platform, which is essentially reconfigured above the beltline to produce very different products.

The modular aspect of GM's Ultralite is emphasized by the strong cut-line around the rear. Alternative cars will undoubtedly incorporate reconfigurable body styles and alternative drive packages that allow the user to select the fuel system according to preference or local fuel policies. A totally modular vehicle might consist of a central passenger pod, a front suspension/steering pod and a rear power pod. Modules could then be interchanged to customize the vehicle for different markets or different consumer preferences. Figure 2.26 shows a totally modular vehicle designed by Jimmy Chow while studying at the Art Center College of Design in Pasadena, CA.

Flexible/Plastic Body Panels

Conversion to plastic body components has been underway for many years and is gaining momentum. Major incentives include:

- Corrosion and damage resistance
- Weight reduction on the order of 30-50%
- A 30-50% reduction in annual tooling costs
- Low-cost facelifts
- Increased styling freedom
- Conducive to just-in-time manufacturing at or near assembly plants

Plastic components account for approximately 10 percent of the mass of today's automobiles and the applications are growing each year. Plastics are especially

Fig. 2.24. The Esoro E301 was developed by Esoro AG, Switzerland. A finished prototype is shown in Chapter Five. Esoro E301 utilizes a basic two-seater platform that can be configured as a pickup or van, or as a coupe, hard-top, or even a four-seat passenger car. (Courtesy: Esoro AG)

Fig. 2.25. Ford's Zig and Zag: A perfect example of design commonality between two very different vehicles. (Courtesy: Ford Motor Co.)

compatible with the concept of ultralight alternative cars. Outer body panels, particularly at the front and rear, and even the fenders, might be flexible polyurethane or a similar compound that will resist damage from minor contacts

Fig. 2.26. An illustration by student Jimmy Chow while at the Art Center College of Design shows an idea for a totally modular automobile. Front suspension, occupant compartment, and rear power pod are self-contained modules with controls that interface electronically, rather than mechanically. The vehicle can be reconfigured for different markets and operating environments by combining a variety of different modules. (Rendering by Jimmy Chow, Courtesy: The Art Center College of Design, Pasadena, CA. Photographic reproduction by Kevin Michael Studio, San Marino, CA)

with other vehicles. Honeycomb structures and rigid foams can create light, energy-absorbing sub-structures for appropriate energy management at key locations. Flexible front-end components that are more kind to pedestrians also are possible. In the future, the use of molded foam/vinyl laminates and integral skin foams will be expanded to include even the seating areas. Conventional fabricated upholstery could be replaced with thermoformed panels made from laminates or from integral skin foams. New materials and processes often reduce manufacturing costs and make the product more aesthetically appealing. The option of inexpensively replacing worn upholstery panels could help promote the product's ecological orientation.

The Electronic Automobile

Once microprocessor technology is introduced, a new world of opportunities becomes available to the designer. Electronics opens the way for new approaches to systems management, driver information, collision avoidance,

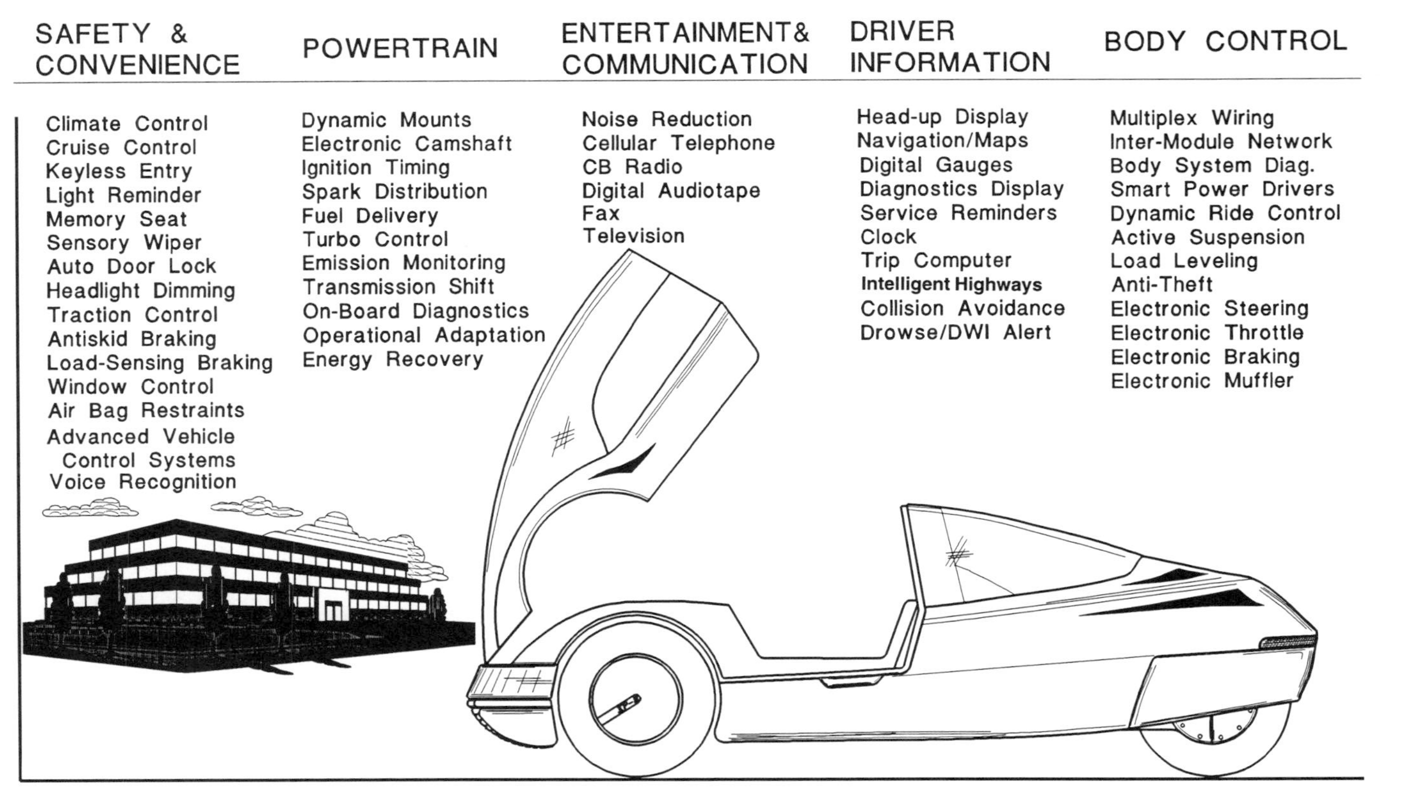

Fig. 2.27. The Electronic Automobile.

and active controls that were not possible before. In addition, significant psychological benefits are intrinsic to modern electronics. Microprocessor technology offers a precise and visually impressive interface with the operator that is uniquely its own. Vehicles now come to life in ways that go far beyond the task of providing information and control. Subtle qualities of sight and sound can create a sense of vehicle competency, adroitness, and even luxury. Audible communication, head-up display, and voice activation and identification are tools for adding perceived value to the product and more fully executing the vehicle theme. Active controls such as steer-by-wire and drive-by-wire replace mechanical connections with electronic terminals making it possible to more fully implement the concept of modular design. Ultimately, vehicle electronics may supplant occupant protection systems as collision avoidance, and advanced vehicle control systems make automobile crashes a thing of the past.

"Active Steering" is an interesting idea that has been around for some time in various forms. It is mentioned here because ultralight designs are especially well suited to electric power steering, and because it speaks so well to the subject of vehicle ambiance. Today's steer-by-wire system is like one being developed at Saab. A joystick takes the place of a conventional steering wheel and steering is accomplished by an electrohydraulic steering servo. Other versions utilize more conventional steering yoke or wheel connected to an electromechanical

Fig. 2.28. Electrohydraulic joystick steering system relies on system redundancy and self-testing and monitoring to provide extreme reliability. (Courtesy: Saab Cars, USA, Inc.)

power steering unit. Perfected steer-by-wire could improve vehicle safety by eliminating the lethal steering column.

Intelligent vehicle highway systems, advanced vehicle control systems, navigation and communication electronics will work together to create attitudes and perceptions that are likely to eliminate any remaining perceptual association between vehicle size and vehicle quality. Like fine watches and personal computers, compact vehicle packaging is more likely to become synonymous with superior engineering and high product quality.

The Potential Market for Alternative Cars

The best that can be done regarding market potential is to speak in terms of estimates and general market trends. Estimating market potential for new products is especially difficult because both buyers and sellers lack experience. There is no sales history on which to base traditional analyses, and consumer surveys can produce inaccurate results when consumers have no product experience to guide them. Moreover, the waters get extremely murky when the product is hypothetical rather than actual. Market potential cannot be determined in the abstract. Much depends on consumer response, not to the idea, but to the actual product.

Overall, trends in the U.S. show a movement toward smaller cars, despite the resurgence of luxury and performance cars. John Hemphill of J.D. Power and Associates suggests that the traditional family automobile may be losing ground because of the growth in multi-car households and declining family size.[11] Trends point toward a greater demand for personal transportation and less demand for family transportation, which appears to open the door for smaller, more specialized vehicles. According to Dr. Charles Lave, professor of Economics, University of California at Irvine, a U.S. study of the demographic factors that influence people's willingness to buy both imported cars and small cars indicated that decisions are overwhelmingly tied to three important factors.[12] The most important factor was family education. More highly educated households are more likely to buy a foreign or a small car. The second most important factor was whether or not the household had multiple cars. As the number of cars in a household increased, members were more likely to buy a foreign or small car. The third most important factor was age. Older people buy larger cars and younger people buy smaller cars. A market study by the Institute of Transportation Studies, University of California at Davis, revealed that the "green market" segment and "moral choices" regarding purchases are becoming increasingly more significant factors. When participants in a product clinic were asked what would motivate them to purchase an alternative fuel vehicle (AFV), 28 percent answered that they "would purchase (an) AFV because it's (the) right thing to do."

Demographic factors and trends, as well as consumer preferences and attitudes regarding conventional cars are probably not directly applicable when radically different transportation products are discussed. Undoubtedly such information can help refine one's judgment. However, traditional market analyses are likely to produce inconclusive and even conflicting results when considering unconventional products. Sub-cars, for example, are so far removed from the traditional automobile product that automobile demographics and indicated market potential may be largely inapplicable.

Vehicle Ownership and Use Trends

Households are becoming smaller and people are driving more. This trend is consistent throughout the industrialized world. In the U.S. from 1969 to 1990, persons-per-household have decreased by 20 percent and vehicle miles traveled have increased by approximately 22 percent. Households have fewer members, they have more cars available to them, they are making more trips, and driving alone more often.[2] In 1990, average vehicle occupancy was 1.6 persons during all trips, and 1.1 persons during trips to work. The NPTS report also reveals that

TABLE 2.4. U.S. HOUSEHOLD TRAVEL RATES IN MILES
(kilometers inserted in parentheses by the author)

	1969[(1)]	1977	1983	1990
Annual Vehicle Trips	1396	1443	1486	1702
Annual Vehicle Miles Traveled	12,414 (19,974)	12,035 (19,364)	11,739 (18,888)	15,096 (24,289)
Annual Person Trips	2322	2808	2628	2707
Annual Person Miles Traveled	22,465 (36,146)	24,919 (40,947)	22,802 (36,688)	24,816 (39,929)
Persons Per Household	3.2	2.8	2.7	2.6

(1) In the '69 survey, only auto and van trips were collected as private vehicle trips. In '77, '83 and '90 surveys, the definition of private vehicle was expanded to include pickups and other light trucks, recreational vehicles, motorcycles and mopeds.
Source: USDOT, Federal Highway Administration

in 1990, 88.1 percent of person miles traveled (PMT) in the U.S. were aboard private vehicles and 67.1 percent of all vehicle miles traveled (VMT) occurred with one occupant aboard. Approximately 83 percent of the nation's driving was done with only two occupants aboard.

The Mission-Specific Car and Multi-Vehicle Households

Most analysts agree that the market for a mission-specific or limited-utility car will come from that segment of households with multiple vehicles. In single-vehicle households, it is presumed that a single car has to serve the broadest possible utility. Therefore, a multi-passenger conventional car is the car of necessity. Today, multi-car households are common and the type of car purchased often depends on preference, rather than utility. Many consumers have different models for different uses and different occasions. When two or more cars are available, criteria other than maximum utility may apply to the second, third and fourth vehicle. Statistics indicate that nearly 60 percent of U.S. households already have at least two vehicles, and about one third of that group have three or more. In Europe, about 20-30 percent of households have two or more vehicles. A mission-specific commuter/urban car would therefore target this universe. Table 2.5 shows the acceleration in multi-vehicle ownership in the U.S.

Potential market is likely to be some percentage of total multi-vehicle households. NPTS figures combine all light-duty vehicles including vans and pickup trucks. There are fewer multi-passenger-car households, and consumers are probably less likely to exchange a personal van or pickup for an ultralight alternative car. In 1990, light trucks accounted for 33 percent of total vehicle sales in the U.S. According to R.L. Polk & Co., approximately 24.2 percent of U.S. households, or about 22.4 million, have two or more passenger cars. Household income also has a large effect on the degree to which households may

TABLE 2.5. HOUSEHOLDS BY NUMBER OF VEHICLES

Vehicles Available	Percent in 1969	Percent in 1977	Percent in 1983	Percent in 1990
Zero	20.6	15.7	13.5	9.2
One	48.8	34.6	33.8	32.8
Two	26.6	34.4	33.5	38.4
Three +	4.6	15.3	19.2	19.6

Source: U.S. Department of Transportation, Federal Highway Administration

be considered as potential new car purchasers. Average household income of new car purchasers is $50,000 per year. R.L. Polk & Co. lists 47.9 percent of all multi-car households as having an annual income of $50,000 per year or more. Potential market is therefore reduced to about 10.72 million, or about 11.5 percent of the total 92.5 million U.S. households.

Potential market for an alternative car is not necessarily limited to new car purchasers. Commenting on the potential U.S. market for micro-mini cars, John Hemphill of J.D. Power and Associates estimated that 25 percent of buyers would be diverted from the used car market.[11] This assumption is based on the idea that a micro-mini car would be less expensive than a standard car. Many second cars are purchased second-hand, primarily because of lower used car prices. The degree to which used car buyers are attracted would then depend on the household's typical budget for a used car, as compared to the price of an alternative car. Research has shown that price is extremely important and consumers will often choose a less expensive model even if long-term operating costs are higher. Charles Lave, Professor of Economics, University of California at Irvine, ran a projection of the potential sales of the 45 mpg Honda City in the U.S. market. In comparison to existing subcompact cars, a simple 25 percent decrease in price increased market share by 14 points. With both a 25 percent lower price and 45 mpg (19 km/L) fuel economy, the combined effect resulted in a full 22 point increase in market share. Introducing a $3500 retail price (roughly equivalent to $4600 at 1993 new-car prices) increased market share by a total of 30 points.[13] Market share of course refers to sales during a given year, and market potential has to do with those consumers who may ultimately purchase the product over time, normally over the car-buying/trade-in cycle.

Other variables include the degree to which household members are willing to swap vehicles to accommodate those days on which an alternative car does not fill a member's driving needs. For example, vehicle image is important to nearly everyone, and in at least some of these households the second car may be a teenager's hot rod or jacked-up truck. Consequently, the idea of swapping vehicles could conflict with someone's self image. In the other direction, potential purchasers are not necessarily limited to two-car households. In my two-adult household, there are two 2-seater cars. Neither of us feels especially limited by a two-passenger vehicle. Vehicle range and freeway capability would be a greater consideration. Still, the most significant limitation of a two-occupant vehicle is undoubtedly that of reduced seating capacity. Although this technical limitation is partially offset by limiting the universe of potential purchasers to that of multi-car households, its effect on potential market must be considered independently.

The Effect of Reduced Seating Capacity

Limited seating capacity is generally believed to have a strong, negative impact on potential market. Although two-seater vehicles will accommodate approximately 90 percent of vehicle trips, consumers prefer four-passenger cars by a large margin. Studies by J.D. Power and Associates place the ratio at about 5 to 1 in favor of four seats.[11] However, other vehicle attributes can combine to outweigh individual preferences and certain attributes may be intrinsic to the vehicle theme—the quality that attracts consumers in the first place. If the 5 to 1 ratio were applied to sales of the Pontiac Fiero, for example, increasing seating capacity to four indicates sales of 467,000 units in 1984. Nissan's "Z" car is offered in both two-seater and four-seater versions, yet two-seater sales far surpass four-seater sales. The "Z" car provides a real-world example of how consumer preferences may not apply in specific cases. With the Fiero, we can only guess at how increased seating capacity might have impacted sales. In both cases, seating capacity is closely associated with vehicle theme.

When considered in the abstract, one must assume that reduced seating capacity will limit the universe of potential alternative car purchasers. One can then move on to ideas that have the potential to counteract the limiting effect and expand the product's market. The most effective tool would be a strong vehicle identity (theme) that separates the limited-capacity vehicle from comparison with the standard car. If consumers perceive the commuter/urban car as essentially the same as a standard car, reduced seating capacity is likely to have a strong negative impact on potential market. If it is perceived as an entirely new type of transportation product, seating capacity may be unlinked from traditional buying habits and become more closely associated with actual driving needs. Seating capacity might then be perceived as one of the defining attributes of the commuter/urban car and reduced capacity may have little effect on sales, beyond the general limitation of the vehicle type itself. Product clinics could determine the effectiveness of this approach in advance of market entry; however, the actual product must first be developed. Studies based on abstract product ideas will invariably reflect traditional preferences and buying habits.

Additional Impact of Alternative Fuels and Electric Power

Two-car households are seen also as the consumer base for electric cars. However, when defining the initial target market, additional criteria are normally applied because of the electric car's limited range and the need for a recharging site. Studying the market for electric vehicles, researchers at the Institute of

Transportation Studies, University of California at Davis, applied the following criteria to define potential purchasers:[14]

1. The family must own their own home.
2. The home must have a garage or secure carport for overnight charging.
3. There must be two or more vehicles in the household.
4. There must be at least one person in the household who drives less than 70 miles (113 km) round trip to work.

Based on the above criteria, the potential market included 28 percent of the households in the U.S. Vehicle occupancy was not considered. This estimate was then reduced by the $50,000 annual income criterion in order to filter out potential new car purchasers. The remaining universe then totaled 13 percent of the households in the U.S. When consumer preferences having to do with the electric car's reduced range, limited baggage space, reduced power, and higher costs were applied, market penetration dropped to approximately one percent. This finding is supported by a 1991 study by Ford Motor Company on the potential market for electric cars. The Ford study indicated a one percent initial market penetration as well. For their study, Ford defined the electric car as being 100 percent emissions-free, having a top speed of 75 mph (121 km/h), 50 percent less space, reduced range, and costing $3000 more than a conventional car.**

Consumer preferences regarding a variety of alternative fuels were measured during the ITS study. Electric cars won hands down with 54 percent of participants favoring electric power over methanol (12%), hybrid/electric (11%), propane (7%), CNG (6%), hydrogen (5%), and ethanol (5%). Participants who picked electric cars most often cited their belief that EVs are "the best technology to solve pollution" (61%) and because they "will end oil dependence" (42%). In the U.S., technical constraint studies (the degree to which trip lengths limit electric cars as a transportation option) indicate that electric cars can serve a high percentage of consumer driving needs. However, the market limitations of electric cars have more to do with consumer perceptions than with vehicle technical capabilities. Consequently, more recent market estimations have emphasized stated preference studies in which participants make hypothetical choices from sets of product attributes. Although many consumers do not know the range of their present cars, consumer reaction is strongly negative to limitations in vehicle range. Disutilities because of limited range and long recharge times on the order of $10,000 to $15,000 have been reported by some researchers.[15]

** It is significant that the product definition includes one societal benefit and four personal sacrifices.

As previously discussed, seating capacity also is closely associated with perceived limitations, rather than actual limitations. One significant difference, however, is the fact that limited range could leave the occupants stranded and without a car at all. Limited seating capacity is an inconvenience, but limited range represents the potential for total failure of the automobile. Consequently, consumer attitudes toward limited range may not yield as easily to the influence of vehicle theme. Pioneers may have to accept a more limited initial market and rely on the combination of consumer familiarity and gradually improving technology to expand the market above this initial level. Additional help can come from strong government support and grass roots support from ecology groups and environmentally conscious consumers.

It is important to understand that estimates of market potential have to do with the initial market wherein consumers have no previous experience with the cars under review. When participants in the ITS study had an opportunity to test drive an electric car, 61 percent reported that their opinion of electric cars was improved, compared to 16 percent who said their opinion was worse. Researchers generally agree that the EV market will quickly grow from the one percent initial estimate. Consumers will learn that range limitations have very little effect on actual vehicle utility, experiences of friends and associates will reinforce the EV's validity as a satisfactory vehicle, and technology will also improve over time. In addition, the traditional bell curve will come into play with early purchasers paving the way for more conservative buyers later on. This of course applies to more conventionally powered, but limited-utility alternative cars as well.

In the final analysis, the green consumer and concerns for the environment could have a powerful effect in rapidly expanding the market above the initial level. Consumer preferences are based on experience with traditional automobiles. Traditional preferences could quickly become less meaningful when alternative vehicles actually enter the market. New vehicles types may not necessarily supplant existing products on a value/function basis, but instead motivate consumers with new benefits and appeals. When consumer attitudes change, statistical relationships also change.

Subliminal Messages and New Consumer Values

Much has been written about new roles for industry, the responsible corporation, and new consumer values. It used to be that a product was just a product. It had no meaning beyond its simple utility. At one time the renowned analyst, Sigmund Freud, affirmed that: "Sometimes a cigar is just a cigar." Today, a cigar is no longer just a cigar. Instead, it is a social blight, a corrupter of the young, and a messenger of illness, suffering, and death. In the eyes of many, a cigar speaks to the smoker's irresponsibility to himself and disregard for those who live and breathe in the same environment. A cigar is now an attitude and a moral code of ethics to a growing new wave of consumers.

Modern consumers see other products in a similar complex perspective. The message a product projects goes beyond simple utility into an abundant world of implied meanings. The world is no longer a simple habitat and the consumer is no longer unaware. Consumers are better educated and more worldly than at any time in history. People experience a level of education by osmosis that often exceeds the formal education of the not-so-distant past. Consumers are exposed in the popular press to political ideas, cultural values, and advanced scientific theories that in some respects outstrips the college curriculum of years past. In this new age, the corporation has new responsibilities and its products have new meanings that were not even considered just twenty or thirty years ago.

The role of the corporation as a responsible member of the world community, and the role of the product as an example of corporate attitude, is a new dynamic in the transaction between business and consumer. Today's consumer is aware of the environmental effects of technology and wants to know about the integrity of the institutions that manage the technology. A product is no longer just a product, but it speaks about the values of the people who create it. When consumers purchase a product, they also subscribe to the product's implied values. Today, when a designer is commissioned to redesign a company logo or to create a new corporate image, a review of company values is the first item on the agenda.

An automobile expresses company values in the language of its design. Its products tell consumers whether the company values quality and durability, or whether they think in terms of throwing things away and making more. Presumably, a well-made product comes from a solid company. Abundant safety features imply a concern for the welfare of consumers. A company that is concerned about the welfare of the human species will likely build products that are kind to the environment. Stereotyping is a powerful by-product of prevailing attitudes. Consumers will increasingly ask: "Does this product represent values to which I want to subscribe, or am I being sold on something that does not match my belief system?"

Likewise, there also are below-the-line meanings associated with green consumer attitudes. Consumers naturally want to protect the environment, but they do not want to sacrifice. People are more affluent and they want to be healthier, happier and more free. Self discipline is not the wave of the future and consumers are not likely to become environmental purists and sacrifice their own lifestyles for the planet. To paraphrase Dormer's characterization, regardless of the emerging green consumer, no one will make a lot of money selling hair shirts. People are concerned about consumption and pollution, but they want manufacturers to resolve problems in ways that merge with lifestyles.

The U.S. automobile market is especially sensitive to the self-indulgent quotient in purchasing habits. A popular theory of the '80s suggested that educating consumers about the costs of energy and the benefits of conservation would ultimately convince them to accept more inconvenience and expect less from their cars. However, econobox cars, ride sharing and public transportation were largely unsuccessful in spite of the popularized energy crisis. Buying habits and studies of psychological motivations indicate that conservation and sacrifice are not especially marketable attributes. Green market appeals must carry a different message. Instead, consumers must be motivated by the inherent desirability of the product. Conservation must add in some way to the product's attractiveness. A green orientation should be an upscale attribute: a chic product theme that in no way speaks to limited utility or reduced aesthetic appeal. Astute manufacturers will listen to consumers' needs and concerns, then package products in ways that appeal to consumers' fantasies. Consumers must be able to satisfy their moral and social concerns in the process of filling their more basic ego needs.

A New Paradigm for Personal Transportation Vehicles

I recently visited a class at the Art Center College of Design in Pasadena, California, where students were working on their ideas for the car of the future. A rendering, a beautiful work-in-progress of an urban car design, caught my attention. When asked about his design, the student released an avalanche of information about an unseen world to come. He talked on for several minutes about future lifestyles, and the new generation's awareness of the environment and the planet's limited resources. He described how the ongoing desire for fun transportation would merge with environmental concerns and ultimately guide the design of vehicles. Instinctively, this young creator was setting the context for his design. He was creating a perspective from which to experience his vehicle concept—a set of conditions and attitudes, or a paradigm within which his design would be appropriate and attractive. This ability to see through different eyes can be a powerful tool for reshaping perceptions and opening new horizons.

The paradigm, once accepted, determines how ideas and designs will be perceived and how they will fit into the world. It determines which ideas are valid and which are not, and it sets guidelines for interpreting data. A new world view can lead creative work in new directions, and it can reveal products and markets that may not have been apparent before. There is a story about two market researchers who were sent, independently, to one of the world's less-developed countries by one of the world's largest manufacturers of shoes. Their instructions were to report back on the potential market. When the first researcher arrived and observed the conditions he immediately faxed the following message to corporate headquarters: "No market here. Nobody wears shoes." Within a few hours, headquarters received word from the second researcher. His message read: "Great market here. Nobody has any shoes."

This humorous story about diametrically opposed conclusions from identical observations lightly illustrates the degree to which perspective influences interpretation. By changing one's focus, interpretations can be turned upside-down and inside-out. Of course, neither researcher had adequate data, and determining which of their conclusions would ultimately prove correct is unimportant to this discussion. The focus here is on individual perceptions, and how a change in perspective can reveal different and even seemingly contradictory facets of the same phenomenon. A paradigm shift, which Joel Barker describes in Future Edge as a switch to a different framework of rules, can sharpen the judgment of the entrenched practitioner, but more importantly it can give birth to electrifying new visions. Einstein's theories introduced a new set of rules that opened new frontiers in human understanding. Before having seen the other side of the mountain, Einstein's brilliant revelation described its treasures in glorifying detail. Singularities, quarks, mesons, and the connectivity of time and space are real phenomena that were first visited by the mind's eye before they were seen in the physical world. That they existed at all flowed from the new scientific paradigm. The objects and relationships that were ultimately revealed through Einstein's paradigm shift would have remained forever invisible through the filters of Newtonian physics.

New designs for markedly downsized transportation products, products that have large market and profit potential for their creators and provide safe transportation to their purchasers, can emerge through new ways of seeing. By adopting new filters, new opportunities and new product ideas can begin to take shape. Through the filters of the existing paradigm, markedly downsized vehicles are generally envisioned as unsafe motorscooter cars that have little market appeal. But such a characterization is already as outmoded as mechanical brakes and the traditional grille. Although the grille is still a widely accepted feature, with just a slight change in perspective the traditional expanse of chrome-plated latticework attached to the front of an otherwise modern automobile can suddenly seem oddly out of place.*** A change in perspective can also reveal the large multi-purpose passenger car as oddly out of place when used primarily for single-occupant trips to work or the corner market. Today, consumers purchase motorcycles for a sense of fun and freedom, four-wheel-drive vehicles for driving in places that are not appropriate for the family sedan, motor homes as a sort of resort on wheels, and trucks and vans for a variety of personal reasons. Mission-specific design in consumer vehicles is already widely accepted. Urban and commuting-specific cars simply expand the concept to include additional vehicle types.

Prevailing attitudes tend to cast urban cars and commuter cars as low-cost, utilitarian versions of full-size cars. Once this premise is accepted, vehicle attributes consistent with the vision naturally emerge and an outline of market potential, profitability, and even vehicle safety then follows suit according to the core idea. These details can quickly change when the vehicle itself is seen from a different perspective—as an essentially new and separate vehicle type with its own personality, its unique place in the transportation system, and its individual potential in the marketplace. By emphasizing safety features, visually impressive driver information systems, advanced vehicle control and crash avoidance systems, and distinctive and attractive vehicle layouts and styling, alternative cars can be positioned as a superior transportation product. An outmoded perspective of the recent past failed to see trucks as widely appealing alternatives to conventional passenger cars, but today, one-third of new-car sales are trucks of a different nature. When alternative cars address environmental concerns with attractive new vehicle themes that satisfy consumers' psychological needs, a powerful matrix will have been formed. A low-end alternative car in a high-end market is likely to produce disappointing sales. When the vehicle theme and the market are correctly matched, both sales and corporate profits are likely to soar.

*** Consider the grille's origin as an ornamentation for the old-fashioned automobile radiator. Grilles belong on industrial products, household vents, and 1920s-era automobiles, not on the front of modern passenger cars. Air can be discretely inducted, rather than through a gaping hole covered by an ornamental derivative of the household vent. If one can accept this characterization, automobile grilles will suddenly stand out as a rather odd design attribute. A shift in perspective can also reveal the oddity of large, multi-passenger cars in bumper-to-bumper traffic, each with its single occupant.

References

1. *The Nissan Report*, Edited by Steve Barnett, 1992.

2. "1990 Nationwide Personal Transportation Study, Early Results," Office of Highway Information Management.

3. F.T. Sparrow, "Automotive Transportation Productivity: Feasibility and Safety Concepts of the Urban Automobile," Final Report to The Lilly Endowment, Inc., December 1984.

4. Results of the studies are unpublished. This figure was obtained during a phone conversation between the author and Al Sobey.

5. The percentage of vehicle trips by number of occupants is based on 1983 NPTS data. At the time of this writing, the tabulation of 1990 NPTS data for this category had not been completed, except for trips with a single occupant only. Early 1990 NTPS single-occupant tabulations indicated that 67.1% of all vehicle trips were made with only a single occupant, versus 65.7% in the '83 study and 59.6% in the '77 study.

6. "1992—The Motor Industry of Japan," Japan Automobile Manufacturers Association, Inc., Washington, D.C.

7. William L. Garrison and Mark E. Pitstick, "Lean Vehicles—Strategies for Introduction Emphasizing Adjustments to Parking and Road Facilities," SAE Paper No. 901485, Society of Automotive Engineers, Warrendale, PA, 1990.

8. Wasielewski, Paul F., "The Effect of Car Size on Headway in Freely Flowing Freeway Traffic," Report No. GMR-3366, General Motors Research Laboratories, 1980.

9. John W. McClenahan and Howard J. Simkowitz, "The Effect of Short Cars on Flow and Speed in Downtown Traffic: A Simulation Model and Some Results," *Transportation Science*, Vol. 3, No. 2, May 1969.

10. Gino Sovran and Mark S. Bohn, "Formulae for the Tractive-Energy Requirements of Vehicles Driving the EPA Schedules," SAE Paper No. 810184, Society of Automotive Engineers, Warrendale, PA, 1981.

11. John Hemphill, "The Market Potential for Micro-Mini Cars in the United States," speech delivered before the "Mini and Microautomobile Forum: Overview and Potential Problems," Transportation Research Board, National Academy of Sciences, Circular Number 264, September 1983.

12. Dr. Charles Lave, "Economic Considerations," a speech delivered before the "Mini and Microautomobile Forum: Overview and Potential Problems," Transportation Research Board, National Academy of Sciences, Circular Number 264, September 1983.

13. "Mini and Microautomobile Forum: Overview and Potential Problems," Transportation Research Board, National Academy of Sciences, Circular No. 264, September 1983.

14. Thomas Turrentine, *et al.*, "Market Potential of Electric and Natural Gas Vehicles: Final Report for One Year," Institute of Transportation Studies, University of California, Davis, CA, (UCD-ITS-RR-92-8).

15. Tom Turrentine, *et al.*, "Household Decision Behavior and Demand for Limited Range Vehicles: Results of PIREG, a Diary Based, Interview Game for Evaluation of the Electric Vehicle Market," unpublished paper submitted for presentation at 1993 TRB Annual Meeting, July 1992.

Chapter Three

The Technology of Fuel Economy

(Courtesy: Ford Motor Co.)

......to myself I seem to have been like a boy playing on the seashore, and diverting myself in now and then finding a smoother pebble or a prettier shell than ordinary, whilst the great ocean of truth lay all undiscovered before me.

Sir Isaac Newton

Automobile manufacturers have produced a number of experimental cars that can run 40-50 kilometers on a liter of fuel (91-118 mpg). An experimental car built by British Ford will cruise at 24 km/h (15 mph) with fuel economy on the order of 892 km/L (2100 mpg). The difference between experimental, super-fuel-efficient cars and those on the showroom floor largely depends on the degree to which fuel economy is emphasized over other vehicle attributes. With an experimental vehicle, designers have free rein to maximize fuel economy without the normal constraints imposed by manufacturing and marketing realities. When it comes to a production car, hundreds of additional criteria become relevant and a different product emerges. Attributes such as reliability, repeatability, costs, creature comforts, and vehicle performance characteristics that may not be essential in an experimental vehicle are crucial in a production design. Ultralight composites give way to more affordable materials, a sleek aerodynamic body sacrifices to styling and room considerations, power-hungry accessories are installed, and high-tech experimental drivetrains forfeit to more buildable, more affordable, and more reliable hardware. In effect, the design gives up one excellence in favor of others.

The cooperative effort between the U.S. government and the auto industry is designed to triple the fuel economy of conventional passenger cars through technological improvements. As much as $1 billion may be committed to research and development efforts over a ten-year period toward the goal of producing a prototype conventional-size car capable of 35 km/L (82 mpg) by the year 2003. Although proponents are cautious about the potential for achieving the program's bold fuel economy goals with a marketable, safe, and affordable design, the technology of passenger car fuel economy will surely be advanced. In addition, the politics of vehicle fuel economy is, by definition, elevated to a cooperative, rather than an adversarial plane.

But achieving triple the present fuel economy with a production design is already possible with vehicles of markedly reduced size and mass. The idea of defining a new vehicle category with a different mass/performance profile was explored in the last chapter. Using existing technology, it is possible to design production cars with fuel economy on the order of 40-50 km/L (94-118 mpg) with performance equal to today's compact car. Challenges have to do primarily with marketing and safety issues, both of which are manageable by applying traditional marketing and engineering techniques.

Minimum fuel consumption per passenger-kilometers traveled occurs when vehicle capabilities most closely match mission requirements. When vehicle capabilities significantly exceed mission requirements, transportation energy intensity is generally greater, regardless of the basic technology. One- and two-occupant local missions are most efficiently served by smaller, limited-capacity vehicles. Although downsizing is not normally considered a technology, significant downsizing requires an increased emphasis on compact packaging, cost-effective and lightweight materials, more efficient processing, smaller and lighter powertrains, and improved strategies for managing the forces of collisions or avoiding them all together. Its application relies primarily on innovative design, rather than improved basic technology. But planners do not have to choose between better technology and vehicle downsizing. Both are interrelated, and significant downsizing will propel a technological re-emphasis of its own.

Mission-specific design is especially important to electric cars because of the limited energy supply available on-board these environmentally friendly vehicles. State-of-the-art battery technologies imply greater costs, higher mass and larger volumes that must be devoted to on-board energy storage. Downsized cars require less energy; therefore, smaller, lighter and less-costly motors, controllers, and battery packs can provide equal transportation. An optimum relationship between energy storage mass/volume, tractive systems, and vehicle size is best achieved when the vehicle mission is limited and more clearly defined. In multi-car households, vehicle missions tend to be segmented with different vehicles serving different missions. Traditionally, the choice of vehicle is usually limited to a choice between different multi-purpose cars. The practice of mission-specific vehicle design simply acknowledges this segmentation in trip requirements and configures the vehicle accordingly. Vehicle choice can then become purposeful rather than happenstance. When vehicles are accepted as having more specific applications, consumers may more easily accept electric cars as viable alternatives to conventional cars.

The Ingredients of Fuel Economy

In the most elementary sense, fuel economy is the product of load times system efficiency. Consequently, there are two ways to reduce vehicle fuel consumption: increase the efficiency of the powertrain in order to deliver more work from the fuel consumed, or reduce the required work (load). Work required to propel the vehicle is defined as *road load*. Road load is normally divided into the three categories of *inertia weight*, *aerodynamic drag*, and *rolling resistance*. Together, they determine the energy required to propel the vehicle throughout the course of the trip.

The vehicle's total energy supply comes from the fuel it consumes. However, only a portion of the fuel is actually transformed into useful work. The percentage of the fuel's latent energy that is delivered to the drive wheels as useful work is determined by *system efficiency*. The primary subsystems that determine overall system efficiency are the *engine*, *transmission*, and *final drive*. Each of these sub-systems is affected to varying degrees by the values of thermal efficiency, friction, and the viscous drag of lubricants, as well as by the operating variables over the course of the driving cycle.

Modifying the details of a vehicle's design affects the balance of energy flow throughout the system. Individual changes have complex effects which are often difficult to analyze. An engine with greater thermal efficiency may actually consume more fuel if its operating characteristics are not well matched to the vehicle's demand for power (operating schedule). Altering the engine's brake specific fuel consumption (bsfc) curve or changing the transmission shift schedule to take advantage of engine operating characteristics can significantly improve fuel economy, even though engine thermal efficiency remains unchanged.

When matched by complementary changes in the powertrain, reducing vehicle mass is a simple and relatively low-tech modification that can have a marked effect on fuel economy. Given the tenacity of powerplants to resist improvements in thermal efficiency, new technology should not be overemphasized in relation to the effects of vehicle downsizing and systems engineering on energy requirements. Mass effects alone are responsible for roughly 82 percent of the net energy consumed over the urban driving cycle. Eliminating load is the most direct route to greater fuel economy. In addition, most, but not all, harmful emissions tend to be proportional to the total fuel consumed, assuming that vehicles of equal technology are compared. In the future, reducing vehicle fuel consumption may be the single most important strategy in further reducing harmful emissions.

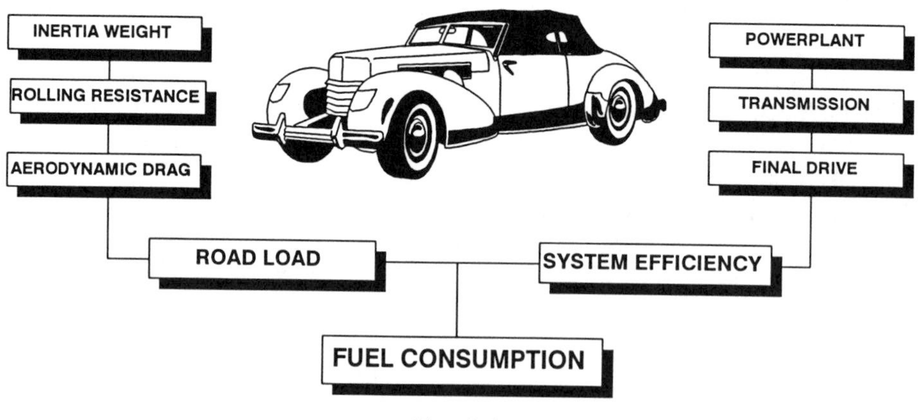

Fig. 3.1.

Fuel economy is normally expressed as distance traveled per quantity of fuel consumed (km/L or mpg). Except under controlled conditions, however, distance/fuel-consumed provides only an approximate measure of system thermal efficiency. In actual practice, fuel consumption is affected by a variety of intermingled conditions, including ambient temperature, trip duration, driving cycle and driver technique. Total fuel consumed throughout a trip is averaged against the total distance traveled to obtain a fuel economy figure. Even though km/L (or mpg) is a highly variable figure, it nevertheless serves as a convenient and easily understood measure.

Urban Driving Cycle Patterns Effect on Fuel Economy

Over the city driving cycle, fuel economy plummets. Early experiments with traffic density and its effect on fuel economy showed that light traffic conditions with an average speed of 29 km/h (18 mph) resulted in fuel economy that was only 60 percent of the steady-state 29 km/h value. When increased traffic density reduced average speed to 19 km/h (12 mph), fuel economy fell to 44 percent of the steady-state 29 km/h value.[1] Degraded urban fuel economy comes primarily from low engine loading, high inertia loads, and from frequent periods of braking and time spent at idle.

During city driving, conditions such as acceleration, engine loading, and time spent braking or at idle are continually changing across a wide range. These variations result in equally wide swings in fuel consumption. Studies by Gino Sovran at General Motors Research Laboratories in the early '80s revealed the following breakdown in times spent under various driving modes.[2]

TABLE 3.1. TIME SPENT IN VARIOUS OPERATING MODES

	Urban		Highway	
Operating Mode	Seconds	Percent Time	Seconds	Percent Time
Acceleration	544	39.7	252.8	33.0
Cruise	104.5	7.6	292.7	38.3
Powered Deceleration	164	12.0	159.2	20.8
Braking	311	22.7	57.3	7.5
Idle	248.5	18.0	3	0.4

Source: Ref. [2]

Vehicle downsizing first reduces fuel consumption by reducing steady-state road load. But over the urban driving cycle, the vehicle and its rotating components reside at a steady-state speed only 7.6 percent of the time. The effect of inertia is a much more significant factor in city driving. Inertia loads are directly proportional to the mass of the vehicle and its rotating components. Nearly 40 percent of urban-cycle driving time is spent under acceleration. Time spent at idle and during closed-throttle deceleration accounts for 41 percent of the total operating time over the urban cycle. Fuel consumption at idle and during closed-throttle deceleration is directly proportional to engine size. Engine size is determined primarily by vehicle acceleration requirements, which in turn is proportional to vehicle mass.

Figure 3.2 shows the relationship between the total energy (instead of total time) required for inertia, aerodynamic drag, and rolling resistance over urban and highway operating cycles. Inertia loads are less significant during highway driving. However, inertia consumes 60 percent of the total vehicle energy over the EPA urban driving cycle.[3] Inertia loads and rolling resistance together account for 82 percent of the total energy required to propel the vehicle over the

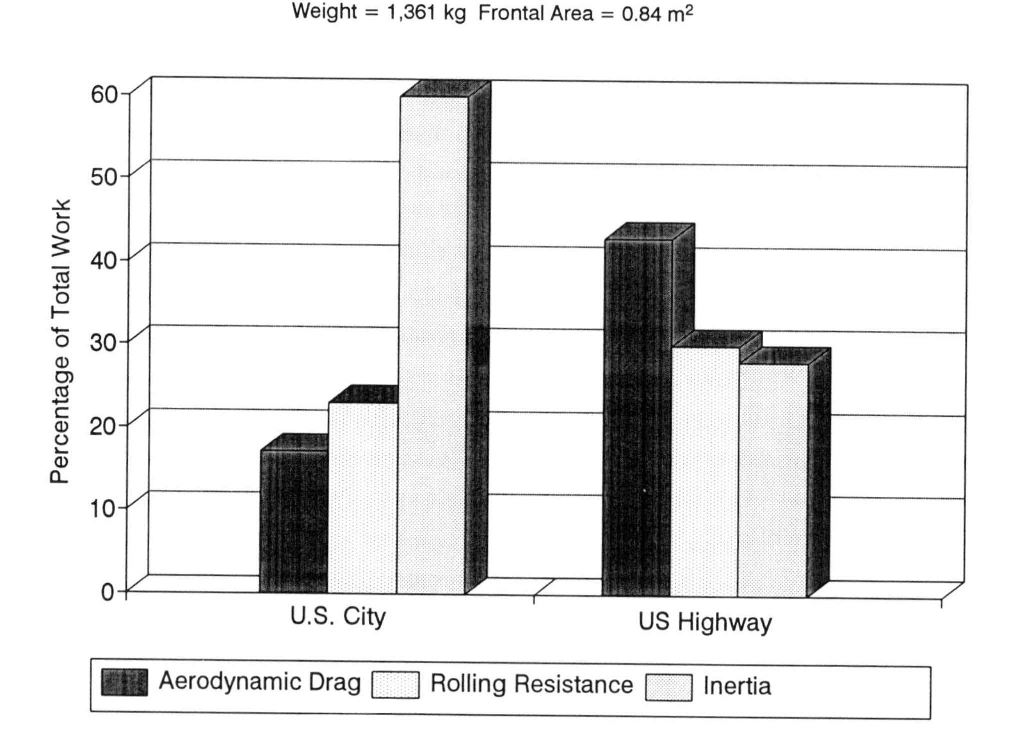

Fig. 3.2. Driving Cycle Load Comparison. (Source: Ref. [3])

urban cycle. A reduction in vehicle mass therefore has a direct effect on 82 percent of the vehicle's total energy requirements in the urban environment.

Idle shut-off systems have been employed to reduce fuel consumption in heavy stop-and-start traffic. With idle shut-off, the engine is automatically shut off when the vehicle stops, then restarted when power is required. This technique has been successfully used in automobile applications by both Audi and VW, and today it is often used in IC golf cars in order to reduce noise and emissions on the golf course. Basically, when the accelerator pedal is released, the engine stops, and when it is depressed the engine starts. An IC-engine-powered urban car might utilize a small electric tractive system to power the car at crawling speeds (up to 15 km/h, for example) where the heat engine is especially inefficient. At a driver-selected kick-in threshold, the IC engine could start and power the vehicle for greater acceleration and speed. The relatively small battery would then be recharged by the IC engine or, more efficiently, at home.

Cold-running characteristics are another important consideration for cars that operate primarily in the urban environment. Most urban trips are of short duration and, consequently, the vehicle may never be fully warmed up. This causes poor fuel economy as well as increased exhaust emissions. Modern IC automobile engines run relatively clean once they are fully warmed up. With today's engines, most HC and CO exhaust emissions occur during the warm-up period while the charge is enriched and before catalyst light-off. Tests at GM several years ago demonstrated that on a cold winter day (–12°C), 6.4-km (4-mile) trip that began with a cold start-up resulted in an average fuel economy that was only 72 percent of a similar, cold start-up 24-km (15-mile) trip. Engineers discovered that even after 24 kilometers (15 miles), the vehicle was not fully warmed up and fuel economy was still approximately 10 percent below optimum (see Figure 3.3).[4]

Data for the curves in Fig. 3.3 were obtained from vehicles that were not equipped with charge-air heaters. Such preheaters utilize exhaust manifold heat to reduce engine warm-up time. However charge-air heaters have no effect on bearings, seals, lubricants, transmission, final drive and tires. Warm-up energy can come only from burning fuel. The amount of energy required to reach a fully warmed-up state depends primarily on the heated mass. Reducing the mass of the powertrain and lowering the quantity of lubricants, as well as utilizing waste heat to augment the warm-up process, can significantly reduce fuel consumption during short-duration trips. Catalyst light-off time also can be significantly reduced.

Another option for reducing warm-up and catalyst light-off time is latent heat storage. With smaller powerplants, latent heat storage devices become more

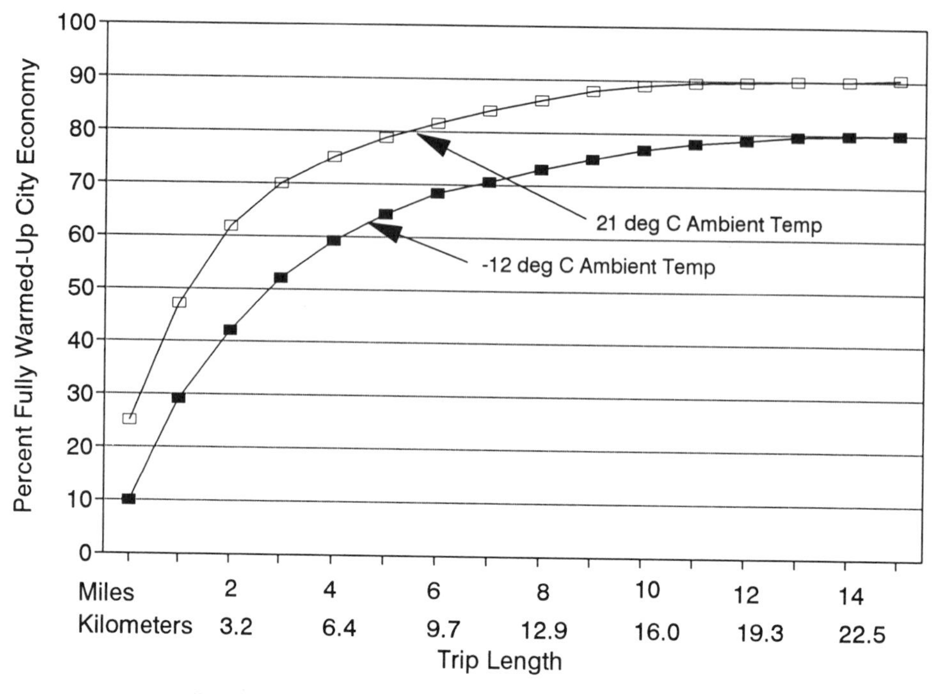

Fig. 3.3. Urban Cycle Warm-up Fuel Economy.

effective tools for rapidly increasing engine temperature after a cold start-up. Due soon on production cars, a heat battery will look like a large tin can tucked away inside the engine compartment. A typical device might measure 170 mm in diameter by 370 mm long, and weigh approximately 10 kg. A heat battery of this size can release 50-100 kW of heat during the first 10 seconds. Heat capacity is on the order of 600 W/h when cooled from 80° to 50°C. The storage medium is a water/salt mixture which is insulated by a conventional vacuum barrier. Heat is delivered to the engine coolant to induce rapid warm-up. Reheating is done with waste heat returned by the coolant, and stored heat can be used several days later.[5] Light-off time also can be reduced by electrically preheating the catalyst.

Accessory Loads

Accessories account for a significant portion of the automobile's total energy requirements. The power consumed by accessories can equal or exceed the power required to actually propel the vehicle. Loads imposed by the air conditioner, power steering, alternator, and cooling system can total as much as 3.5-7.5 kW in a full-size sedan. The Quincy-Lynn prototypes, Urbacar and

Trimuter, were each powered by a 12 kW (16 hp) engine and could reach a maximum speed of 96 km/h (60 mph). On a hot day, the tractive power of Trimuter or Urbacar would barely run the accessories of two full-size passenger cars.

Part of the technology of fuel economy therefore includes a close scrutiny of energy-robbing convenience items. With an alternative vehicle, power-assist features such as power steering and the host of electrically operated convenience items might be eliminated as part of the vehicle theme with little effect on perceived value. The reduced thermal loads of a smaller powerplant would naturally result in less demand on the cooling and lubrication systems. Air conditioning, however, may not yield as easily to simple solutions.

Eighty percent of new cars sold in the U.S. are equipped with air conditioning. Today, air conditioning is perceived as a necessity and consumers are not likely to give it up in order to save fuel. However, the air conditioning system is the largest single energy parasite aboard the automobile. Power required to cool the interior of a full-size automobile can equal the power required for all other accessories combined. Thermal load on the air conditioner depends on the thermal characteristics of the cabin area. A cabin with high surface-to-volume ratio (large surface area in relation to interior volume) is more difficult to cool. An appropriate heat barrier between the passenger compartment and the engine room and exhaust system is crucial. Contrary to popular belief, experiments have indicated that exterior color makes little difference in vehicle heat gain. Lighter interiors do, however, help reduce the demand on space conditioning. Glazing is undoubtedly the largest culprit of all. A solar heater consists of a flat box (high surface-to-volume ratio) with a darkened interior covered on one side with a large expanse of glass. Air conditioning a modern automobile with its large expanse of glazing is somewhat like air conditioning a solar heater. Solar control glazing can reduce heat gain by approximately 30 percent, but it cannot solve the problem. Efficient space conditioning may be one of the greatest challenges to designers of tomorrow's energy-efficient vehicles.

Road Load

Road load defines the power required to move the vehicle. It has a direct effect on fuel consumption. The greater the road load, the greater the appetite for fuel. A road load graph will typically show steady-state demand over a range of driving speeds. In actual practice, however, road load is much less consistent than the image portrayed by the smooth lines of a road load curve. Instead, it swings widely according to the variations in vehicle payload and speed, weather

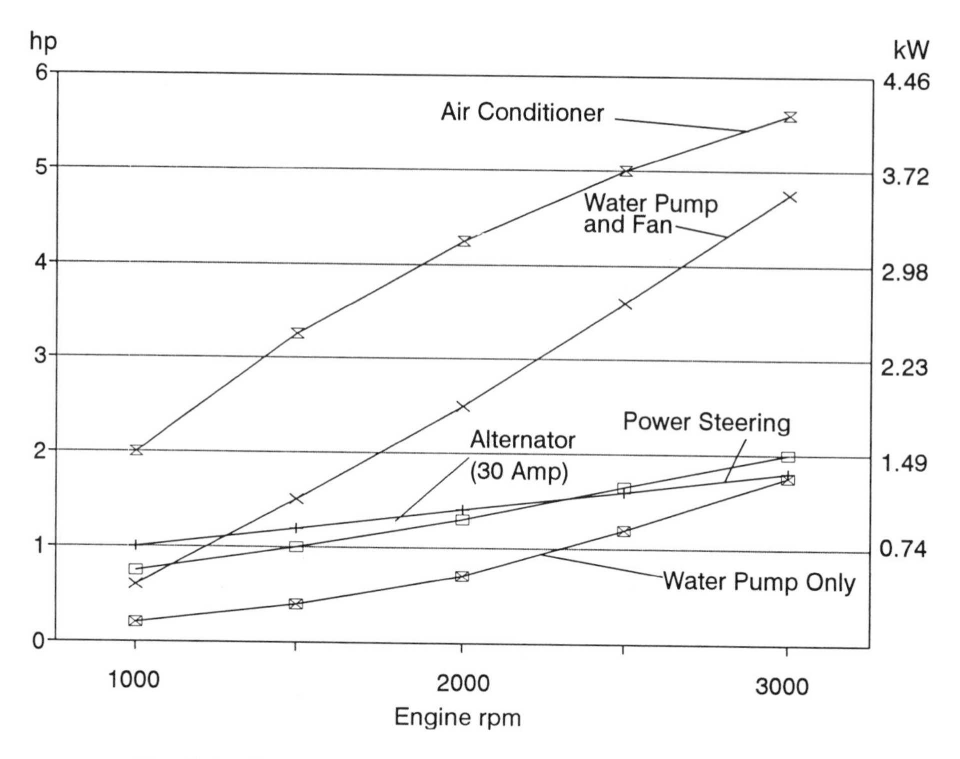

Fig. 3.4. Accessory Power Requirements (Full-Size Sedan).

conditions, and the grade and condition of the road surface. The engine converts fuel into work to meet the continuously changing demands of the road load. Reducing the road load places less demand on the engine and thereby reduces fuel consumption.*

During steady-state cruise, road load is comprised of rolling resistance and aerodynamic drag. Rolling resistance is directly proportional to the weight and speed of the vehicle. Consequently, if vehicle weight or speed doubles, rolling resistance also doubles. The power required to overcome aerodynamic drag is directly proportional to the frontal area of the vehicle, but increases to the cube of vehicle speed. Figures 3.5 and 3.6 compare the steady-state aerodynamic drag and rolling resistance of a hypothetical large car to those of a hypothetical commuter vehicle of about the same size and weight as Quincy-Lynn's Trimuter. Both graphs use the same criteria and neither graph takes into account the drivetrain losses or the power required for parasite systems.

* Fuel consumption will be reduced if the engine is appropriately downsized.

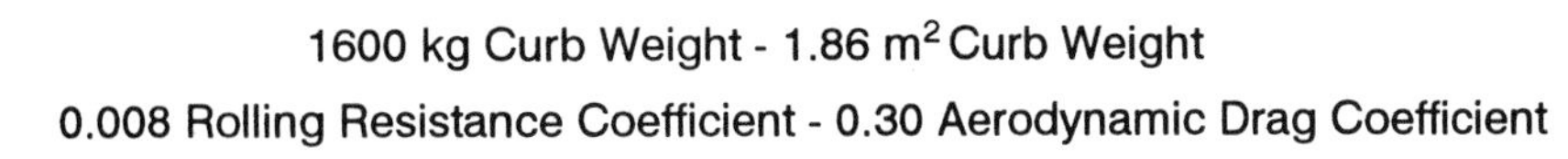

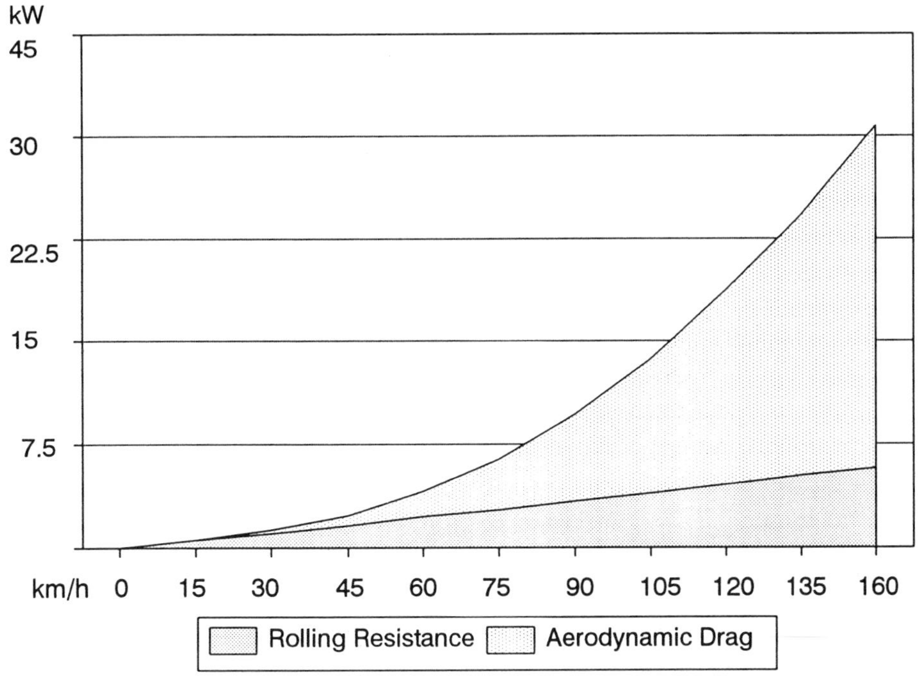

Fig. 3.5. Steady-State Road Load of Full-Size Car.

Developing a Road Load Graph

A road load graph can be developed for an existing vehicle by performing a coastdown or a drawbar test. A coastdown test is performed by allowing the vehicle to coast to a stop from a predetermined speed. If the inertia weight of the vehicle is known, road load can be calculated from the deceleration curve.[6] A drawbar test is performed by towing the test vehicle with a rigid bar equipped to measure the pulling force. Readings are taken in equally spaced velocity increments across a relevant range of speeds. These points, which indicate velocity and force values, are used to a develop the road load curve.

More often a road load graph is developed for a hypothetical vehicle as an engineering aid. Such a graph provides the basis for selecting the correct engine, configuring the drivetrain, and for predicting performance and fuel economy. In order to develop an engineering road load graph, the velocity/force values must be calculated on the basis of the target vehicle weight, the projected rolling resistance, and the estimated aerodynamic drag.

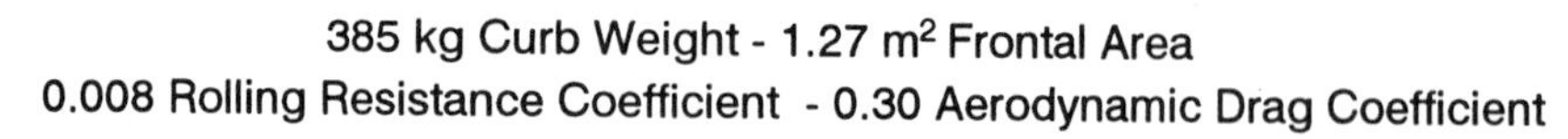

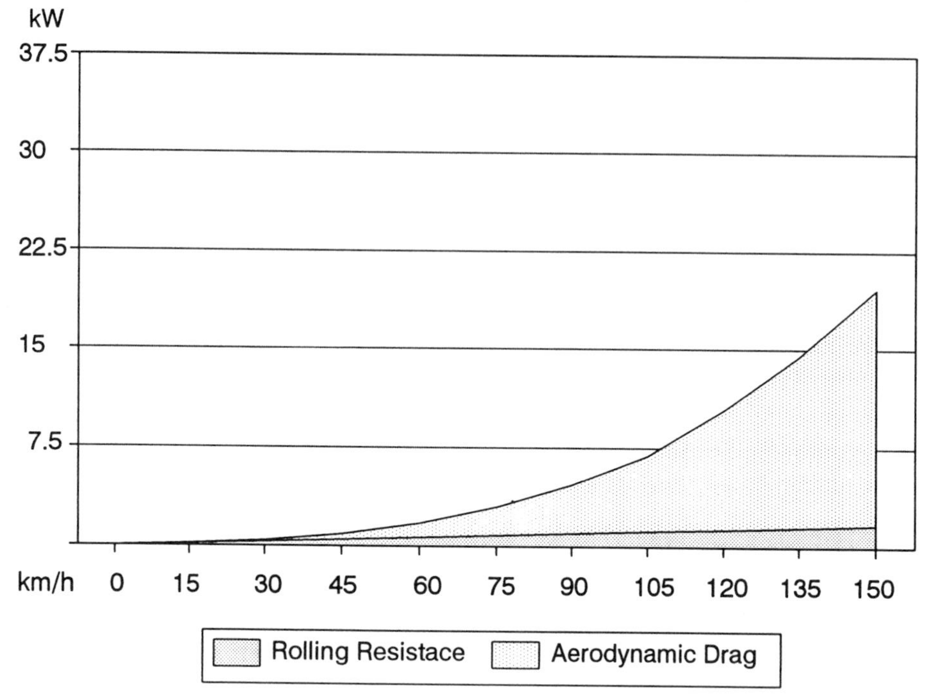

Fig. 3.6. Steady-State Road Load of Hypothetical Commuter Vehicle.

Rolling Resistance

Rolling resistance is the force it takes to make the vehicle roll. It is a linear function that is directly proportional to vehicle weight and velocity. The rolling resistance coefficient can vary widely depending on the chassis design, the surface conditions of the road and, most importantly, the design and inflation pressure of the tires. Although pneumatic tires are essential for good handling and ride, they are responsible for approximately 85 percent of the rolling resistance of the vehicle. The remaining 15 percent comes from bearings, seals, gear friction, and the viscous drag of lubricants.

Tires—The Balance Between Ride, Handling and Drag

A steel wheel running on a steel rail provides the least rolling resistance of any wheel design. Steel wheels, however, have no inherent damping and very poor adhesion. Unlike a pneumatic tire, a steel wheel does not flatten across the area of contact with the ground (contact patch). The effective radius remains unchanged throughout the circumference, and as a result, rolling resistance is minimal. By deflecting on contact with an irregularity, a pneumatic tire acts as a sort of air spring to smooth out the ride. This same flexibility also allows the tire to flatten at the contact patch, which places a greater portion of the tire in contact with the road. The tire's "pneumatic resilience effect" is one of the sources of its excellent adhesion, ride, and handling characteristics. Pneumatic resilience is also the primary source of its high rolling resistance. Much depends on sidewall flexibility and materials compounding.

At 90 km/h (55 mph) each tire on the vehicle is forced though a complete radial distortion at the rate of nearly 800 times per minute. Such flexing requires a significant input of energy which, as always, comes from fuel consumed by the engine. The energy required for tire distortion is normally expressed as the coefficient of rolling resistance. A full-size passenger car equipped with soft-riding bias-ply tires might have a rolling resistance coefficient as high as 0.015. A chassis equipped with more efficient state-of-the-art radial tires could exhibit a rolling resistance coefficient as low as 0.008. Rolling resistance can be significantly reduced by increasing the inflation pressure, but at the expense of ride. Improved tire design offers the greatest potential for reducing vehicle rolling resistance.

Tire manufacturers have been hard at work for many years to develop low-rolling-resistance tires that still maintain their superior adhesion and riding qualities. During the past two decades, tire rolling resistance has fallen by approximately one third and prototype tires have been demonstrated that promise even greater improvements for the future. Goodyear's most recent contribution is a concept tire for GM's new electric car, Impact. The experimental tire has a rolling resistance coefficient of 0.0048 which is about 55 percent less than today's conventional highway tire. Tires for Impact are inflated to 448 kPa (65 psig), or about twice the inflation pressure of a conventional tire.

Other Components of Rolling Resistance

Approximately 15 percent of rolling resistance is comprised of the drag of seals, gear friction, and the drag caused by churning the high-viscosity lubricants. Special seals with a reduced contact pressure have been shown to reduce drag. The tradeoff is in the reduced ability of the seal to maintain seal integrity over an equal life span. An area of greater concern is the viscous drag of lubricants.

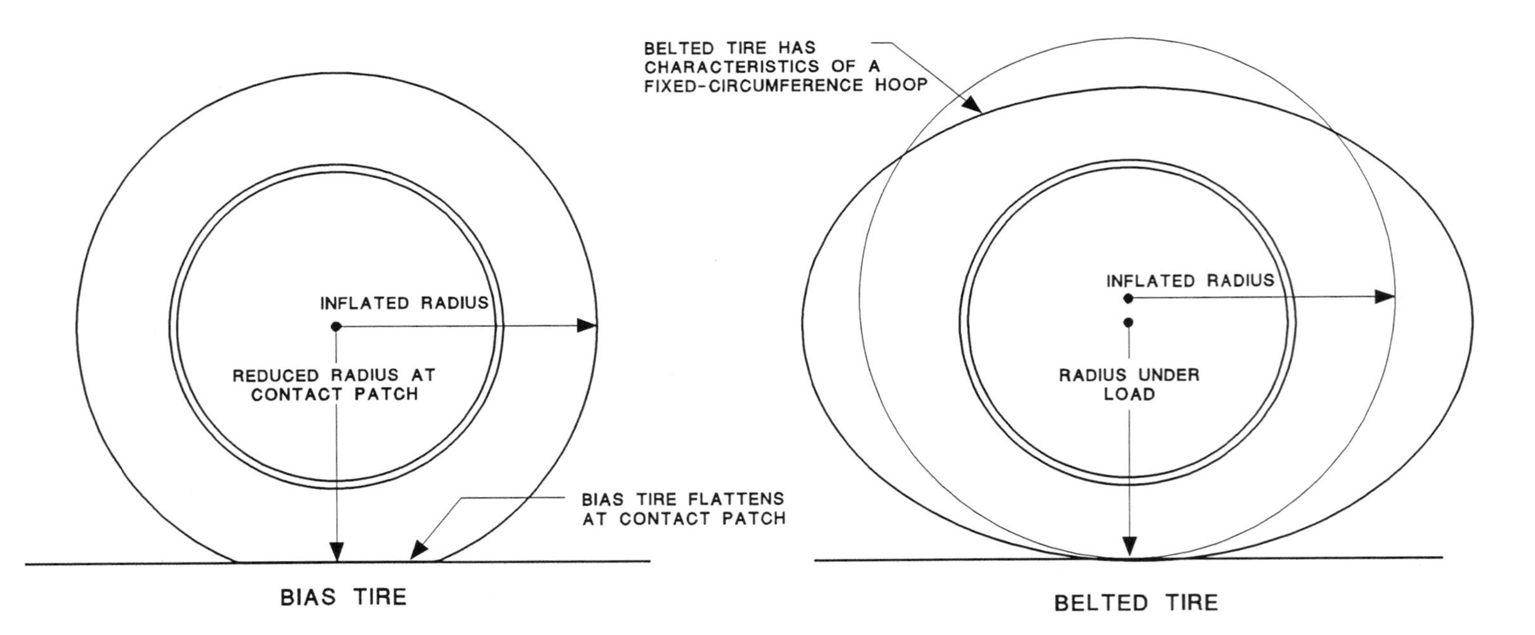

Fig. 3.7. Tire Distortion Characteristics.

Fig. 3.8. Low-Rolling-Resistance Tire from Goodyear. (Courtesy: Goodyear Tire and Rubber Co.)

For many years, engineers have been aware of the drag on transmission and final drive components caused by the high viscosity of lubricants. The viscosity must be high to prevent gear pressure from squeezing lubricant away and allowing the teeth to make metal-to-metal contact. However, switching to lower-viscosity lubricants has produced mixed results. When the lubricating film can be maintained at a lower viscosity, drag is reduced. If lubricant is squeezed away, friction increases and more than offsets any reduction in viscous drag. In this respect, the lower gear pressures of a relatively low-power commuter or urban vehicle present new opportunities for reducing drag with low-viscosity lubricants.

Automobile Aerodynamics

Nearly every aspect of vehicle design, including creature comforts, vision, systems cooling, interior ventilation, noise, and styling, are affected by, or have an effect on, the aerodynamics of the design. Aerodynamic drag alone is a major component in the vehicle's appetite for energy, and it becomes increasingly so at higher speeds. Aerodynamic drag is a product of the vehicle's frontal area, its drag coefficient, and the cube of its speed. The cubic increase of drag in relation to speed makes aerodynamics a major consideration when designing high-speed highway vehicles. At the lower speeds typical of urban traffic, aerodynamic drag is not nearly as significant.

Features that determine the drag profile of the vehicle include:

- Appropriate Curvatures for Maintaining Laminar Flow
- Shape of the Nose
- Shape of Pillars
- Ground Clearance
- Rear Body Shape
- Cooling System
- Spoilers & Dams
- Antenna
- Door Handles
- Slope of the Hood
- Rain Gutters
- Body Inclination
- Wheel House Openings
- Internal Air Flow
- Mirrors
- Recess of Glazing
- The Effect of Air Leaks
- Windshield Angle
- Gap Profiles
- Backlight Angle
- Wheels
- Underbody Flow
- Headlights
- Wipers

Drag is not the only aerodynamic effect the designer must consider. Passenger compartment flow and climate control, mud spray and particle deposits, systems cooling, exhaust dispersion, and noise are all affected by a vehicle's aerodynamic characteristics.

Working with the Relative Wind

The velocity and direction in which the vehicle and air encounter each other is referred to as the relative wind. Automobile aerodynamics is concerned with the effects of the relative wind as it flows around and through the vehicle. Relative wind has a number of effects on the vehicle in addition to the drag it imposes. It can increase or decrease the pressure at the contact patch through positive or negative lifting forces, it can stabilize or destabilize the vehicle with yaw inputs, and it can cool and ventilate or destroy cooling and ventilation flow due to pressure differentials around the body.

At highway speeds the forces of induced lift can reach a magnitude of 1300 N (292 lb) or more on a full-size sedan. These high lifting forces can have a

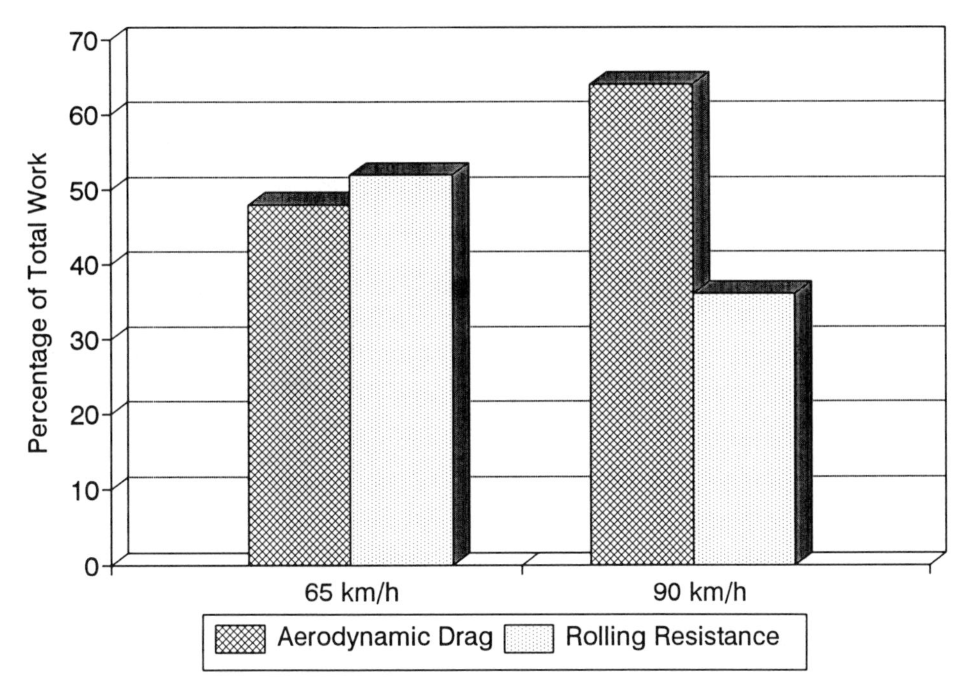

Fig. 3.9. The Effects of Speed on the Relationship Between Aerodynamic Drag and Rolling Resistance.

significant effect on the vehicle's stability. Cross-winds also impart destabilizing forces that cause the vehicle to yaw. High lift and yaw inputs, combined with the reduced adhesion of a rain-slicked road, can cause loss of control under extreme conditions. Low-mass vehicles imply greater attention to aerodynamic effects. At freeway speeds a lightweight car also could develop high lifting forces. As a percentage of the vehicle's gross weight, lifting forces could be proportionally higher than in vehicles of greater mass. The destabilizing effects of inertial forces in combination with cross-winds and side-blasts from passing vehicles may therefore become more significant as vehicle mass decreases.

Streamlining and Aerodynamic Drag

The techniques for reducing aerodynamic drag are collectively referred to as *streamlining*. Air resistance has been studied for over two hundred years. Early on it became clear that the shape of the body has a large effect on the magnitude of the drag it produces. However, as recently as the wind tunnel experiments by the Wright Brothers in 1902, experimenters were unsure about which shapes actually produced the lowest drag. It was soon discovered that the ideal aerodynamic shape was the teardrop with the large end facing into the relative wind. The trailing end of the teardrop should taper at a shallow angle of 15 degrees or less. Figure 3.10 compares the actual drag on 30.5-meter (100-ft) lengths of rod of various profiles when exposed to a 160 km/h (100 mph) wind. Each length of rod presents an identical frontal area to the relative wind. Shape (profile) is the only difference.

The objective of streamlining is to shape the body in a way that will smooth out the flow of air and thereby reduce the drag coefficient. The coefficient of drag (Cd) can therefore be used to compare the ability of various shapes to slip through the air with minimal resistance. Cars have become increasingly more slippery over the years. A modern, aero-styled car has a Cd on the order of 0.25 to 0.35. A 1930 Model A Ford had a Cd of about 0.85, and the Cd of a 1955 Ford was on the order of 0.50. A 1977 Porsche 924 had a Cd of 0.36.

Sources of aerodynamic drag are normally divided into the categories shown in Table 3.2.

TABLE 3.2. AERODYNAMIC DRAG BY SOURCE

Source of Drag	Percentage of Total Drag
Profile Drag	55%
Parasitic Drag	17%
Internal Flow	12%
Skin Friction	9%
Induced Drag (lift)	7%

Profile Drag

Profile drag is a function of shape. The ideal aerodynamic shape is the teardrop with the large end toward the relative wind. Unfortunately, the teardrop does not accommodate passengers and cargo well, nor is it very attractive. A box-like shape makes better utilization of space but has poor aerodynamic characteristics. Modifications to the basic shape of the box can result in large improvements in flow characteristics.

Profile drag is minimized by maintaining a laminar flow as far rearward as possible. Laminar flow refers to a condition in which the air stream follows smoothly along the body without breaking away into turbulence. However,

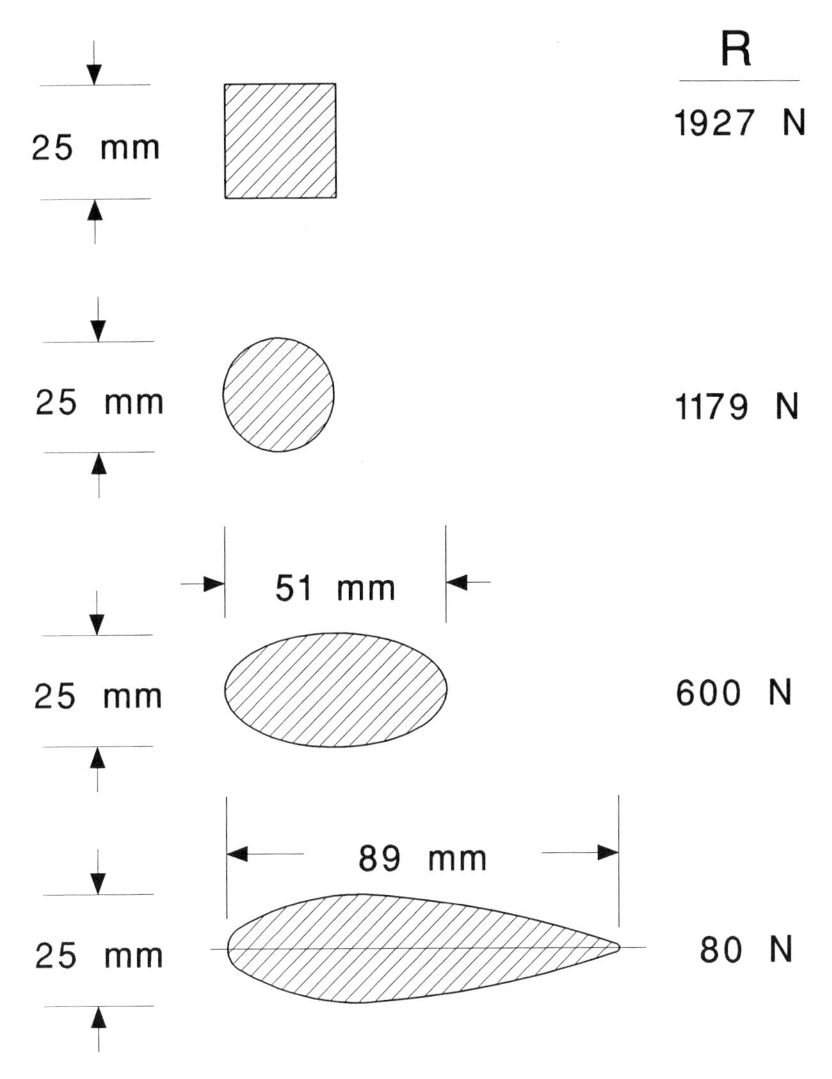

Fig. 3.10. The Effect of Shape on Aerodynamic Drag.

laminar flow is difficult to maintain. As long as the airstream is accelerating across a smooth, convex surface it has a good chance of remaining laminar. Experiments indicate that separation is most effectively delayed when the body surface's radii of curvature vary continuously according to an equation of the second order. This implies that all curves of the body surface must be of the third

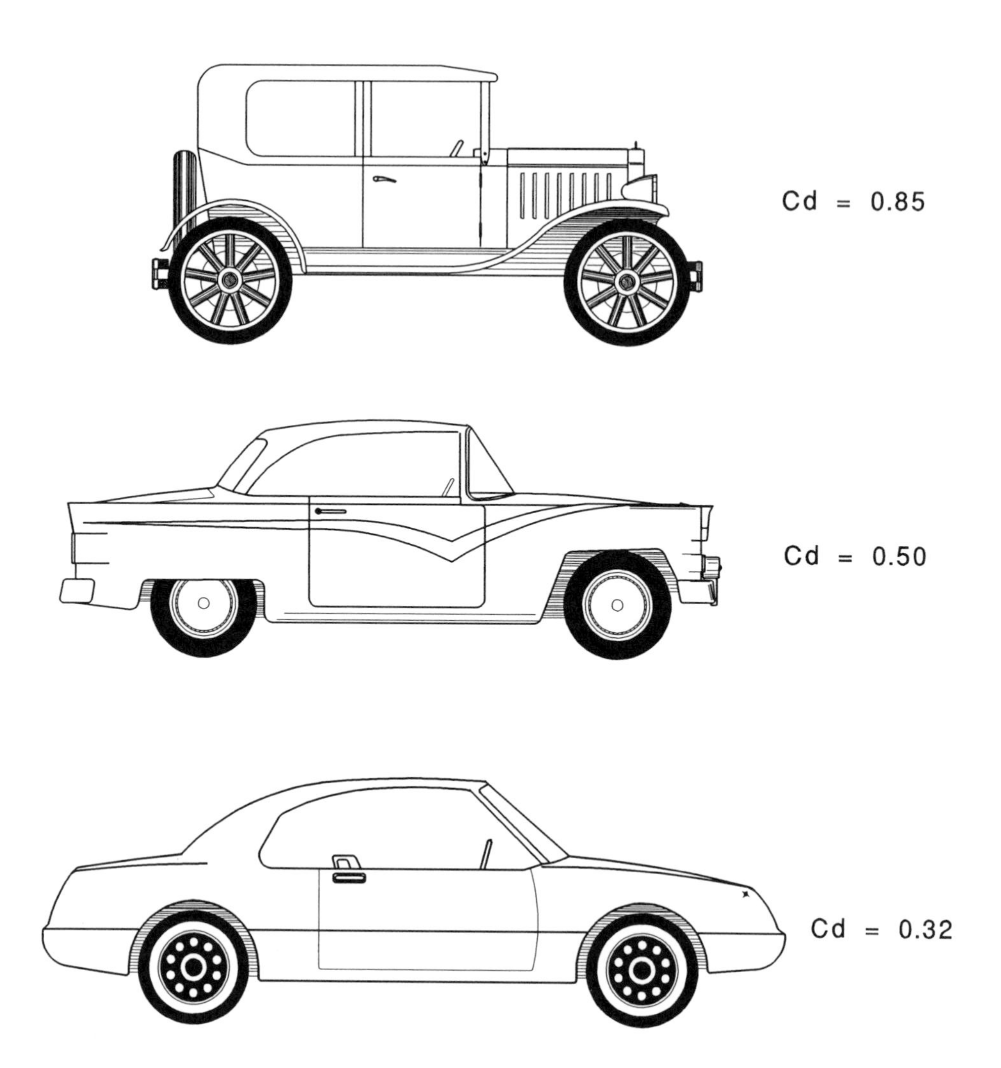

Fig. 3.11. Declining Aerodynamic Drag of Automobiles.

order. Cubic equations applied to test model body surfaces demonstrated very low drag.[7] As the flow moves downstream, the boundary layer (a thin layer of slow-moving air next to the surface) begins to thicken, form rollers and eddies, and finally break away into vortices. At this point, separation is complete and drag sharply increases.

Any sudden break in a smoothly contoured surface can result in flow separation. Wheel wells have always been problem areas for engineers because they destroy laminar flow along body sides. Recessed window glazing is another design attribute that is known to destroy laminar flow across the sides of the greenhouse.

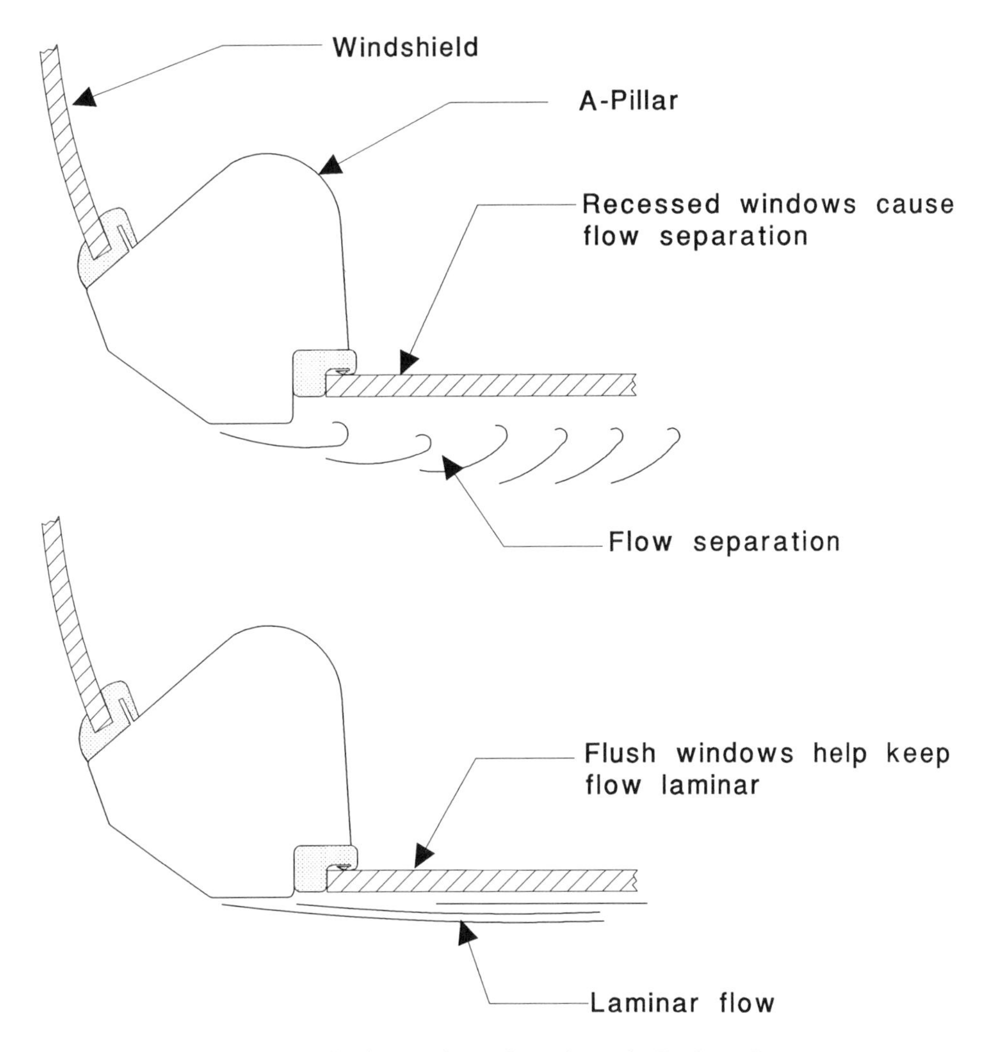

Fig. 3.12. Flush Windows Significantly Reduce Drag.

Once separation occurs, it is usually difficult to re-establish a laminar flow. Protuberances such as mirrors, handles, bug deflectors and antennae produce drag of their own, but they also can induce separation.

Regardless of how well the body encourages the flow to remain laminar, separation must ultimately occur at the rear of the vehicle. Separation creates a standing low-pressure area behind the vehicle. The breakaway cross-section at the rear has an effect similar to that of a flat plate of equal size presented to the air stream. Consequently, the body should taper toward the rear, similar in effect to the reverse teardrop, to form an aft section that is as small as possible

(Figure 3.13.). Another technique is to direct some of the relative wind into the low-pressure zone in order to reduce the magnitude of the drop in pressure behind the vehicle.

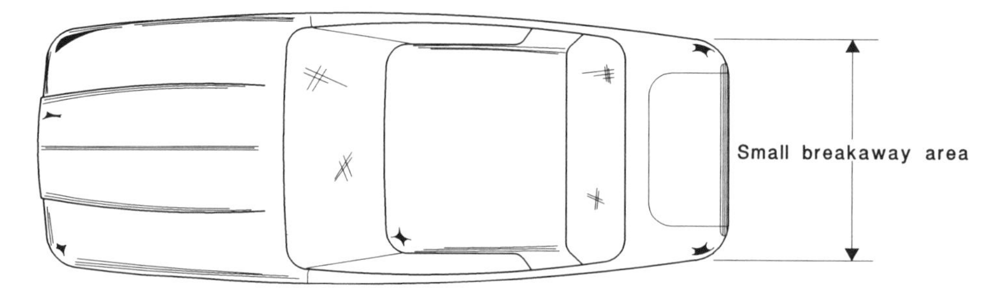

Fig. 3.13. Aft Tapering Reduces Separation Area.

Induced Drag

An automobile is relatively flat along the bottom and curved across the top. That is the basic shape of an airplane wing (see Figure 3.14). As a result, an automobile is normally subject to significant lifting forces at higher speeds. As mentioned before, a standard sedan can generate lift on the order of 1300 N (292 lb) at highway speeds. Drag is a necessary component of lift. A design that minimizes lift will also reduce the resulting drag, provided that flow separation or turbulence is not induced.

Parasitic Drag

Parasitic drag refers to the drag induced by components that project into the airstream. This includes items such as ornaments, bug deflectors, wipers, antenna, rearview mirrors, license plates, door handles, air scoops, rain gutters, bezels and trim, and luggage racks. Exposed mechanical components underneath the vehicle also contribute to parasitic drag. The magnitude of parasitic drag can be much greater than the simple aerodynamic drag of the item itself. For example, a rearview mirror is typically located on the side of the body just aft of the windshield. In this location it may be directly in line with the high-speed air spilling from the sides of the windshield. This high-speed airstream can induce drag that is more than 1-1/2 times greater than the drag of the same mirror in free air. Further, if the mirror induces flow separation, the combined effects become even greater. Exposed wipers cause similar effects. Wipers induce drag of their own, plus they encourage flow separation across the windshield.

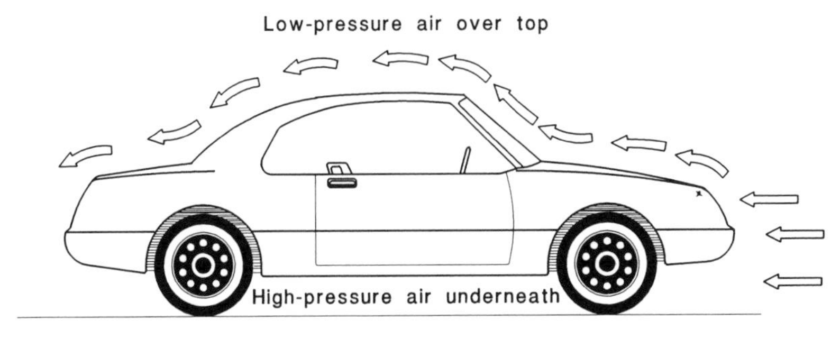

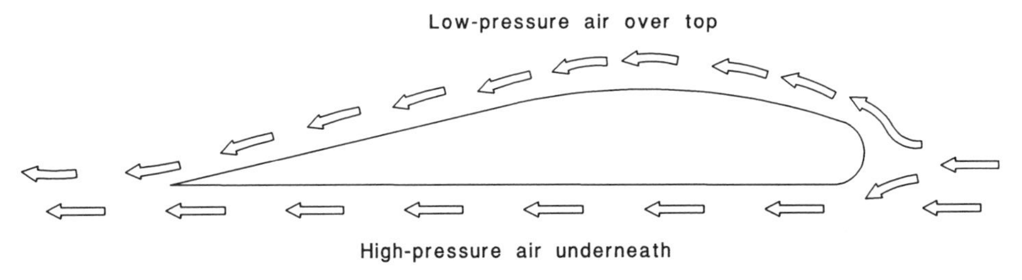

Fig. 3.14. Lift is Developed Across Top of Both Shapes.

Enclosing the underbody with a flush pan can reduce drag by as much as 17 percent. Flush windows will reduce drag by 5-10 percent, depending on the overall body design. Raising the headlights of a Corvette can increase drag by approximately 3 percent. The drag coefficient of a Porsche 924 is more than 5 percent greater with the headlights raised.[8] Parasitic drag becomes much more significant on an aerodynamically clean shape. Extra effort taken to establish laminar flow may be to no avail if mirrors, door handles, or trim moldings destroy laminar flow and increase drag.

Skin Friction

The effect of viscosity produces a thin layer of relatively slow-moving air next to the surface of the body. This is called the boundary layer and the friction it produces is called "skin friction." At high speeds, a very polished surface will reduce skin friction by presenting a "slippery" surface to the air. However, at the relatively low speeds of an automobile, the polish of the surface has little effect on drag. The primary culprits of automotive skin friction are raised fasteners, exposed hinges, trim moldings and the configuration of gaps.

Internal Flow

Drag caused by air flowing through the passenger compartment and cooling system can amount to as much as 12 percent of the total drag of the vehicle. A properly designed cooling and ventilation system can keep drag to a minimum and reduce energy demands by working in unison with the pressure differentials on the body surface. For example, air that is drawn into the passenger compartment from the base of the windshield will help maintain laminar flow across the windshield by relieving the high-pressure area. Allowing air to exit into the low-pressure area behind the backlight can also reduce drag. However, the airflow through the cabin normally has minimal effect compared to that of the cooling system.

Flow through the radiator and engine compartment produces the greatest drag from internal flow. Appropriate ducting can make a large difference. In his book, Designing Tomorrow's Cars, Walter Korff suggests that the principles of efficient ducting developed for the aircraft industry can be used to design more aerodynamically efficient automobiles.[8] These principles require the air outlet to be as carefully controlled as the air inlet. Also, ducting must be sealed or loss of energy will occur. Korff suggested that the size of the cooling air intake is much too large on most automobiles. A fast-moving vehicle with the radiator located behind the entrance of the duct at a distance equal to its own height needs an opening only 17 percent as tall as the radiator. If the vehicle spends much of the time in slow-moving urban traffic, the duct size should be increased by approximately 50 percent, or to approximately 25 percent of the height of the radiator. The closer the radiator is moved toward the intake, the taller the opening must be. Also, cooling system flow should be controlled by restricting the outlet flow, not by blocking the inlet (see Figure 3.15).

Frontal Area

Reducing vehicle size is one of the most straightforward ways to reduce air drag. The frontal area, the silhouette the vehicle presents to the relative wind, has a direct effect on the drag it produces. Reduce the frontal area by half and aerodynamic drag also drops by half. This technique is limited because vehicles must have room for occupants. Frontal area can be reduced only by reducing height and width. Reducing length has very little effect on aerodynamic drag.

Minimum vehicle height is established by the sum of the ground clearance, the seat height, the seated height of the 98th-percentile male, and the required

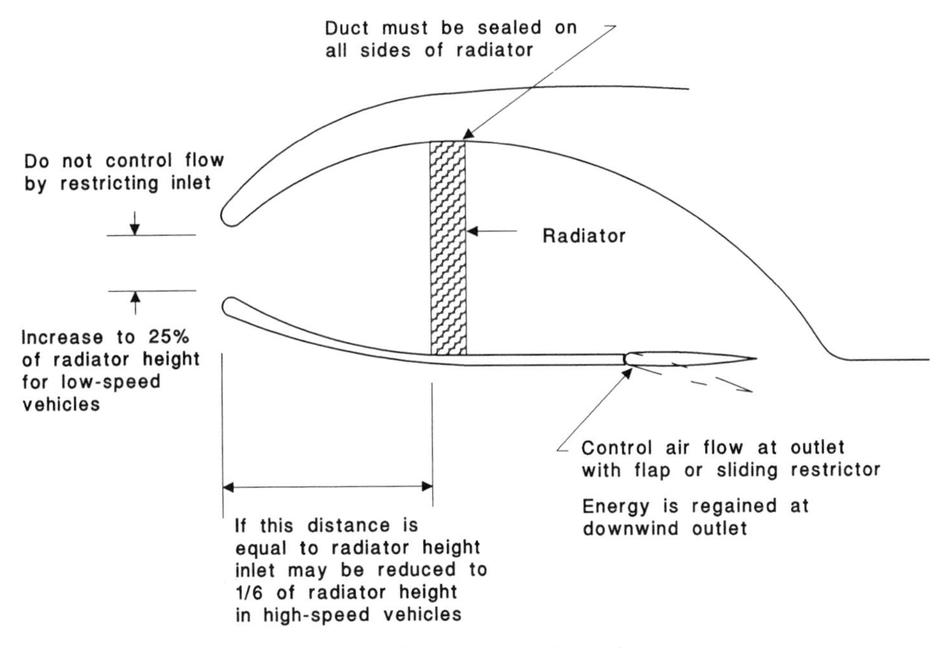

Fig. 3.15. Minimize Air Inlet.

headroom. These values are based on the mechanical requirements of the chassis, along with interior dimensions established through prior research into human space requirements and comfort levels. The cabin height of a typical sedan will be on the order of 1150-1200 mm (45-47 in.). A sports car might be limited to slightly over 1000 mm (40 in.) inside. Placing the occupant in a semi-reclining position can reduce the cabin height to approximately 900-950 mm (36-37 in.) (see Figure 3.16).

Shoulder width establishes the minimum cabin width. The cabin may slant inward at the top and bottom because of the inward displacement of the occupants' head and hips. However, a close-fitting roof line will create a crowded feeling and can be hazardous in a collision. Crowding across the shoulders also tends to be uncomfortable. A simple method of reducing frontal area of a two-place vehicle is to place the occupants in tandem. By sitting one behind the other, vehicle frontal area can be reduced by as much as 40 percent (see Figure 3.17). The Honda EP-X concept car is an excellent example of the tandem seating arrangement.

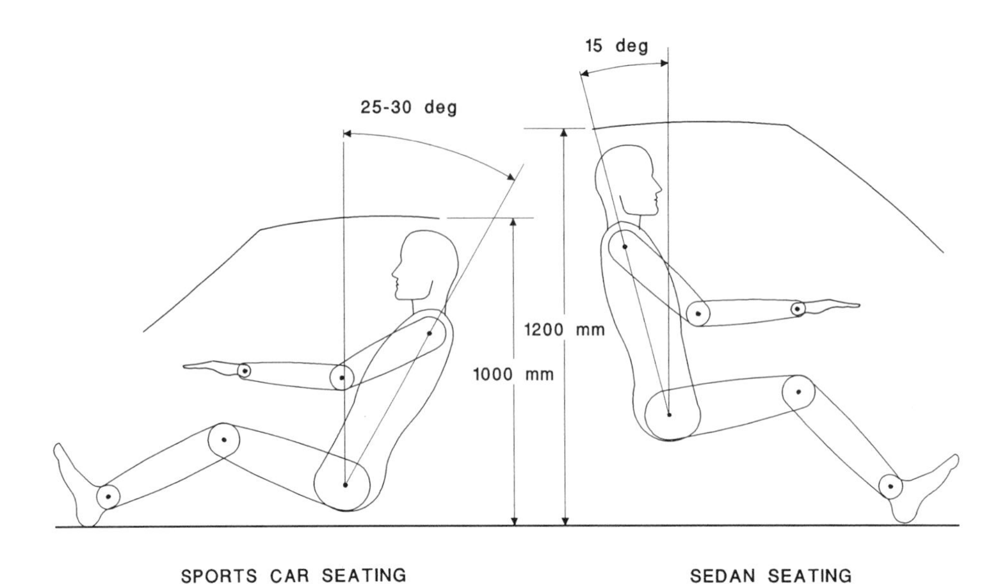

Fig. 3.16. Seating Position Affects Vehicle Height.

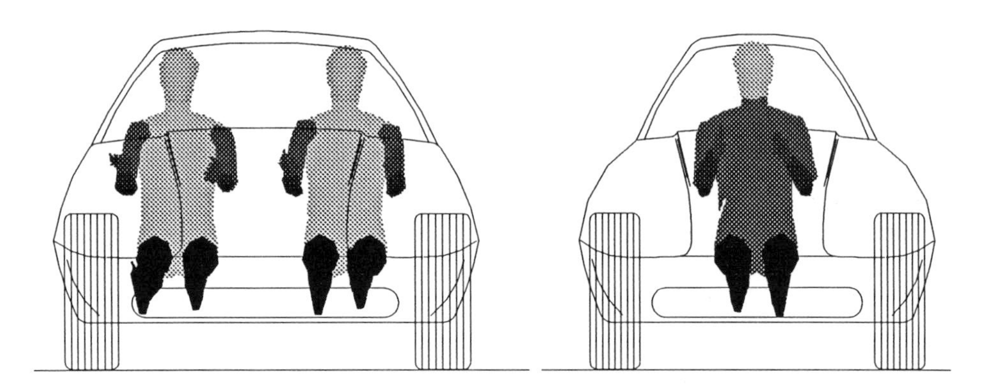

Fig. 3.17. Tandem and Side-by-Side Seating.

The Fuel-Efficient Powertrain

The powertrain consists of the engine, transmission, and final drive. Ideally, the powertrain will be light, small, and simple to manufacture, it will have high specific power, and it will consume as little fuel as possible and produce few harmful emissions. The quality of the vehicle's performance, its fuel efficiency, and emissions profile are determined by the balance between hundreds of intricately intertwined variables in the design. A vehicle's performance, economy, and emissions depend on the interaction of the various subsystems as they respond to the demands of the operating schedule. Configuring a powertrain for maximum efficiency consists primarily of capitalizing on desirable subsystem characteristics and minimizing characteristics that are undesirable. Advancements in powertrain design are normally based on technologies that increase control of events and relationships, and thereby more precisely match powertrain performance to operating conditions.

The Prime Mover

In Newton's time the prime movers were the beasts of burden and the water wheel. The prime mover of the twentieth century's personal transportation vehicle has been the internal combustion engine. Today's IC engine, however, bears little resemblance to its early predecessors. The design of engines has reached a level of refinement that no doubt was unimaginable to early pioneers. The experience of the Wright Brothers is a case in point. In 1903 the Wrights wrote to several engine manufacturers with specifications for an engine to power their soon-to-be-built flying machine. To the Brothers' dismay, not a single manufacturer could meet their requirement for an engine that would develop 7.4 kW (10 hp), yet weigh no more than 68 kg (150 lb). Such great power from an engine of such minimal weight was not within the capability of existing manufacturers. The Wrights ultimately built their own: a 91 kg (200 lb) machine that developed 11.7 kW (15.75 hp) at 1200 rpm. Today, after a century of refinements, a high-performance automobile engine can develop power on the order of 1 kW/kg (1 hp/1.5 lb), and a two-stroke engine can develop nearly twice the power per weight.

The IC engine is often maligned because of its harmful exhaust emissions and poor thermal efficiency. But only in the last 15 years, since the widespread use of electronic engine management, has it been possible to gain control of the events inside the engine. The future will surely unveil an even more power-intensive, fuel-efficient, and clean-running machine as improved designs, materials, and engine management systems reach the production line.

The Losses of Converting Fuel to Mechanical Power

Any discussion of fuel efficiency will ultimately involve concepts that enable the engine to develop more useful work for the fuel it consumes. Fuel, regardless of its form, contains a specific energy value (specific heating value). Only a percentage of this energy is converted to useful work by the engine. The difference between the heat-energy work-equivalent of the fuel and the mechanical work available at the output shaft represents the thermal efficiency of the engine. The thermodynamically ideal combustion cycle for the internal combustion reciprocating engine is the so-called "constant-volume process," which is based on adiabatic and frictionless conditions and the behavior of ideal gases. Unfortunately, real engines do not operate according to theoretical ideals. Instead, they operate with real gas, incomplete combustion, and with a number of losses to flow, heat dissipation and discharge, and mechanical friction and parasitic loads. Of the energy that is released during combustion, much is either discarded through the cooling and exhaust systems (waste heat), used to move the fuel/air charge through the engine (pumping losses), or is sacrificed to internal friction and parasitic loads (oil pump, water pump, fan, alternator, etc.). A primary objective of advanced engine technologies is to put some of this lost energy to work.

Figure 3.18 compares the fuel that is converted into useful work by a typical Otto engine against the various losses that occur. Brake horsepower is the power available to run the accessories and propel the vehicle. As the chart indicates, the largest losses are due to rejected heat. When fuel is converted into heat energy and then into mechanical power, energy losses tend to be inversely proportional to the combustion temperature (the higher the temperature, the lower the loss). By raising the operating temperature of the engine, it is possible to recover some of the rejected heat and put it to work. However, there are practical limitations to the ability of materials to operate at extremely high temperatures. Ultimately, low heat rejection technologies may be perfected and engines of greater fuel-efficiency may become available. It is theoretically

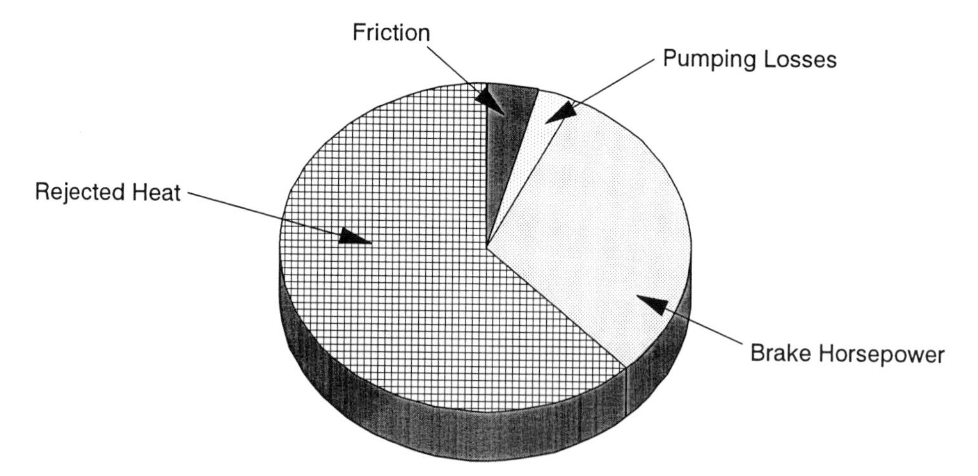

Fig. 3.18. Distribution of Fuel Energy at Wide Open Throttle.

possible to develop adiabatic diesels (low heat rejection) that operate at twice the fuel efficiency of today's designs. Goals, however, are much more modest with expectations in the range of 20-30 percent improvement in efficiency.

Power and Fuel Efficiency Characteristics of the Reciprocating IC Engine

Figure 3.19 shows power and fuel consumption curves of a spark ignition gasoline engine. The values expressed are those of the engine operating at wide open throttle (WOT). They show the characteristics of the engine under its most efficient operating conditions. In actual practice, the engine will rarely operate according to the conditions represented in the chart. Instead, fuel economy will be degraded by the effects of part-load operation.

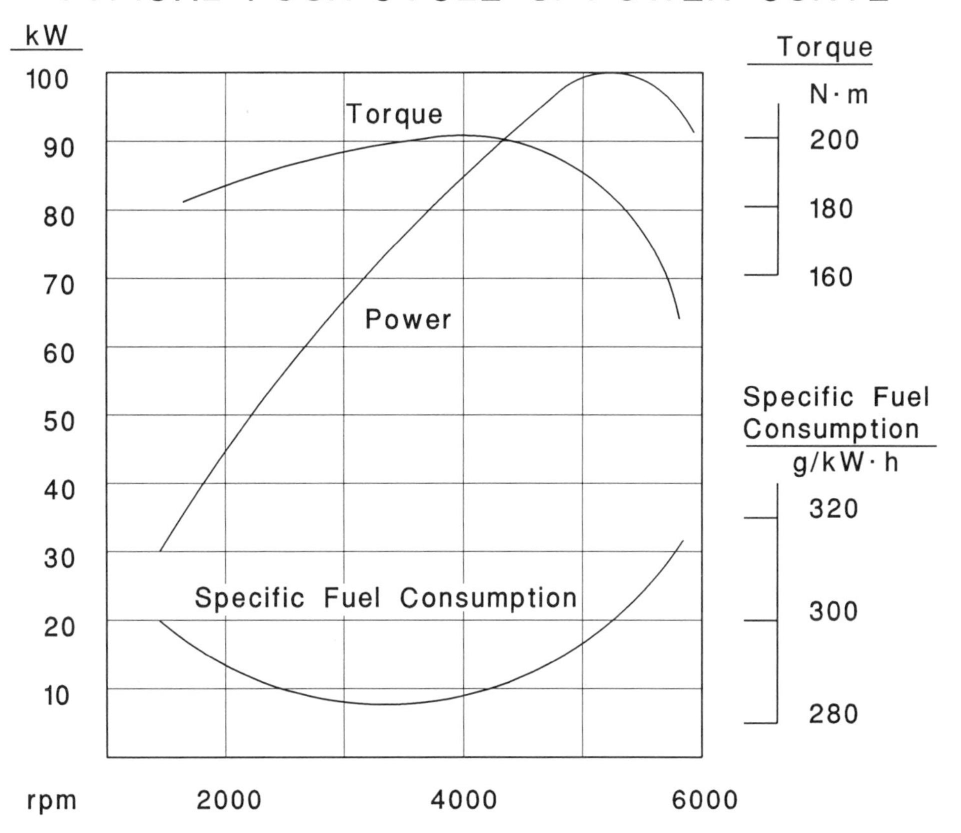

Fig. 3.19. Power and Fuel Consumption at WOT.

The power band of the engine extends from slightly above idle to the rpm at which power begins to drop. Depending on the design, a spark ignition engine will idle at 800-1000 rpm and develop peak power at 3500-6000 rpm. Power, torque and fuel consumption curves are typically established at wide open throttle.

An important characteristic is that the engine can produce a given power across a wide rpm range by regulating the throttle. For example, in Fig. 3.19, 30 kW (40 hp) is developed at approximately 1500 rpm at wide open throttle. Output can also be limited to 30 kW at any speed above 1500 rpm by appropriately throttling the engine. As far as the vehicle is concerned, engine speed is unimportant as long as output matches demand. However, the rpm at which the engine delivers power has a significant effect on fuel consumption.

The Challenge and Opportunity of Part-Load Fuel Consumption

A typical four-stroke, spark ignition gasoline engine operates most efficiently near wide open throttle (WOT). That is, the engine consumes the least fuel per given output when it is developing close to maximum power across its rpm band. With diesel and two-cycle gasoline engines, peak fuel efficiency occurs at slightly lower loads. A diesel operates most efficiently at approximately 60-70 percent load, and a two-cycle engine develops peak efficiency at approximately 50 percent load.

Over the course of a trip, engine efficiency is continually changing according to load. Unfortunately, it changes in a way that is essentially opposite of what is needed for good fuel economy. When engine load is reduced at cruising speeds, efficiency plummets. When engine load increases, such as during periods of acceleration, efficiency increases. When the engine is operating under a light load it may consume two or three times the fuel for a given output as it does when loaded into its maximum efficiency range. Most automobiles operate most of the time with the engine loaded to only a fraction of its capability.

Part-load fuel consumption characteristics have been responsible for a popular wives' tale: namely, that reducing vehicle weight does not do much to improve fuel economy. There is some truth at the basis of the idea. If weight is removed from a vehicle without a corresponding reduction in engine size, fuel consumption will not necessarily decrease. This is because the engine is more lightly loaded in the lighter vehicle and therefore is operating in a region of reduced fuel efficiency. The same result is produced by increasing engine size and leaving vehicle weight unchanged. As a general rule, doubling engine size in a given vehicle will result in approximately 50 percent greater fuel consumption per given distance traveled.

Reducing vehicle weight has a direct effect on fuel consumption if the engine is downsized to maintain an equivalent power-to-weight ratio. By the same dynamics, fuel economy is improved when the vehicle weight is left the same, and the engine is downsized to increase engine loading. To obtain maximum fuel economy, vehicle weight should be reduced for less rolling resistance and inertia loads, vehicle size should be reduced for minimum aerodynamic drag, and the power-to-weight ratio should be limited for increased efficiency.

Poor part-load fuel economy normally results from excessive installed power, which tends to translate into light engine loading at cruising speeds. Under lightly loaded conditions, most of the energy (fuel) is expended to keep the engine itself working, rather than to provide motive power. This results primarily from the fact that pumping and friction losses are speed dependent and function independently of power output. Fig. 3.18 shows friction and pumping losses of an engine at WOT. At high output levels friction and pumping losses are relatively insignificant compared to the energy lost to rejected heat. However, if a given rpm is maintained and the engine is throttled to reduce output, the portion indicating friction and pumping losses will expand into the brake horsepower region virtually in step with the degree to which the engine is throttled. Carried to the extreme, an engine revved under no load will exhibit friction and pumping losses that entirely replace the brake horsepower band. Figure 3.20 shows how load affects the various losses.

The relationship between engine load and fuel efficiency is most clearly illustrated by a fuel consumption map. Such a map indicates fuel consumption in islands, which represent rpm/load regions of operation. Cruising power should be taken at the lowest possible rpm where the engine will be loaded into its region of greatest fuel efficiency. Peak power demands are then met by transmission downshifting which lets the engine accelerate into its region of greatest output.

Part-load fuel consumption characteristics offer both challenges and opportunities. At urban speeds, the typical automobile operates at very light loads where losses to rejected heat, friction and pumping loads can account for 90 percent or more of the fuel consumed. Operating at greater engine loads is therefore the most straightforward method of increasing vehicle fuel economy. However, the vehicle's ability to respond to a change in throttle setting suffers when the engine is more heavily loaded, which has traditionally limited the practice of operating at extremely high engine loads. In addition, consumers are accus-

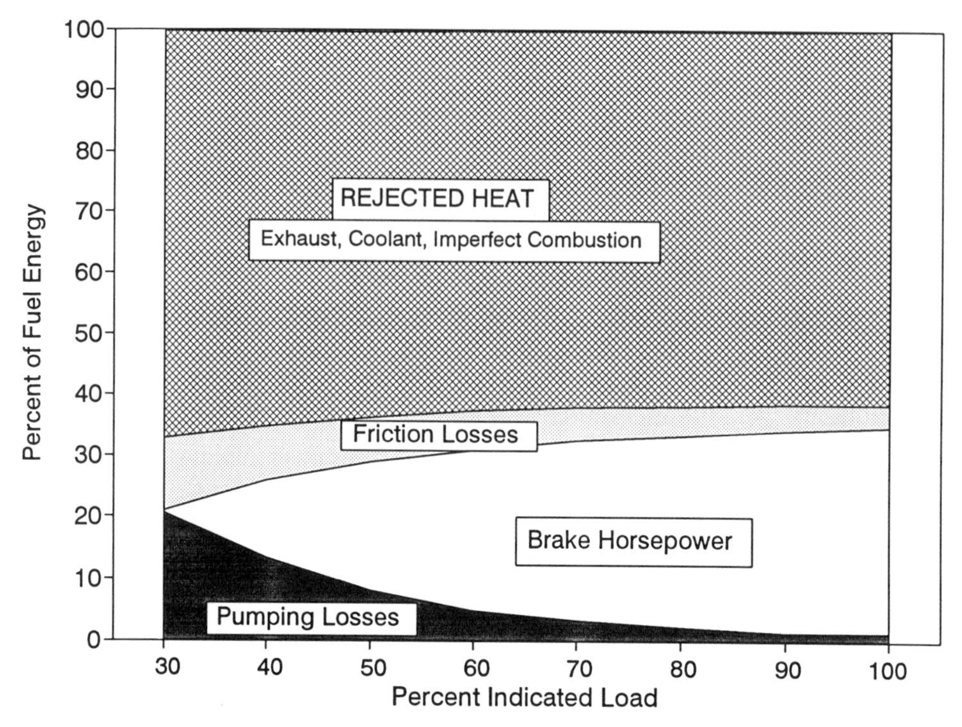

Fig. 3.20. Distribution of Fuel Energy (Spark Ignition Otto Cycle Engine).

tomed to a certain feel, and a vehicle that feels as though it is laboring is not appealing. Engine loading is sensed by sound, vibration, and vehicle responsiveness. A highly loaded powertrain must therefore be engineered to maintain the smooth, lively feel that consumers have learned to expect. Part of the technology for greater vehicle fuel economy therefore centers on small, low-inertia engines that can rapidly respond to changing demands, supercharging systems that enable smaller engines to provide greater peak power yet operate more heavily loaded at cruising speeds, and continuously variable transmissions that more precisely match the engine speed to operating conditions.

The strategy of reducing vehicle road load and operating the engine in a region of greater load is an effective, but relatively low-technology approach to greater vehicle fuel economy. Additional tools in the form of improved engine technology will make it possible to deal more directly with the causes of poor part-load fuel efficiency. The energy lost to rejected heat can be reduced by higher operating temperatures and exhaust recovery systems such as turbocharging. Engine pumping losses will decline with better valving systems. Improved valving, more efficient supercharging, better fuel delivery systems,

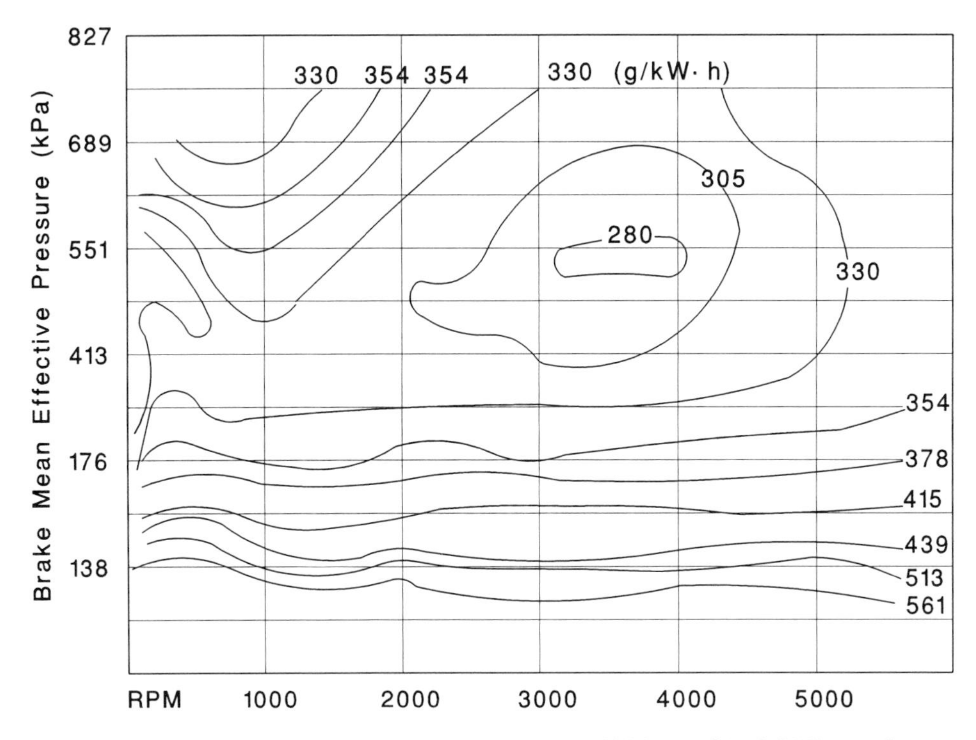

Fig. 3.21. Specific Fuel Consumption Map (4920 cm³ V-8 SI Engine).

variable compression, and even full expansion concepts offer a number of ways to put more of the fuel's energy to work.

Rotary Valves

Poppet valves were adopted by early engine designers because they provided a quick and simple solution to valving needs. Poppet valves are easy to lubricate, they seal well, and they operate with relatively little load on their linear bearing surfaces. In addition, the sealing surface of the valve lifts away from, rather than slides across its stationary mating surface. Consequently, there is no need to lubricate surfaces that are directly exposed to the hot gases of combustion. But poppet valves also have inherent disadvantages. Their characteristically high surface area and the feature of opening into the combustion chamber tend to place limitations on combustion chamber and piston design. Hot spots can develop on the exhaust valve and cause pre-ignition. In addition, the reciprocating action of poppet valves often limits engine rpm due to their tendency to "float" at high speeds. These characteristics have been an ongoing problem for engineers since the poppet valve's inception. As a potential replacement, the rotary valve has been an intriguing alternative.

Rotary valves have been around almost since the inception of the Otto engine. The design is actually a derivative of the steam engine's sliding valve. The fact that the valve rotates, rather than reciprocates, makes it inherently appealing. Because of its rotary action, the valve can operate at much greater speeds. A rotary valve also runs cooler because the same surface is not continuously exposed to hot combustion gases. This characteristic reduces the tendency for pre-ignition, and it allows the engine to operate at higher compression ratios on lower octane fuel. The rotary valve exhibits better breathing characteristics, and therefore less pumping loss, even though the aperture opening does not provide the average aperture size (over cycle time) of a poppet valve of equal maximum aperture. Rotary-valve engines tend to be more fuel efficient, are quieter, and offer a lower engine profile. Reduced valvetrain friction is also claimed by proponents, although comparisons often are made with older poppet valvetrain layouts. As early as 1946, Marcus Hunter, author of Rotary Valve Engines, calculated that rotary valves actually develop greater friction loads than poppet valves.[9]

Unfortunately, rotary valve designs have been plagued with significant problems of their own. Typical problems include thermal distortion, excessive wear, imperfect sealing, and excessively high oil consumption due to the need to lubricate surfaces that are exposed to combustion. This, combined with the fact that poppet valves are proven and reliable, has tended to discourage research by major car companies. The field has therefore been left primarily to independent experimenters and to companies in related industries. A perfected rotary valve design may ultimately result in a more simplified and more fuel-efficient engine.

The most notable early rotary valve designs are probably those produced by Rowland Cross. Cross began developing rotary valve engines in 1922. These early experiments produced impressive results that even Cross himself did not fully appreciate at the time. Speaking in London at a meeting of the Institution of Mechanical Engineers in 1957, Cross said: "Apparently I did not realize at the time that I was setting standards only met by the finest poppet engines then in existence, and which were highly developed in comparison with my crude and very badly made rotary-valve engine." In the 1970s, Cross Manufacturing, in conjunction with Esso Research Co., developed a unique four-cylinder radial engine that used a single rotating valve at each cylinder to control both the intake and exhaust ports. The compact 1.6-liter engine is reported to have been extremely fuel efficient and would operate on low-octane gasoline.

Work over the past 15 years at GV Engine Research PTY., Ltd., in Cronulla, Australia, has produced a design that developers claim eliminates many of the valve's difficulties (see Figures 3.22 through 3.25). The key component is a specially designed sleeve that controls lubrication and provides a positive seal. Most of the work at GV Research has been with small-displacement four-cycle engines. Part-load specific fuel consumption of engines equipped with the new valve is lower at all points of the operating range. Most importantly, the rotary valve has the greatest impact on fuel efficiency at low rpm and light loads typical of the urban environment (see Table 3.3).[10] Improved part-load fuel consumption is characteristic of rotary valves.

TABLE 3.3. ROTARY VALVE EFFICIENCY IMPROVEMENTS
Based on 500 cm³ SI Engine

RPM	Percentage of Full Load Operation			
	15%	25%	40%	60%
	Percent Reduction in Brake Specific Fuel Consumption			
1500	30%	11.8%	8.0%	9.3%
2500	21%	12.3%	11.5%	8.0%
3500	18%	13.7%	8.6%	5.2%

George Coates, an Irish inventor and mechanical engineer, has taken a different approach by entirely eliminating oil from the valve's environment. Unlike the GV cylindrical valve design, Coates uses a spherical valve made of Nitralloy that rides on a floating carbon-ceramic pressure-activated seal. The valve itself is unlubricated, except for the minimal lubrication provided by the fuel. Higher compression ratios and lower valve temperatures are claimed. By eliminating the hot poppet exhaust valve, both pre-ignition and NO_x emissions are reduced. Each cylinder is equipped with two valves, one for intake and one for exhaust. Variable phasing is therefore possible. Coates claims an overall 12 percent savings in fuel and a 20 percent increase in output.

A perfected rotary valve might be especially applicable to the valved, direct-injection, two-stroke engine. A power stroke at each revolution, along with a characteristically short duration for exchange of gases, triples the speed at which gases must be cycled through the combustion chamber. The rotary valve's improved breathing and greater speed capability could resolve some of the

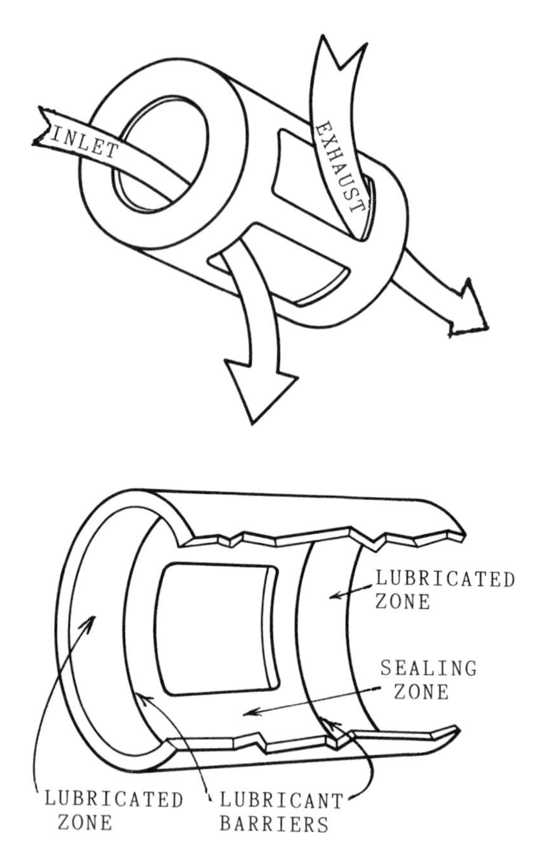

Fig. 3.22. Rotary Valve Gas Path and Seal Details. Top drawing shows gas path through rotor. Bottom drawing shows floating seal that eliminates oil consumption problems typical of previous designs. (Courtesy: GV Engine Research, Cronulla, Australia)

difficulties of the valved two-stroke engine. Rotary valves might also reduce manufacturing costs. One manufacturer of lightweight industrial engines has implemented a program with one of their models to eliminate the camshaft, pushrods and rocker arms, and replace them with a simple, self-contained, low-profile rotary valve cylinder head. Lower costs are especially important in automotive two-stroke applications where conventional valving may increase efficiency, but may also increase costs and thereby diminish the cost advantage of the two-stroke engine. Although the rotary valve has a history of designs that have fallen short of a practical replacement for the poppet valve, that does not preclude the rotary's ultimate success.

Fig. 3.23. Floating Valve Seal (Courtesy: GV Engine Research, Cronulla, Australia)

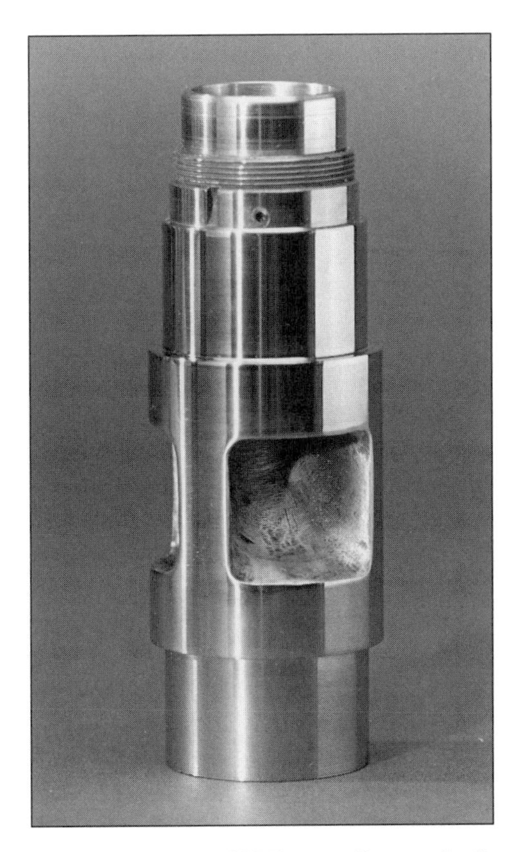

Fig. 3.24. Valve Rotor. (Courtesy: GV Engine Research, Cronulla, Australia)

Fig. 3.25. Low-profile rotary valve assembly eliminates complicated valvetrain and allows extra freedom in combustion chamber design. (Courtesy: GV Engine Research, Cronulla, Australia)

Fig. 3.26. Coates Retrofit Rotary-Valve Cylinder Head. (Courtesy: Coates Enterprises, Ltd.)

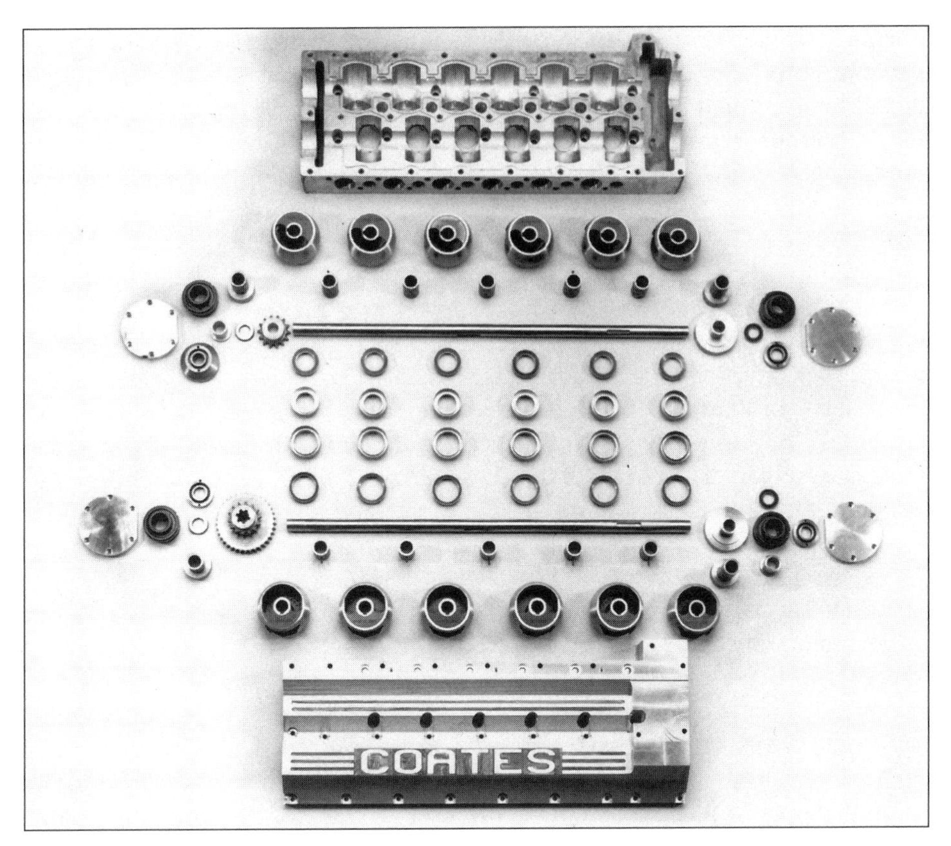

Fig. 3.27. Coates Rotary-Valve Head Assembly. (Courtesy: Coates Enterprises, Ltd.)

Variable Compression Ratio

The compression ratio of an engine is typically fixed at a specific value. Of necessity, compression ratio is generally a compromise value in order to account for operational extremes. As a result, the engine is often operating at less than peak efficiency throughout the rest of its operating band. The ability to adjust the compression ratio according to operating conditions results in a more fuel-efficient engine of greater output.

Higher compression lets the engine run with leaner mixtures, and it improves the engine's thermal efficiency. The limiting factor is that high compression ratios lead to overheating and knocking at higher engine loads, especially in the lower speed ranges. Ideally, the compression ratio should be increased when the engine is lightly loaded, and reduced when load increases. A continuously variable compression ratio that changes according to operating conditions would increase fuel efficiency throughout the operating range. More importantly, fuel efficiency would be especially enhanced in the lower load ranges that are typical of city driving. Variable compression ratio (VCR) technologies give engineers the ability to selectively adjust compression according to operating conditions.

Variable compression ratio can be achieved in a number of ways. Several engines have been designed with features such as variable height pistons, vertically sliding cylinder heads, adjustable length connecting rods, and a secondary combustion chamber with an adjustable volume. A VCR engine developed by Volkswagen utilizes a secondary chamber with a small piston that is adjusted to vary the volume of the chamber and thereby control compression. Piston depth is controlled by an electric motor and worm gear arrangement.

Another VCR idea utilizes a spring against a secondary-chamber piston to automatically vary the compression ratio according to cylinder pressure.[11] This design automatically reduces compression as load increases, and increases compression as load is reduced. The prototype shown in Figures 3.28 and 3.29 is equipped with a large adjustable tensioning device for experimentation purposes. Dynamometer tests conducted on a single-cylinder two-stroke engine at the laboratory of TVS-Suzuki, Ltd., in India, demonstrated significant improvements in fuel consumption throughout the engine's operating band. Power output also was improved.

Active Thermo-Atmosphere Combustion (ATAC) in two-stroke cycle applications is described by NICE and others, and is another concept that may ultimately benefit from VCR technology. ATAC is initiated by controlled auto-ignition. It is a different combustion process than occurs during either spark-ignition or diesel combustion. In a spark-ignition engine, combustion begins when a kernel is established by the spark plug. A flame front is thereby established and the flame then propagates outward from the kernel and quenches against the cooler surfaces of the cylinder, piston, and cylinder head. Under this scenario, combustion begins and ends at a precisely defined point in time. During ATAC there is no traditional flame front and the onset of combustion is much more gradual. Instead of a traditionally propagating flame front, ATAC is a more homogeneous process with combustion initiating at multiple points throughout the charge and progressing at a slower rate.[12]

Once initiated, ATAC continues thereafter regardless of the occurrence or absence of an electric spark. The shift from spark-controlled combustion to

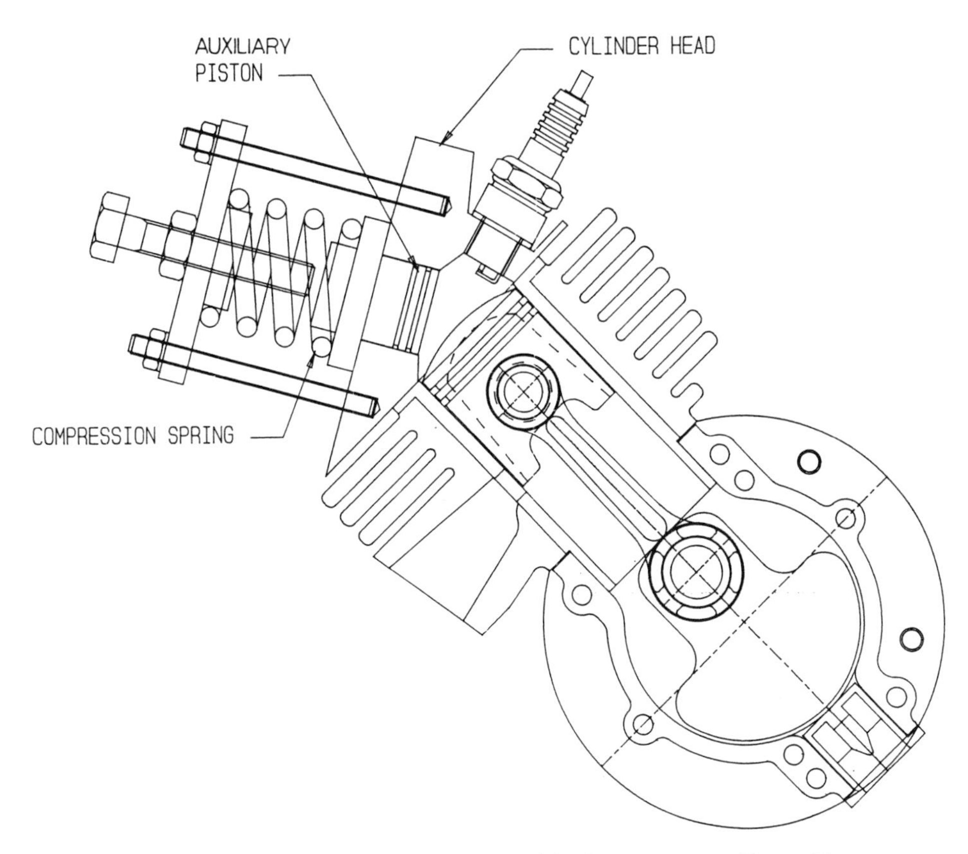

Fig. 3.28. Layout of Automatic Variable Compression Ratio Engine. (Courtesy: TVS-Suzuki Limited, Tamilnadu, India)

ATAC is smooth. During ATAC, cycle-to-cycle P_{max} becomes much more uniform, and the engine runs more smoothly and quietly, and without knock. During light to moderate loads, ATAC stabilizes the lean-burn charge, essentially eliminating the misfires typical of the two-stroke cycle, and thereby significantly reducing fuel consumption and HC emissions. A diagram of cylinder pressure curves reveals a uniform rise and fall in pressure over multiple cycles that is unequaled by the traditional combustion process. VCR technology utilizing continuously variable and electronic-feedback-timed compression may provide additional control over ATAC and open the way for improved engine performance, as well as reduced emissions, and lower fuel consumption at cruising speeds.

Fig. 3.29. AVCR Engine Cylinder Head Components and Components Assembled on Test Stand. (Courtesy: TVS-Suzuki Limited, Tamilnadu, India)

Variable Valve Actuation

Variable valve actuation (VVA) is another emerging technology that is already producing significant improvements in engine performance. Historically, the point at which valves open and close, the dwell time, and the overlap between intake and exhaust valves (phasing) are fixed to specific crankshaft rotational points (crank angles). As a result, the timing of these events is compromised throughout much of the operating band because ideal valve timing varies according to engine speed and load. Control over valve lift allows the engine to operate more efficiently over the entire speed/load range. The torque curve, as well as emissions and fuel consumption characteristics, become more fluid and controllable attributes. They are no longer fixed by the geometry of the engine.

Early versions of VVA systems enhance specific areas of engine performance through a limited number of valve timing and phasing options. The more sophisticated the system the greater the benefits and the higher the costs. A simple two-position system of control over intake valve lift is relatively inexpensive and can still provide improved idle and a flatter torque curve. Such systems are already incorporated into engines of production automobiles by Nissan, Mercedes-Benz, Alpha Romeo, and Porsche. Figure 3.30 shows the operating principle of the variable cam phasing (VCP) system used in the Porsche 968. It controls the angular relationship between intake and exhaust cams by moving the chain tensioning device.

Continuous and non-incremental control over valve timing and phasing, and controlling intake and exhaust valves independently of each other and independently of the crank angle, opens new possibilities in engine design. VVA provides engineers with control of events not possible before. VVA can control compression ratio and change the effective displacement of the engine. Advanced VVA systems are expected to result in more power from smaller engines, greater low-rpm torque, a broader torque band, greatly improved part-load fuel efficiency, and lower emissions. Figure 3.31 shows the torque improvements possible with an optimized variable valve actuation system.[13]

Several versions of a VVA system have been built and tested at Clemson University. Professor Alvon C. Elrod, along with the help of engineering students, has built a number of prototypes of a variable timing camshaft system that in laboratory tests produced a 20 percent improvement in fuel efficiency. The system can be retrofitted to existing engines or incorporated in an existing line of engines without redesigning the engine or the production line. The device takes the place of the existing camshaft and, on engines with dual overhead cams, it provides control over timing as well as phasing.[14]

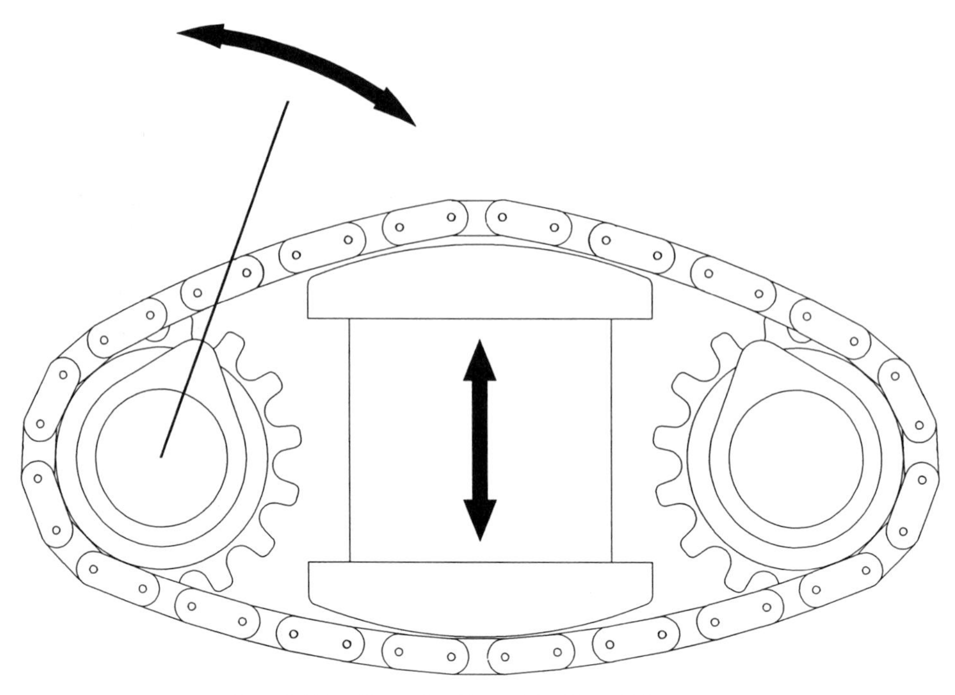

Fig. 3.30. Porsche Variable Cam Phasing.

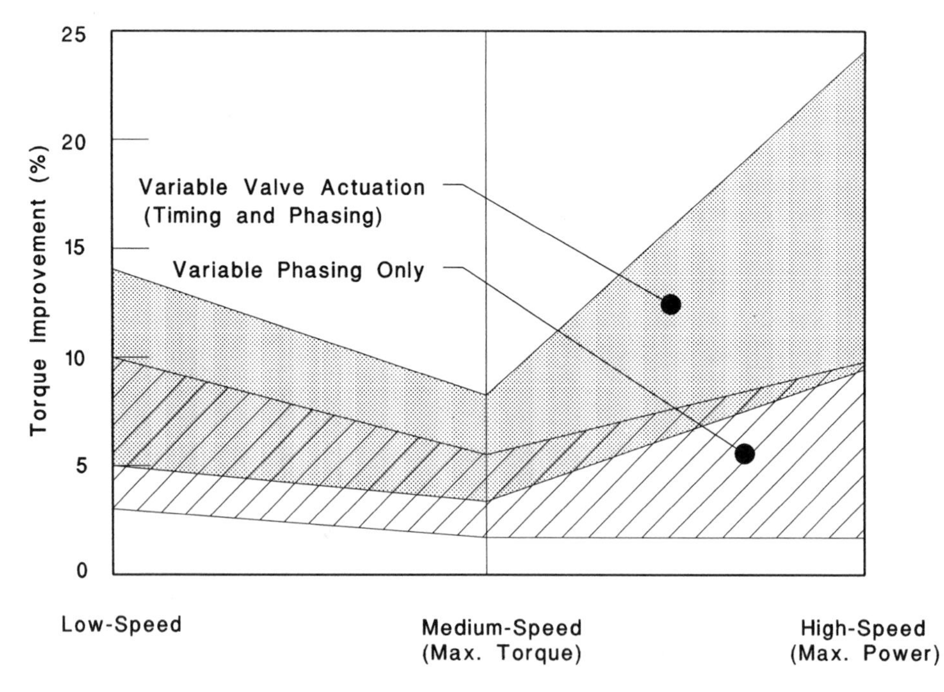

Fig. 3.31. Potential Torque Improvement with Variable Valve Actuation.

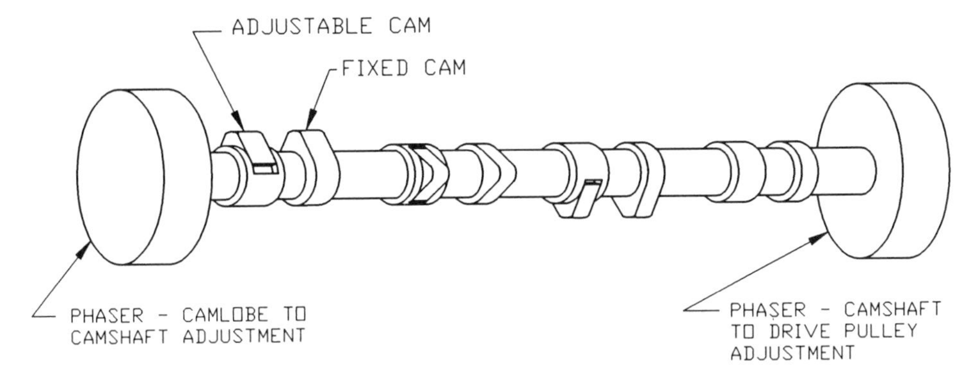

Fig. 3.32. The Clemson Variable Timing and Phasing Camshaft can be assembled in a variety of configurations. The design above is for a dual overhead cam application. It allows for optimization of all valving events. (Courtesy: Clemson University)

Development began with the premise that poor part-load fuel economy is basically caused by the throttling process in the spark ignition engine. Part-load throttling losses occur when charge airflow is restricted (throttled) to control engine output. Restricting intake creates a vacuum which places an additional load on the engine and reduces compression. At wide open throttle, manifold vacuum is decreased because charge air can flow more freely into the engine. The innovative camshaft can control engine speed by varying valve timing instead of throttling charge-air. As a result, the engine runs under conditions that more closely emulate WOT. Twin camshafts provide independent control of intake and exhaust valves.

A more recent VVA system is the Electromagnetic Valve Actuator (EVA) developed by Aura Systems of El Segundo, CA. Aura's relatively new line of high force actuators appears destined for a variety of automotive applications. The company's exceptionally precise and efficient linear motors can produce smaller audio speakers that outperform larger speakers of traditional design, more effective active suspension systems, noise and vibration cancellation devices, and precisely controlled and variable fuel delivery systems. Aura's EVA is configured as a complete drop-in module that replaces the entire valve and seat assembly, as well as the rest of the valvetrain (Figure 3.34). Valve lift is then controlled by the car's ECU. Dwell, timing, and phasing are infinitely variable. Moreover, the operating schedule can be easily switched to accommodate different fuels. In combination with similarly controlled fuel metering, flexible fuel vehicles can operate in essentially a dedicated mode with different fuels. Specifications are shown in Table 3.4.

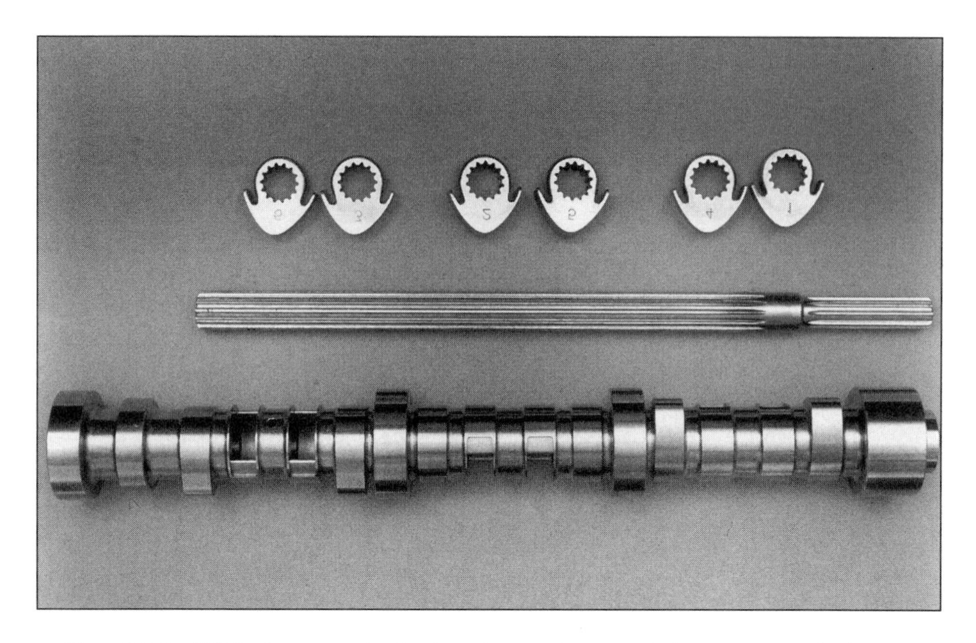

Fig. 3.33. With this simple design the entire cam lobe rotates. A design utilizing split cam lobes in which movable cam lobe halves rotate axially out of plane with their mating fixed halves may be used to control dwell, phasing, and timing.

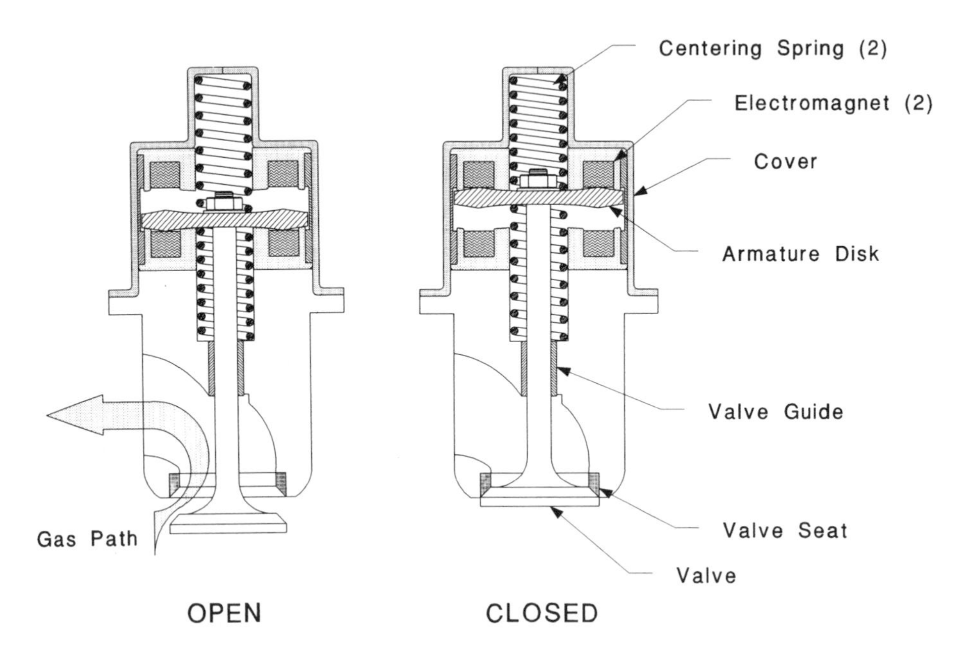

Fig. 3.34. Layout of Aura Systems Electromagnetic Valve Actuator (EVA).

TABLE 3.4. AURA SYSTEMS ELECTROMAGNETIC VALVE ACTUATOR (EVA) SPECIFICATIONS AND PERFORMANCE

Specification	2-Valve Cylinder[1]		4-Valve Cylinder[2]	
Mechanical:				
Maximum rpm	6250		7500	
Transition Time	2.42 ms		3.3 ms	
Lift	7.62 mm		9.5 mm	
Valve Opening Force	500 N		500 N	
Valve Seating Force (adjustable)[3]	250 N		250 N	
Touch-Down Velocity (setable)	0.15 m/s		0.15 m/s	
Diameter of EVA Body	62 mm		40 mm	
Length (surface of head to top of body)	72 mm		85 mm	
Valve Mass	90 g		40 g	
Overall EVA Weight (includes valve)	0.85 kg		0.35 kg	
Electrical:				
Required Voltage	12 or 24 V Nominal		12 or 24 V Nominal	
Current Draw @ 24 Volt:				
Initialization (on start of engine)	32 amp, 8 ms		22.8 amp, 5.5 ms	
Transition (once per transition)	18.4 amp, 5.2 ms		13.1 amp, 5.5 ms	
Holding (between lifts)	4.5 amp capture to release		3.2 amp capture to release	
Power Consumption in Watts (incl. 50% alternator efficiency loss):	Ea. Valve	(8 Valves)	Ea. Valve	(16 Valves)
800 rpm	14.4	(115.2)	9.3	(148.9)
2500 rpm	31.4	(251.2)	20.3	(325.2)
6000 rpm	66.4	(531.2)	43.1	(689.3)
7500 rpm	-		52.8	(845.0)

1. Verified by tests. Tested on a 2.3-L Ford Ranger.
2. Estimated.
3. The force at which the valve is held against the valve seat.

Source: Aura Systems, Inc.

The Diesel Engine

The design for a compression ignition (CI) engine was conceived by Dr. Rudolph Diesel in an attempt to improve upon the poor fuel efficiency of the spark ignition engine. A paper on Diesel's theories was published in 1893 and stimulated additional research among German contemporaries. However, early designs were cantankerous and some prototype engines actually exploded during tests. While experiments were being conducted in Germany on Diesel's design, an English engineer by the name of Herbert Akroyd-Stuart developed and patented a more successful version. Stuart's design utilized an intake of uncharged air that was ignited by injecting a liquid fuel directly into the combustion chamber, which is how the modern diesel engine operates. In a diesel engine, ignition takes place when the fuel is injected. Ignition timing is controlled by fuel delivery, rather than by a spark. Engine rpm is controlled by regulating injection duration. Air intake is unthrottled.

Diesel engines are either direct injection (DI) or indirect injection (IDI) designs. The DI engine is more fuel efficient, requires more exacting tolerances, is less tolerant of fuel inconsistencies and is generally more noisy and rough than an IDI engine. In the early '80s, a typical diesel-powered automobile would consume about 30 percent less fuel than a similar gasoline-fueled car. Today, improvements in SI engine technology have reduced the gap to about 10-15 percent in the diesel's favor. Diesel engines are still more fuel efficient, and efficiency is likely to improve with new developments in turbocharging and low-heat-rejection technologies. Part of the diesel's greater fuel economy is due to the slightly greater energy content of diesel fuel, but most of it comes from the inherently greater thermal efficiency of the cycle. The high compression ratio and the reduced pumping losses of the unthrottled intake are generally accepted as the primary contributors to the diesel's high fuel efficiency. Both power and fuel economy are significantly improved by turbocharging.

Regulated emissions are also inherently lower with the diesel engine. Without aftertreatment, diesel exhaust emissions are as low as those of a typical gasoline engine equipped with a three-way catalytic converter. The diesel's inherently clean combustion has therefore allowed diesels to operate without exhaust aftertreatment. However, the emissions profile is not all positive and more stringent regulations are under review. Diesel engines produce a large quantity of particulate emissions, primarily in the form of soot that is laden with liquid hydrocarbons. Animal studies have indicated that some of these hydrocarbons may be carcinogenic. Consequently, different components of the diesel's exhaust may require greater emphasis in emissions control.

A diesel engine is typically more costly to manufacture and more expensive to service. Also, a diesel's high-pressure fuel pump is more easily damaged by contaminated or water-laden fuel. However, proper filtration and routine maintenance can avoid problems attributed to fuel contamination. And higher manufacturing costs are ultimately returned to the customer in the form of increased engine life. To match the engine's long service life, the rest of the vehicle should also be designed for longer life. A diesel engine will typically run 320,000 km (200,000 miles) or more.

Regardless of its inherent advantages, diesel engines have yet to win favor with U.S. consumers. In 1990, just slightly over 0.5 percent of new cars sold in the U.S. were equipped with diesel engines. In Europe, the diesel is much more popular and far more prevalent. One-third of new cars sold in Belgium are equipped with diesel engines. In France diesels account for nearly one-quarter of new car sales, and in Germany, Spain and the UK diesel sales average 11, 14, and 11 percent, respectively. Although the engine is inherently more noisy and less smooth than its gasoline counterpart, proper design can result in a quiet and smooth-running vehicle. In addition, the engine also offers greater opportunity

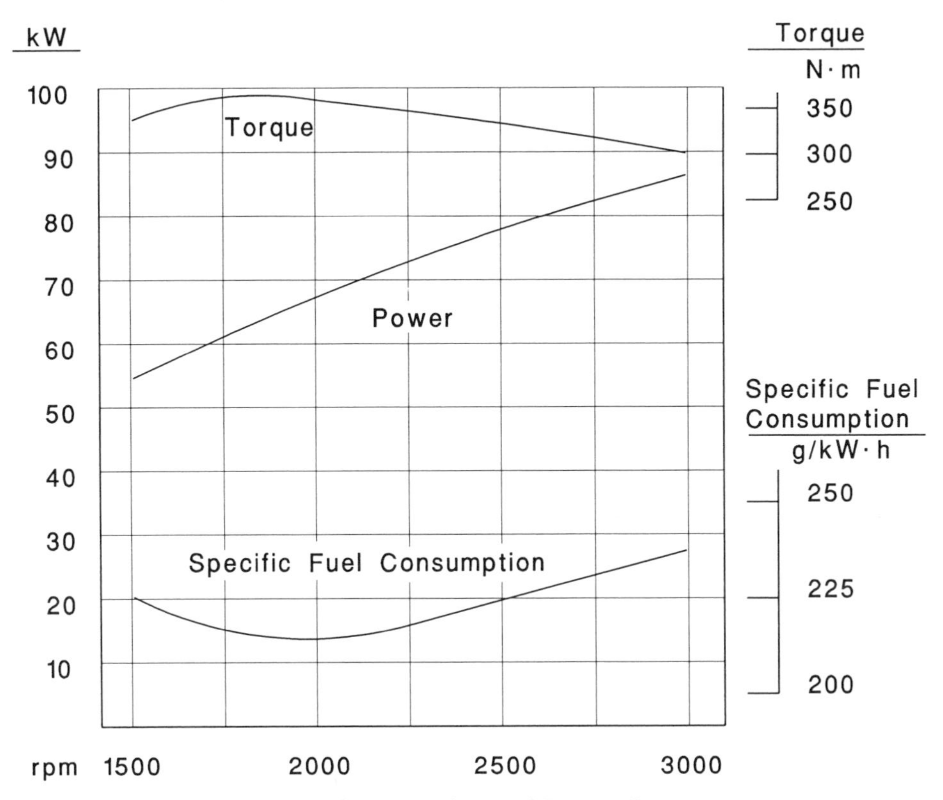

Fig. 3.35. Typical Diesel Power Curve.

for significant improvements in operating efficiency. Improvements are likely to flow from improved fuel delivery systems, more precisely controlled combustion, and low-heat-rejection technologies.

Supercharging

In the past, supercharging has been associated with increased power for high-performance cars, and often at the expense of fuel economy. Supercharging also can be used to increase engine efficiency and improve vehicle fuel economy. The difference in results depends on the orientation of the application. If a given vehicle is supercharged for increased power and performance, fuel consumption will also increase. However, if the goal is to improve fuel economy, the designer can use supercharging to achieve equal power and performance from a smaller, lighter engine and thereby improve fuel economy. Fuel economy gains as high as 20 percent or more are typical for a supercharged diesel. Traditionally, the spark ignition engine is less responsive to supercharger-induced improvements in fuel economy. However, operating a small engine at relatively high loading and utilizing the supercharger to increase output in order to meet peak demands can result in significantly improved fuel economy.

The supercharger operates by pressurizing the incoming air. Pressurization reduces the internal pumping losses associated with air intake, and it raises combustion chamber pressure for increased combustion efficiency. A supercharger therefore increases the power output of the engine. A supercharged engine of a given size burns more fuel and produces more power than a naturally aspirated engine. However, the supercharger also allows a smaller engine to produce equal power, and it improves low-rpm torque characteristics.

The traditional supercharger is comprised of a mechanically driven compressor or blower. A supercharger is normally belt-driven from the engine output shaft. A more widely used method of pressurizing charge air is turbocharging. The turbocharger works on the same principle, but it is driven by exhaust gases passing through a turbine assembly. The turbocharger is therefore a type of waste heat recovery system. Another exhaust heat recovery system is the pressure-wave "Comprex" unit. Each of these systems has its advantages and disadvantages.

An interesting concept called *air-injection supercharging* was explored many years ago at General Motors Research Labs.[15] Instead of a conventional pump set to pressurize charge air, compressed air from high-pressure tanks was introduced through a third valve directly into the cylinders during the compression stroke. An engine-driven compressor then recharged the compressed air

TABLE 3.5. COMPARISON OF SUPERCHARGE SYSTEMS
(+ signifies advantage over other systems)

Characteristics	Exhaust Turbocharger	Comprex Wave Charger	Positive Displacement Supercharger
Nominal Power Output	++	+	
Torque at Low Engine rpm		++	+
Response		+	++
Improved Fuel Consumption in Consumer Driving Cycles		++	+
Noise Radiation	++		
Pollutant Emissions	+	++	
Function With Additional Exhaust Cleanup Devices	+		++
Space Requirement and Free Choice of Placement	++		+
Technology Close to Production Release	++	+	
Manufacturing and Installation Cost	+		++

Source: Automobile Technology of the Future, SAE R-107, Society of Automotive Engineers, Warrendale, PA, 1991.

reservoirs. Tests of a converted 2.7-L engine showed a 250 percent increase in power output at 2000 rpm. With the compressor inoperative, fuel economy was 47 percent greater at 32 km/h (20 mph) and 25 percent greater at 97 km/h (60 mph). With the compressor engaged, gains dropped to about 35 percent and 15 percent, respectively. Disadvantages included the large volume of air required and the complications of the high-pressure delivery and metering system which required an extra valve at each of the eight cylinders.

Low Heat Rejection Engines

The low heat rejection (adiabatic) diesel engine is based on the concept of retaining the waste heat that is now discarded through the cooling and exhaust systems. Experimental designs concentrate on the heat lost to the coolant system, leaving exhaust recovery to various forms of supercharging. With an adiabatic engine, the cooling system is eliminated and the engine is insulated against radiation heat losses. Ceramic surfaces on pistons, valves and cylinder heads are normally required to withstand the extremely high temperatures.

Operating temperature of the cast iron itself increases from 200°C to about 875°C; even higher temperatures are possible. Theoretically, much of the heat that is normally lost through the coolant should be retained within the engine, resulting in a corresponding increase in thermal efficiency.

Experimental adiabatic engines have been built by a number of automobile manufacturers. In the early '80s, an experimental Ford DI diesel, with a 93 percent reduction in cooling system heat loss, produced fuel economy gains of 2 to 25 percent (depending on engine load), and 38 to 56 percent lower particulate emissions in the light to medium load ranges.[16] Fuel efficiency gains as great as 40-50 percent are theoretically possible if waste heat can be retained. However, early experiments have indicated that much of the rejected heat is not necessarily retained by simply eliminating the cooling system. Instead, heat is redirected to the exhaust system where it is rejected in the form of higher exhaust temperatures.

The Two-Stroke Cycle Engine

Although some automobiles have been equipped with two-stroke engines, traditionally the two-stroke cycle has not been considered a satisfactory automobile powerplant. The most limiting characteristics of the engine have been its high fuel consumption and high exhaust emissions. Irregular part-load combustion, relatively short life, and reduced low-rpm torque are other disadvantages. For a while, it looked as though its characteristically high emissions might actually force the two-stroke engine into oblivion. Today, electronic controls and new fuel delivery and scavenging techniques seem to point to a brighter future.

Theoretically, the two-stroke cycle is more fuel-efficient than the four-stroke cycle. This results primarily from the power stroke at each revolution, which essentially reduces friction losses by half. Another advantage includes improved part-load fuel consumption characteristics. Lowest specific fuel consumption occurs at approximately 50 percent load, which is much closer to the typical loads experienced during urban driving. The engine also produces more power per given displacement, and it is lighter, mechanically more simple and less costly to manufacture than a four-stroke engine of equal power. The engine's two primary deficits, poor fuel consumption and high exhaust emissions, have been its undoing. These shortfalls arise primarily from the difficulty in controlling the exchange of gases.

In its most basic form the two-stroke engine has no valves, at least not in the traditional sense. Gases flow into and out of the cylinder when ports in the cylinder wall are uncovered as the piston nears the bottom of the stroke. By uncovering the exhaust port first, cylinder pressure drops and much of the burned charge is expelled into the exhaust header. Slightly afterwards, the intake port is uncovered. At this point, charge air, which has been pressurized in the crankcase by the downward migrating piston, is injected into the cylinder. As the piston moves upward after bottom dead center, the ports are closed off and the fresh charge is trapped and compressed where it is ignited near top dead center, and the process begins anew (see Figure 3.36).

The task of clearing out exhaust gases and introducing a fresh charge takes place at essentially the same time, during approximately 250-270 degrees of crankshaft rotation. As a result, some of the exhaust gases mix with the fresh charge, and some of the charge air ends up being discharged through the exhaust port, unburned. Short-circuiting losses alone can amount to as much as 10-30 percent of the total fuel consumed. The engine's characteristically high emissions of unburned hydrocarbons results primarily from the unburned fuel

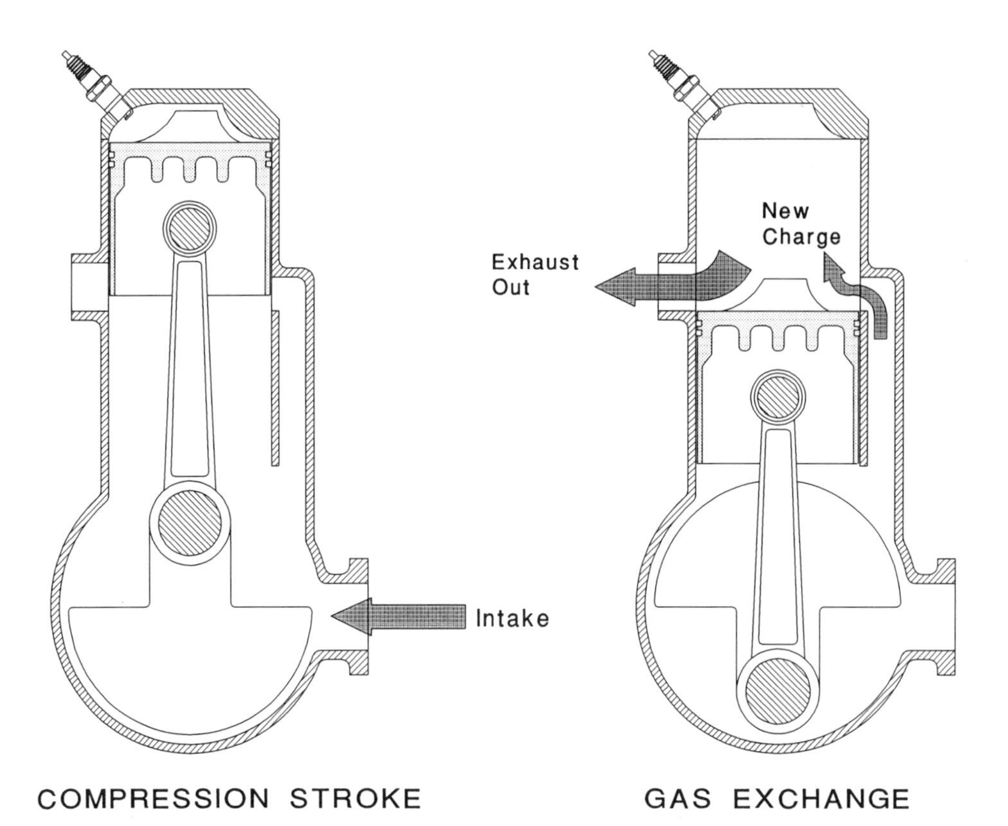

Fig. 3.36. Two-Stroke Cycle.

that escapes into the exhaust header. Since the four-stroke engine dedicates one entire cycle to the exchange of gases, short-circuiting losses do not occur, and exhaust and intake gases typically do not mix.

The limited time available for exchange of gases is another negative characteristic. Lacking a separate stroke for gas exchange, one might assume that the two-stroke must complete the process in half the time. Actually, it must be accomplished in one-third or less of the time available in a four-stroke cycle engine. Engine efficiency significantly improves when uncharged air is used for scavenging and the fuel is injected directly into the cylinder after the ports have been covered by the upward-migrating piston. This requires an injector nozzle pressure of approximately 10 MPa (100 bar), and events must take place over an extremely condensed time budget. Table 3.6 shows the time allotted to events in a direct injection two-stroke engine at 6000 rpm.[17]

TABLE 3.6 DIRECT INJECTION TWO-STROKE CYCLE SCHEDULE OF EVENTS

Event	Time Available	Degrees Crank Angle
Exhaust and Induction	4 ms	144
Fuel Injection	1 ms	36
Mixing and Vaporization	2 ms	72
Combustion and Expansion	3 ms	108

Source: Ref. [17]

Injection and vaporization requires a spray velocity on the order of 100 m/s, and droplets must be no larger than 10-20 microns immediately after leaving the nozzle. Complete vaporization and precise control of events were not possible before the advent of electronic controls and high-pressure injectors. It is this technology that is responsible for the resurgence of interest in the two-stroke cycle engine.

The Transmission as a Tool for Reducing Fuel Consumption

Unfortunately, IC engines characteristically develop power in a way that is not very well matched to the requirements of the typical automobile operating schedule. Modern automobile engines can deliver little power at very low rpm. Engine output capability climbs steadily from idle speed and typically reaches maximum levels in the range of 3500-6000 rpm. In general, the vehicle's demand for power is diametrically opposed to these output characteristics (see Figure 3.37). A vehicle typically needs maximum power (for acceleration) near its minimum speed. Road load at cruising speeds may demand as little as 5-10 percent of the engine's capability at a given rpm. Passing and negotiating grades impose high but temporary power demands with moderate to zero change in vehicle speed. The component that makes it possible for these two mismatched curves to work together is the transmission.

In the case of a manual-shift transmission, the task of selecting an appropriate ratio is delegated to the driver. In the 1940s the automatic transmission emerged and ultimately assumed the job of managing ratios, at least for most U.S. drivers. Early versions were sluggish, troublesome, and inefficient devices that could not match the performance of an adroit human at the controls of a manual-shift gearbox. Today, automatic transmissions are much more efficient and they typically run 160,000 km (100,000 miles) or more with very little service. In addition, they have actually become a tool for increasing vehicle performance and fuel economy.

The ability of the transmission to improve fuel economy results from the part-load fuel efficiency characteristics of the internal combustion engine. When

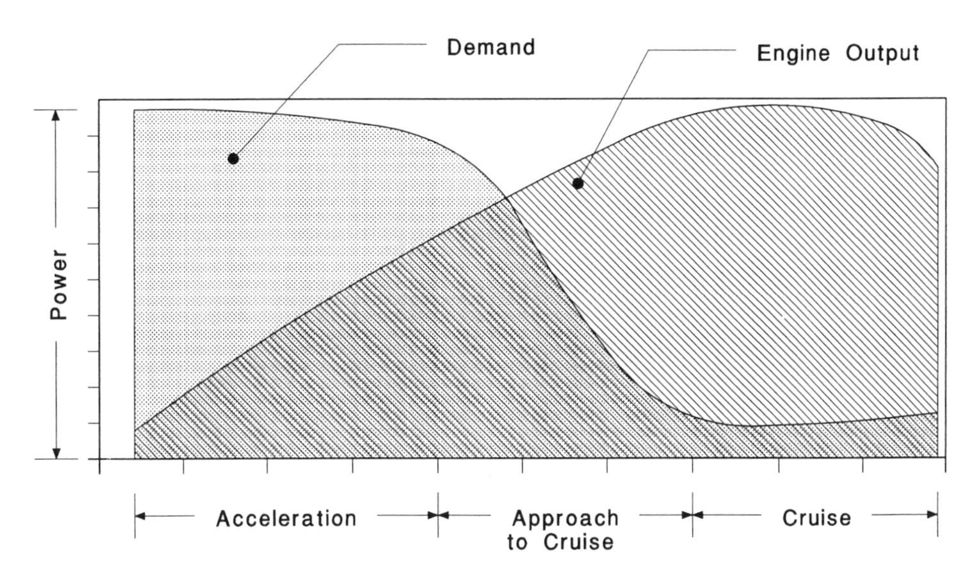

Fig. 3.37. Vehicle Power Demand Compared to Engine Output Characteristics.

acceleration power is no longer required, the transmission can upshift to a ratio that will load the engine into its region of lowest brake specific fuel consumption. A properly designed shift schedule can have a significant effect on fuel economy. Unfortunately, the match between vehicle speed and engine speed is usually a compromise value because of the limited number of ratios that are feasible with a conventional multi-ratio transmission. Multi-ratio transmissions are normally limited to three, four, or five discrete ratios. However, precise engine load management requires a greater degree of ratio selectivity. A promising alternative is the continuously variable transmission.

The Continuously Variable Transmission

The continuously variable transmission (CVT) has been around as long as the automobile itself. Engineers have always recognized its theoretical advantage over the multi-ratio gearbox. A CVT enables the engine to run at its most fuel-efficient or most power-intensive speed while driving the vehicle at any speed desired. With a CVT, engine speed and vehicle speed are no longer connected by a series of discrete ratios. Instead, they can function independently across a wide and stepless band according to engine characteristics and performance requirements. The advantages of this capability are enormous. Most obvious in the IC engine application is that the engine can be loaded into its most fuel-efficient region at cruising speeds, then allowed to accelerate into its region of greatest output when peak power is needed, regardless of vehicle speed. Unfortunately, practical problems have consistently plagued the design. More recently, the CVT appears to be coming of age.

There are essentially two types of continuously variable transmissions: belt CVTs and traction drives. A belt CVT utilizes a belt that connects two pulleys of changeable pitch diameters. The most familiar variety is the unlubricated composite V-belt type normally found on snowmobiles and all-terrain vehicles. Quincy-Lynn's Urbacar and Trimuter, as well as a number of European micro-cars, have utilized transmissions of this type. Perhaps the most notable automotive application is the DAF Variomatic developed by Van Doorne. Unfortunately, transmissions of this type work less well as engine size is increased. Excessive wear and the inability to transmit high torque loads have been consistent problems. In an effort to improve power transfer capability and reduce wear, engineers at Van Doorne developed an improved belt-drive CVT that utilizes a steel belt running in an oil bath. Several versions have been based on this principle, including the production ECVT (Electro Continuously Variable Transmission) used in the Subaru Justy. Another type of CVT—the traction drive—operates differently. These designs employ rolling elements of various configurations and vary the ratio by moving the power transfer points (roller contact points) either closer to, or farther from, the axes of the input and output discs. Traction drives have also been plagued with problems, primarily as a result of inadequate lubricants. However, new lubricants are on the horizon.

With the simple composite V-belt CVT, centrifugal force generated by engine speed causes weights to fly outward and force the movable face against the fixed face of the driver pulley. The greater the engine speed, the greater the force squeezing the two pulley faces together. This causes the belt to ride farther out on the rim as the gap between the fixed and movable faces closes. Concurrently, the belt is pulled deeper into the driven pulley, forcing its two faces apart. Torque-sensitive units employ a cam on the driven pulley designed to increase belt squeeze in response to increased torque (see Figure 3.38). Torque increases when the vehicle encounters greater resistance or when the operator applies power. Consequently, the actual shift position is a result of the continuous balance between the centrifugal forces of the driver pulley and the counteracting forces of the driven pulley.

Transmission efficiency depends on how the unit is tuned to the application. If the belt is subjected to too much squeeze, friction increases and efficiency suffers. If there is too little squeeze, the belt slips and efficiency also suffers. Efficient operation therefore relies on a precise balance of forces throughout the operating range. Belt squeeze must be continuously matched to torque. This ideal operating profile can be difficult to maintain. However, when correctly tuned in a relatively low-horsepower application, efficiency can be quite high (see Figure 3.39). More sophisticated automotive versions were successfully used by DAF and Volvo in Europe.

Van Doorne's lubricated steel-belt CVT works on a similar principle. But instead of a composite belt, the design relies on a lubricated steel belt to transfer power and employs hydraulic pressure to provide belt squeeze and effect changes in shift position. Also, the unit can utilize electronic feedback and logic to analyze load and operating conditions and modify the shift schedule accordingly. In CVT terminology, the pulley and belt assembly comprises the variator. The variator is then combined with other components such as a fixed-ratio reverse gear, hydraulic pumps, governors and actuators, and an automatic clutch or a hydraulic torque converter. A steel-belt CVT is capable of transferring much greater power with much improved wear characteristics. Such designs also provide more precise control of belt squeeze and shift position. Unfortunately, parasitic losses to subsystems and the torque converter are essentially on par with a conventional automatic transmission. However, because of design's ability to take advantage of engine operating characteristics, vehicle fuel economy can be

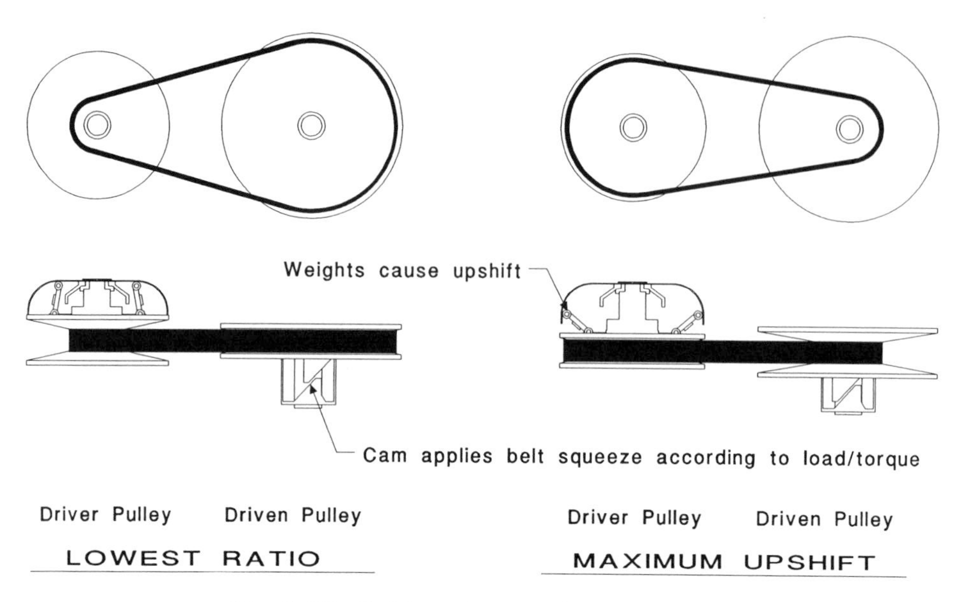

Fig. 3.38. Salsbury Torque-Sensitive CVT.

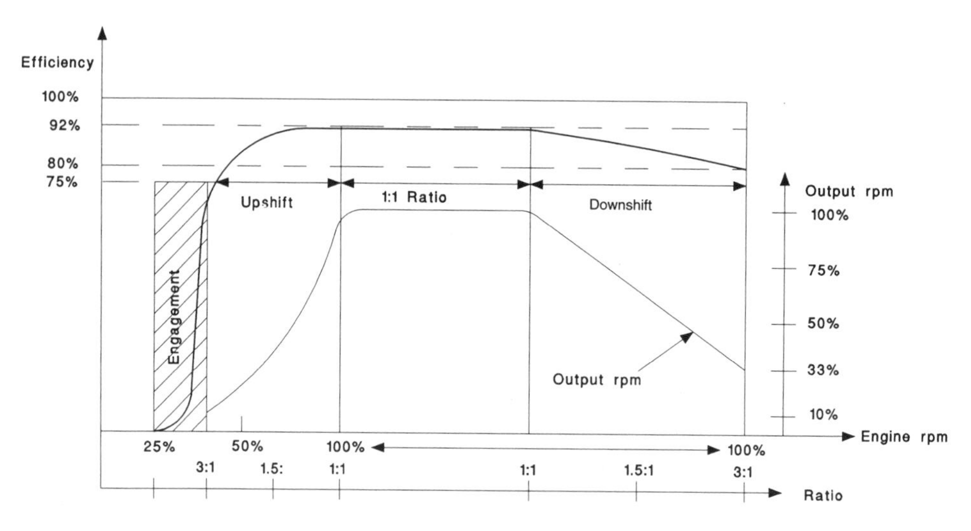

Fig. 3.39. Salsbury CVT Efficiency Curve.

significantly improved. With an optimized shift schedule, the Subaru Justy ECVT achieved fuel economy that was 15-20 percent greater than the same vehicle equipped with a standard three-speed automatic transmission.[18]

Traction drives are also continuously lubricated by oil. They transfer power and control transmission ratio in a manner similar to gears, except the rolling elements have no teeth. Instead, they rely on the hydroelastic properties of the lubricating fluid to transfer shear loads at the contact points between the rollers and discs. As can be imagined, maintaining a lubricant film while preventing roller slippage has been one of the system's greatest challenges. Lubricants that maintain the lubricant barrier, yet transfer high shearing forces, are just now becoming available. New lubricants can transfer approximately 6 percent of the roller contact pressure in the shear direction, compared to about 1 percent for conventional transmission oil. Traction drives are simple, responsive, and compact.

Designs may be based either on the toroidal or the half-toroidal layout (see Figure 3.40). Characteristically, half-toroidal variators exhibit lower spin losses. However, the axial load on rollers and input and output discs are extremely high. Consequently, excessive internal loads and reduced bearing life are common difficulties. A full toroidal layout virtually eliminates axial loads on the rollers, and better lubricants are beginning to solve the traditional disadvantage in load transfer capability.

The full-toroidal Torotrak transmission under development at British Technology Group may be the precursor of tomorrow's production automotive CVTs (see Figures 3.41 and 3.42). The Torotrak CVT utilizes a two-cavity layout with the output disc in the center and an input disc at each end. Roller pressure is maintained by the hydraulically actuated clamping action of the two input discs which traps the rollers between them. A dual-cavity Torotrak variator has been tested at 94-percent efficiency. Half-toroidal designs are reported to operate on the order of 90-percent efficiency. Shift rate is extremely rapid and smooth. Rollers can migrate across the discs in as little as 1/2 revolution of the engine, and changes in ratio are virtually unfelt by the operator.

Another unique feature of the Torotrak unit is what British Technology Group calls the "geared neutral." The geared neutral allows the vehicle to remain at rest, to creep forward, or pull away in a burst of acceleration utilizing an infinitely variable start-up gear that circumvents the need for a traditional torque converter or clutch. The heart of the system is a planetary gearset. To visualize how the geared neutral works, consider the behavior of the gear as illustrated in Figure 3.43. When the sun gear is driven at speed Na, and the planet carrier is "grounded" with the planet gears free to rotate; then the annulus gear rotates

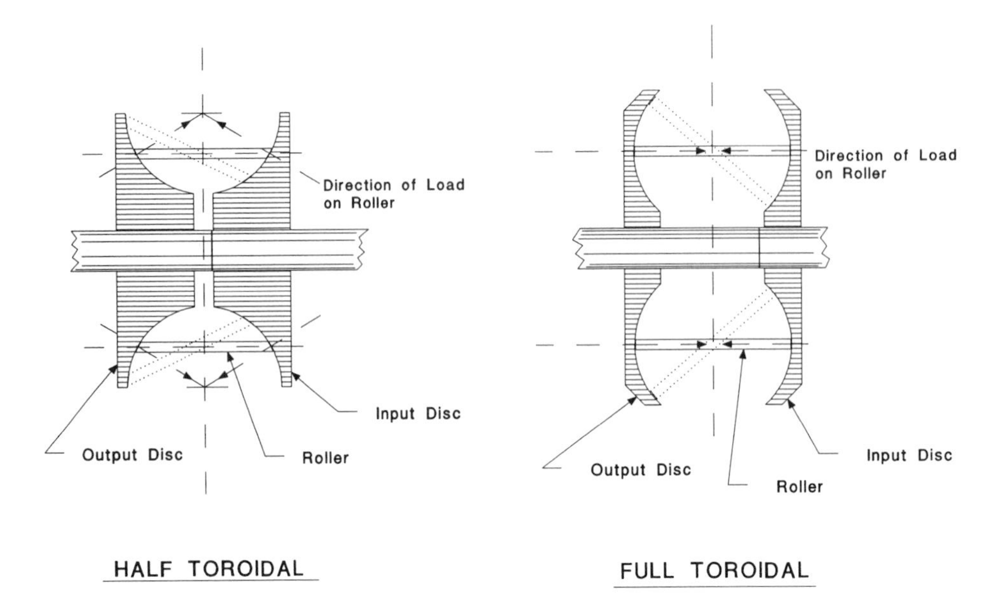

Fig. 3.40. Toroidal and Half-Toroidal Variator.

Fig. 3.41. Torotrak CVT.

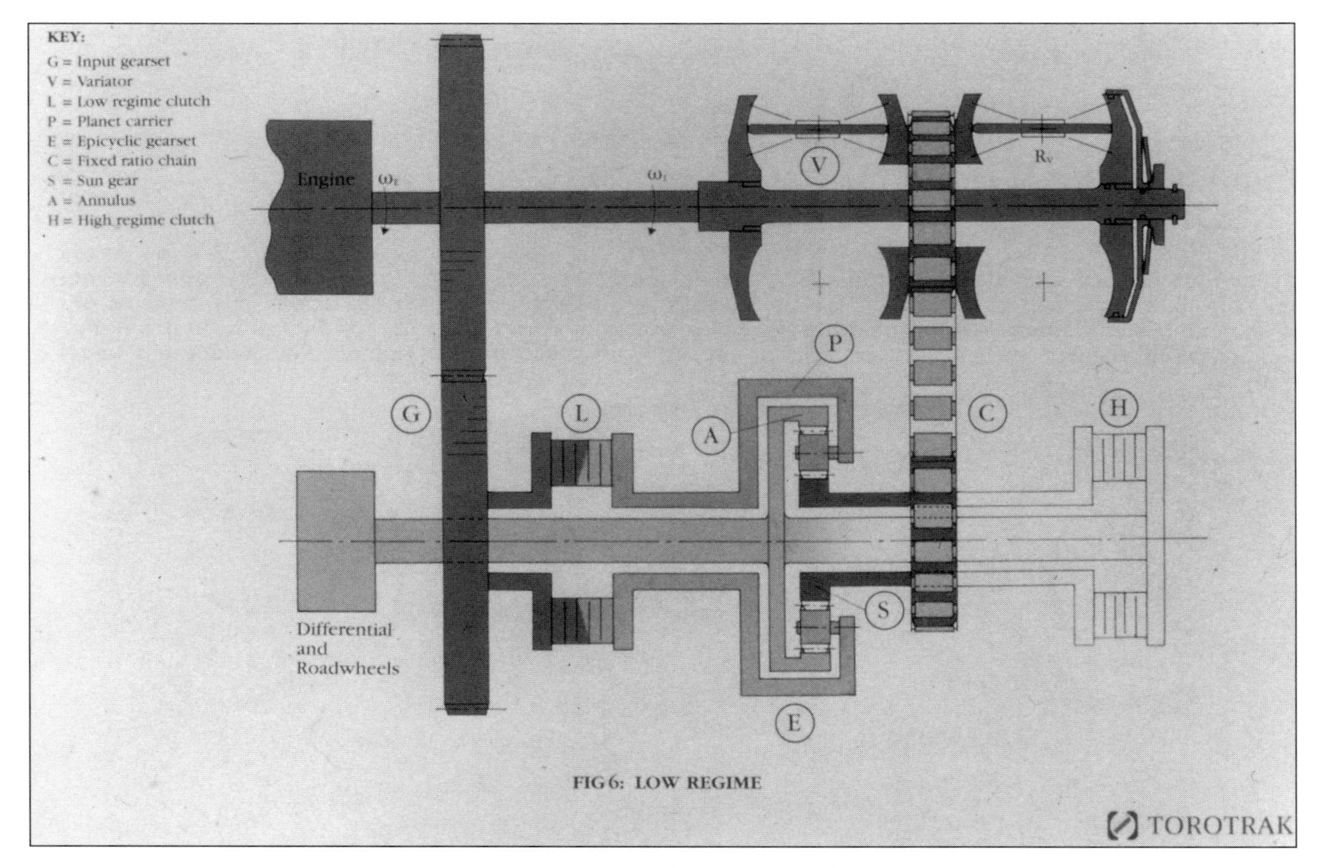

Fig. 3.42. Torotrak CVT General Layout.

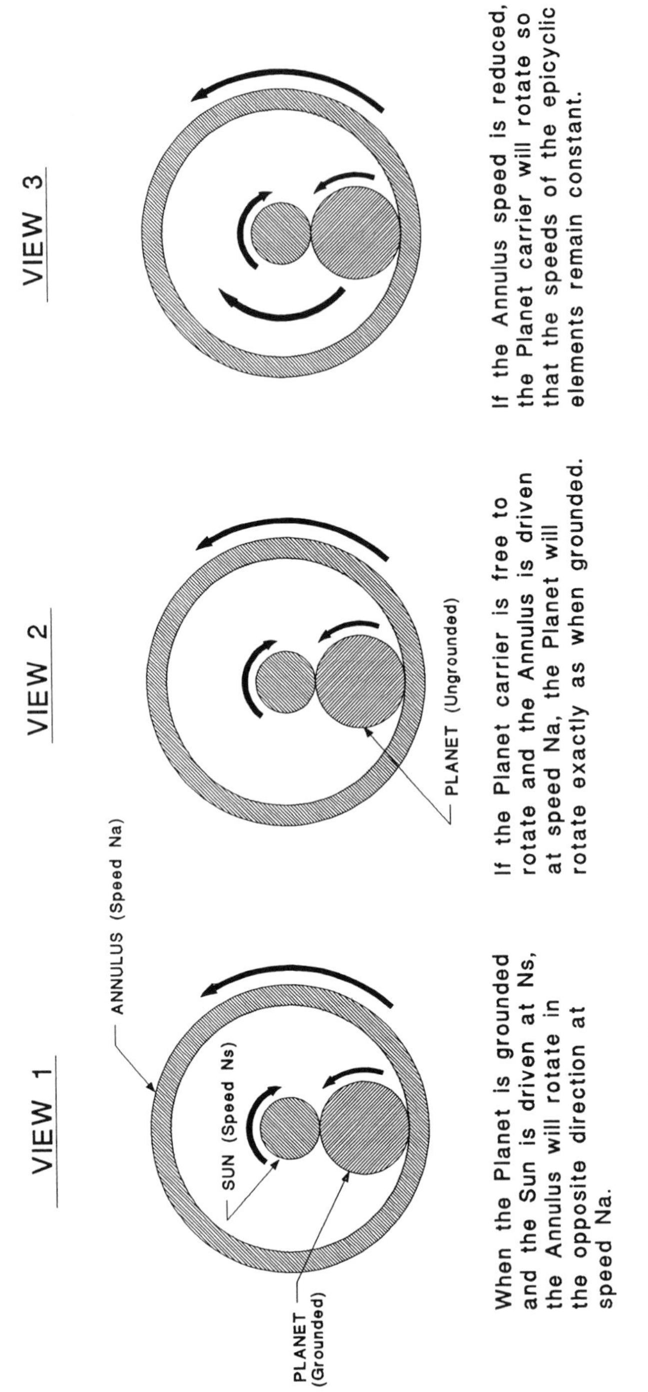

Fig. 3.43. Torotrak's Epicyclic Gearset Provides Geared Neutral.

in the opposite direction at a speed that is determined by the relative size of the gears. Likewise, if the planet carrier is free to rotate and the annulus is driven at the same speed as in View 1, the planet gears will rotate at the same speed as before and the carrier will still remain stationary. However, if the speed of the annulus is reduced or increased, the planet carrier will start to rotate in either the forward or reverse direction so that the sum of the speeds of the epicyclic elements remain constant. By connecting the variator input shaft to the sun gear, the variator output disc to the annulus, and the vehicle wheels to the planet carrier, a slight change in variator ratio will begin to move the vehicle in either the forward or reverse direction. If the variator ratios are matched, the vehicle will not move, although it is still "in gear."

A geared neutral in combination with a perfected variator could provide a new level of freedom to powerplant engineers. With a CVT, the shape of the torque curve is no longer limited by vehicle dynamics, but instead, is free to comply with the optimum characteristics of the engine. The engine and transmission can form a much more fluid and synergistic package when their individual management schedules are more completely integrated. Vehicles can be tuned for maximum performance, maximum fuel efficiency, and minimum emissions within a dynamic system of integrated controls and continuous feedback.

New Technology and the Twenty-First Century Alternative Car

Most of the options for improving vehicle fuel economy can be applied to cars of any size. Even the basic concept of weight reduction is already being applied to conventional multi-passenger cars. The primary difference has to do with magnitude and expedience. Significantly smaller cars are naturally lighter, and as a result, require less energy. Much greater fuel economy is therefore possible with existing technology. Quincy-Lynn's Urbacar (see Appendix) achieved fuel economy on the order of 23 km/L (55 mpg) and was powered by an industrial engine that barely equated to 1930's automobile technology. A modern automotive powertrain could easily deliver double the fuel economy in a vehicle of similar weight, and provide better performance as well. When framed in terms of a choice between reducing work or doing equivalent work more efficiently with better technology, the point can easily become obscured. The best approach is of course to do both. Smaller special-purpose vehicles can result in a large and relatively immediate improvement in transportation energy intensity, and as better technology becomes available the benefits will be compounded.

Although vehicle downsizing is not normally considered a technology, in actual practice, ultralight vehicle design efforts will ultimately produce a technology of their own. Ultralight vehicle designs offer a variety of opportunities to utilize emerging technologies more rapidly and more efficiently. Structural composites can significantly reduce vehicle weight. However, the higher loads typical of conventional high-mass cars can often exceed a material's capability and make a particular application infeasible. Lightweight plastic composites are especially well-suited to the structural and weight requirements of ultralight designs. Vehicle reconfiguration can take advantage of new ideas in powertrain packaging and smaller engines of greater specific power in ways that may not be appropriate for the family sedan. Combination starter/generators, electromagnetic valve lift, composite engine components and miniature drives become more mechanically and/or economically feasible with smaller engines of fewer cylinders. At reduced power levels, traction drives and belt CVTs become less costly and more immediately applicable.

Integrated electronic powertrain control is another emerging technology that can produce a new level of benefits in the fuel-efficient alternative car. Different aspects of existing technologies may be emphasized when they are applied to a new product category. In the process of developing designs for significantly smaller vehicles, new technologies are likely to emerge, and existing technologies may produce new benefits when utilized in new ways. In this respect, a vast reservoir of existing technology is already awaiting a new orientation in systems engineering to create a new level of product benefits.

References

1. R.L. Bechtold, "Ingredients of Fuel Economy," SAE Paper No. 790928, Society of Automotive Engineers, Warrendale, PA, 1979.

2. Gino Sovran and Mark S. Bohn, "Formulae for the Tractive-Energy Requirements of Vehicles Driving the EPA Schedules," SAE Paper No. 810184, Society of Automotive Engineers, Warrendale, PA, 1981.

3. Thomas D. Gillespie, Fundamentals of Vehicle Dynamics, SAE R-114, Society of Automotive Engineers, Warrendale, PA, 1992.

4. Charles E. Scheffler and George W. Niepoth, "Customer Fuel Economy Estimated from Engineering Tests," SAE Paper No. 650861, Society of Automotive Engineers, Warrendale, PA, 1965.

5. "Latent Heat Storage," *Automotive Engineering*, Feb. 1992, Society of Automotive Engineers, Warrendale, PA.

6. Thomas P. Yasin, "The Analytical Basis of Automobile Coastdown Testing," SAE Paper No. 780334, Society of Automotive Engineers, Warrendale, PA, 1978.

7. David M. Tenniswood, Helmut A. Graetzel, "Minimum Road Load for Electric Cars," SAE Paper No. 670177, Society of Automotive Engineers, Warrendale, PA, 1967.

8. Walter Korff, Designing Tomorrow's Cars, M-C Publications, 1980, p. 190.

9. Marcus C. Inman Hunter, Rotary Valve Engines, John Wiley & Sons, Inc., NY.

10. Peter W. Gabelish, Albany R. Vial, and Philip E. Irving, "Rotary Valves for Small Four-Cycle IC Engines," SAE Paper No. 891793, Society of Automotive Engineers, Warrendale, PA, 1989.

11. Vinay Harne and S.R. Marathe, "Variable Compression Ratio Two-Stroke Engine," SAE Paper No. 891750, Society of Automotive Engineers, Warrendale, PA, 1989.

12. Shigeru Onishi, *et al.*, "Active Thermo-Atmosphere Combustion (ATAC)—A New Combustion Process for Internal Combustion Engines," SAE Paper No. 790501, Society of Automotive Engineers, Warrendale, PA, 1979. Also see: Paul M. Najt and David E. Foster, "Compression-Ignited Homogeneous Charge Combustion," SAE Paper No. 830264, Society of Automotive Engineers, Warrendale, PA, 1983.

13. Wolfgang Demmelbauer-Ebner, Alois Dachs, and Hans Peter Lenz, "Variable Valve Actuation," *Automotive Engineering*, October 1991, Society of Automotive Engineers, Warrendale, PA.

14. Michael T. Nelson, Alvon C. Elrod, "Continuous-Camlobe Phasing—An Advanced Valve-Timing Approach," SAE Paper No. 870612, Society of Automotive Engineers, Warrendale, PA, 1987.

15. "Air-Injection Supercharging—A Page from History," *Automotive Engineering*, July 1992, Society of Automotive Engineers, Warrendale, PA.

16. W.R. Wade, *et al.*, “Fuel Economy Opportunities with an Uncooled DI Diesel Engine,” SAE Paper No. 841286, Society of Automotive Engineers, Warrendale, PA, 1984.

17. N. John Beck, *et al.*, “Electronic Fuel Injection for Two-Stroke Cycle Gasoline Engines,” SAE Paper No. 861242, Society of Automotive Engineers, Warrendale, PA, 1986.

18. Yasuhito Sakai, “The ‘ECVT’ Electro Continuously Variable Transmission,” SAE Paper No. 880481, Society of Automotive Engineers, Warrendale, PA, 1988.

Chapter Four

Alternative Fuels

Russian supersonic transport fueled with LNG.
(Courtesy: American Gas Association)

Following the '73-'74 OPEC oil embargo, alternative fuels emerged as a popular and immediate answer to concerns about future motor fuel supplies. Experimental vehicles that run on methanol, natural gas, and hydrogen have since regularly emerged from the labs of automobile manufacturers worldwide. More recently, United States Clean Air Act provisions calling for a phased introduction of alternative-fueled and clean-fueled vehicles, and new legislation led by California, have placed increased pressure on manufacturers to move more quickly with new designs for low-emissions and zero-emissions vehicles. On a more modest level, alternative fuels have already arrived in the form of reformulated gasoline in which additives reduce emissions and lower the fuel's petroleum content. But gasoline is still the world's primary source of transportation energy and may remain so well into the twenty-first century. In general, industry has remained essentially uncommitted to any particular new technology. Once industry commits to a new fuel or a new technology, the direction of vehicle design could be determined for a century into the future. A decision in favor of one alternative fuel may, by default, be a decision against other technologies. A large-scale commitment to methanol, for example, could discourage future development efforts with ethanol, natural gas, and hydrogen, and thereby close off opportunities that may lie just around the corner.

As for gasoline, if it did not exist we might be inclined to invent it. Gasoline is an excellent motor fuel and the gasoline engine runs well even when built according to relatively primitive standards. Undesirable attributes of the technology arise primarily from gasoline's harmful emissions and limited supplies. The value of an alternative fuel therefore begins with its ability to resolve the liabilities of gasoline while avoiding new ones of its own. There should be plentiful and preferably domestic supplies, and the fuel should be environmentally friendly. Further, a viable alternative fuel must be at least potentially inexpensive, non-toxic, and easily contained and dispensed. It should provide a vehicle range and power profile similar to that of gasoline, and it must be adaptable to the existing transportation fuel infrastructure. Its adaptability to flexible fuel operation also is important during the transitional phase.

Even if the perfect alternative were available, substituting another fuel for gasoline could have ramifications that extend far beyond the merits of the fuel itself. Large-scale change will impact the design, manufacture and service of vehicles. Ill-conceived change could create disruptions within the refining and supply infrastructure and place unwarranted financial demands across the spectrum of the transportation and energy supply industries. A forced or too-rapid shift to an alternative fuel may open windows of opportunity for entrepreneurial companies, as well as create windows of vulnerability for companies that are heavily invested in existing technology.

In the final analysis, there is no simple answer in the search for an alternative to gasoline. The twenty-first century will probably require new relationships between the machine, the fuel, and the people who use them. Meeting the challenge of an ever-expanding appetite for personal transportation, confronted with an ever-decreasing supply of natural resources, may require alternative automobiles as well as alternative fuels.

The most promising alternatives to gasoline appear to be methanol, ethanol, natural gas, hydrogen, and electricity. Most of what we know about these fuels (or energy sources) comes from controlled laboratory experiments, experimental vehicles, and comparatively limited in-the-field fleet experience. Consequently, it is difficult to forecast with certainty the environmental and economic impact of large-scale use of any alternative fuel, or the ultimate effect it will have on the design of vehicles that consume it.

Advantages and Disadvantages of Alternative Fuels

Table 4.1 presents the advantages and disadvantages of the most promising alternative fuels.

TABLE 4.1. ADVANTAGES AND DISADVANTAGES OF ALTERNATIVE FUELS

Fuel	Advantages	Disadvantages
Methanol	Familiar liquid fuel Vehicle development relatively advanced Organic emissions (ozone precursors) will have lower reactivity than gasoline Lower emissions of toxic pollutants, except formaldehyde Engine efficiency should be greater Abundant natural gas feedstock Less flammable than gasoline Can be made from coal or wood (as can gasoline), though at higher cost Flexfuel "transition" vehicle available	Range as much as 1/2 less, or larger fuel tanks Would likely be imported from overseas Formaldehyde emissions a potential problem, esp. at higher mileage, requires improved controls More toxic than gasoline M100 has non-visible flame, explosive in enclosed tanks Costs likely somewhat higher than gasoline, esp. during transition period Cold starts a problem for M100 Greenhouse problem if made from coal

Continued

TABLE 4.1. ADVANTAGES AND DISADVANTAGES OF ALTERNATIVE FUELS (CONT.)

Fuel	Advantages	Disadvantages
Ethanol	Familiar liquid fuel Organic emissions will have lower reactivity than gasoline emissions (but higher than methanol) Lower emissions of toxic pollutants Engine efficiency should be greater Produced from domestic sources Flexfuel "transition" vehicle available Lower CO with gasohol (10 percent ethanol blend) Enzyme-based production from wood being developed	Much higher cost than gasoline Food\fuel competition at high production levels Supply is limited, esp. if made from corn Range as much as 1/3 less, or larger fuel tanks Cold starts a problem for E100
Natural Gas	Though imported, likely North American source for moderate supply (1 mmbd or more gasoline displaced) Excellent emission characteristics except for potential of somewhat higher NO_x emissions Abundant worldwide Can be made from coal	Dedicated vehicles have remaining development needs Retail fuel distribution system must be built Range quite limited, need large fuel tanks w/added costs, reduced space (LNG range not as limited, comparable to methanol) Dual fuel "transition" vehicle has moderate performance, space penalties Slower refueling Greenhouse problems if made from coal

Continued

TABLE 4.1. ADVANTAGES AND DISADVANTAGES OF ALTERNATIVE FUELS (CONT.)

Fuel	Advantages	Disadvantages
Electric	Fuel is domestically produced and widely available Minimal vehicular emissions Fuel capacity available (for nighttime recharging) Big greenhouse advantage if powered by nuclear or solar Wide variety of feedstocks in regular commercial use	Range, power very limited Much battery development required Slow refueling Batteries are heavy, bulky, have high replacement costs Vehicle space conditioning difficult Potential battery disposal problem Emissions for power generation can be significant
Hydrogen	Excellent emission characteristics—minimal hydrocarbons Would be domestically produced Big greenhouse advantage if derived from photovoltaic energy Possible fuel cell use	Range very limited, need heavy, bulky fuel storage Vehicle and total costs high Extensive research and development effort required Needs new infrastructure
Reformulated Gasoline	No infrastructure change except refineries Probable small to moderate emission reduction Engine modifications not required May be available for use by entire fleet, not just new vehicles	Emission benefits remain highly uncertain Costs uncertain, but will be significant No energy security or greenhouse advantage

Source: U.S. Office of Technology Assessment, 1990

The comparative energy density of alternative energy sources is as follows:[1]

TABLE 4.2

Energy Storage Medium	Specific Energy* MJ/kg
Gasoline	42-44
Diesel	42.5
Methanol	19.7
Ethanol	26.8
Hydrogen (liquid)	120.0
Hydrogen (gas)**	2.34
Methane	50.0
Lead-Acid Battery	0.19
Regen. Fuel Cell (H_2/Cl_2)	0.44

* Based on lower heating value of fuels
** Includes weight of magnesium-based hydride storage

The Alcohol Fuels

The alcohol fuels are methanol and ethanol. Alcohol is an excellent motor fuel and it can be made from a variety of feedstocks, including renewable biomasses. Ethyl alcohol, or ethanol, is most economically made by fermenting a biomass such as corn. The other alcohol, the one receiving the most attention as a potential replacement for gasoline, is methanol. However, ethanol is in many respects the superior motor fuel. Its major drawbacks are that it is more costly to produce than methanol, and when derived from agricultural feedstocks it ends up as another competitor in the already overstressed food chain.

TABLE 4.3. COMPARISON OF METHANOL, ETHANOL AND GASOLINE

	Methanol	Ethanol	Gasoline
Oxygen Content, wt%	50.0	34.8	0
Boiling Point, K	338	351	308-483
Lower Heating Value, MJ/kg	19.7	26.8	42-44
Heat of Vaporization, MJ/kg	1.11	0.90	0.40
Stoichiometric Air-Fuel Mass Ratio	6.45:1	9.0:1	14.6:1
Specific Energy, MJ/kg per Air-Fuel Ratio	3.08	3.00	2.92
Research Octane Number	109	109	90-100

Ethanol

Ethanol is grain alcohol, like the "white lightning" that people made during prohibition. It is most economically produced by fermenting starch or sugar crops. As a fuel its volumetric energy density is slightly less than two-thirds that of gasoline. Consequently, an ethanol fuel tank must be 1.5 times the volume of a gasoline tank to contain enough ethanol for equal range. This can be slightly offset if the engine is dedicated and tuned to take advantage of the higher octane rating of "neat" ethanol.

The most extensive experience with ethanol as a motor fuel has occurred in the country of Brazil. In an effort to stem oil imports, the Brazilian government initiated a conversion-to-ethanol program by providing disincentives for using gasoline. As a result, Brazil is currently the world's leading producer and consumer of ethanol motor fuel. Sixty-two percent of that nation's total automobile fuel supply is currently ethanol. Unfortunately, the Brazilian experience is a prime example of the unforeseen repercussions that can occur with a major shift to an alternative fuel. In the late '70s it made perfect sense to policymakers to utilize Brazil's vast sugar cane production capability to support the nation's switch to domestically produced ethanol, and away from the ever more costly petroleum. Years later, after a massive, costly and largely successful conversion effort, Brazilians were then confronted with new conditions: a world of cheap oil and expensive sugar, which made it unprofitable to use their valuable crop of sugar cane to produce ethanol motor fuel.

Reservations about a switch to ethanol in other countries are based primarily on the economic unsoundness of ethanol as an alternative fuel, and on the possible negative effects of dedicating huge crops to the production of motor fuel. In addition, with present technology, more energy is spent making bio-ethanol than is obtained from the fuel. The U.S. Office of Technology Assessment (OTA) has estimated that competition for grain and corn to produce large quantities of ethanol might raise that nation's food bill by billions of dollars annually. OTA has also expressed concern about the environmental effects of increasing corn production to meet the demands of ethanol production. Corn is a very energy-intensive, agricultural-chemical-intensive, and erosive crop. Large-scale use of ethanol as a motor fuel could result in an overall negative effect on the environment and the economy.

Even the often-cited benefits associated with the greenhouse effect are questionable. Proponents claim that crops to produce ethanol will compensate for much of the CO_2 by-products of combustion in vehicles. However, substantial quantities of CO_2 are produced by corn growing, harvesting, distillation and other parts of the ethanol production cycle. OTA concludes that, "it is unlikely

TABLE 4.4. ENVIRONMENTAL IMPACTS OF AGRICULTURAL ETHANOL

Water:	Water use (irrigated only) that can conflict with other uses or cause ground water mining. Leaching of salts and nutrients into surface and ground waters (and runoff into surface waters), which can cause pollution of drinking water supplies for animals and humans, excessive algae growth in streams and ponds, damage to aquatic habitats, and odors. Flow of sediments into surface waters, causing increased turbidity, obstruction of streams, filling of reservoirs, destruction of aquatic habitat, and increase of flood potential. Flow of pesticides into surface and ground waters, potential build-up in food chain causing both aquatic and terrestrial effects such as thinning of egg shells of birds. Thermal pollution of streams caused by land clearing on stream banks, loss of shade, and thus greater solar heating.
Air:	Dust from decreased cover on land, operation of heavy farm machinery. Pesticides from aerial spraying or as a component of dust. Changed pollen count, human health effects. Exhaust emissions from farm machinery.
Land:	Erosion and loss of topsoil, decreased cover, plowing, increased water flow because of lower retention; degradation of productivity. Displacement of alternative land uses—wilderness, wildlife, aesthetics, etc. Change in water-retention capabilities of land, increased flooding potential. Build-up of pesticide residue in soil, potential damage to soil microbial populations. Increased soil salinity (especially form irrigated agriculture), degradation of soil productivity. Depletion of nutrients and organic matter from soil.
Other:	Promotion of plant diseases by monoculture cropping practices. Occupational health and safety problems associated with pesticide residues and involvement in spraying operations.

Source: U.S. Office of Technology Assessment, 1990.

that ethanol production and use with current technology and fuel use patterns will create any significant greenhouse benefits."

Ethanol is a major contender in the search for an alternative transportation fuel. A number of programs are under way to develop technology for economically producing ethanol from wood and lignocellulosic materials. New technology could change ethanol's environmental and economical profile and eliminate the food-versus-fuel competition that currently plagues the food-to-ethanol production cycle. If development goals are realized, biomass ethanol could become economically competitive with gasoline by the year 2000.

Methanol

Methanol is a close chemical cousin to ethanol. Its primary advantage over ethanol is that it can be economically produced from abundant resources of natural gas. Methanol can also be produced from coal or wood, but at a higher cost. As a blend of 85 percent methanol and 15 percent gasoline (M85), it is a fuel to which manufacturers can adapt motor vehicles with only minor design changes.

Flexible fuel vehicles (FFVs) that can run on either gasoline or gasoline/methanol blends are also easily designed. A FFV will provide consumers with a familiar vehicle, a familiar liquid fuel, and performance that could surpass their dedicated gasoline-fueled automobile. Such vehicles would also ease consumer resistance to an unfamiliar fuel. With FFVs, consumers could conceivably be unaware of, and even indifferent to, which fuel was actually being dispensed into their vehicles.

Methanol also has disadvantages that make an early switch unlikely as long as gasoline supplies remain high and prices remain low. The primary disadvantages of methanol include low volumetric energy density (about 1/2 that of gasoline), a non-visible flame with M100 (neat methanol), significant cold-starting difficulties, increased formaldehyde emissions, and the likelihood that significant supplies would come from oil-exporting nations. In addition, methanol is also toxic and corrosive. Spills can damage clothes, shoes and automobile paint, and prolonged skin contact can result in poisoning.

Researchers have made significant progress toward resolving many of the technological problems associated with a methanol-based fuel. Consequently, many believe that methanol, most likely in the form of a gasoline/methanol blend, will be the predominant motor fuel of the twenty-first century. However, methanol's fate is still uncertain. Much depends on political, economic and technological developments that take place over the next few years.

Methanol Source and Abundance

At least initially, methanol could be economically produced from the world's plentiful resources of natural gas. Steam reforming of methanol from natural gas is illustrated in Figure 4.1. The U.S. has large reserves of natural gas, but even now some domestic gas needs are being supplied by imports. However, the present resource base of natural gas is much broader and less regional than that of petroleum. As a result, the potential for political difficulties arising from regional monopolies is not nearly as great over the near term. But ultimately, the rapid development of natural gas resources would return market power to the holders of the largest reserves. The world's largest reserves of natural gas lie under the Middle Eastern OPEC countries.

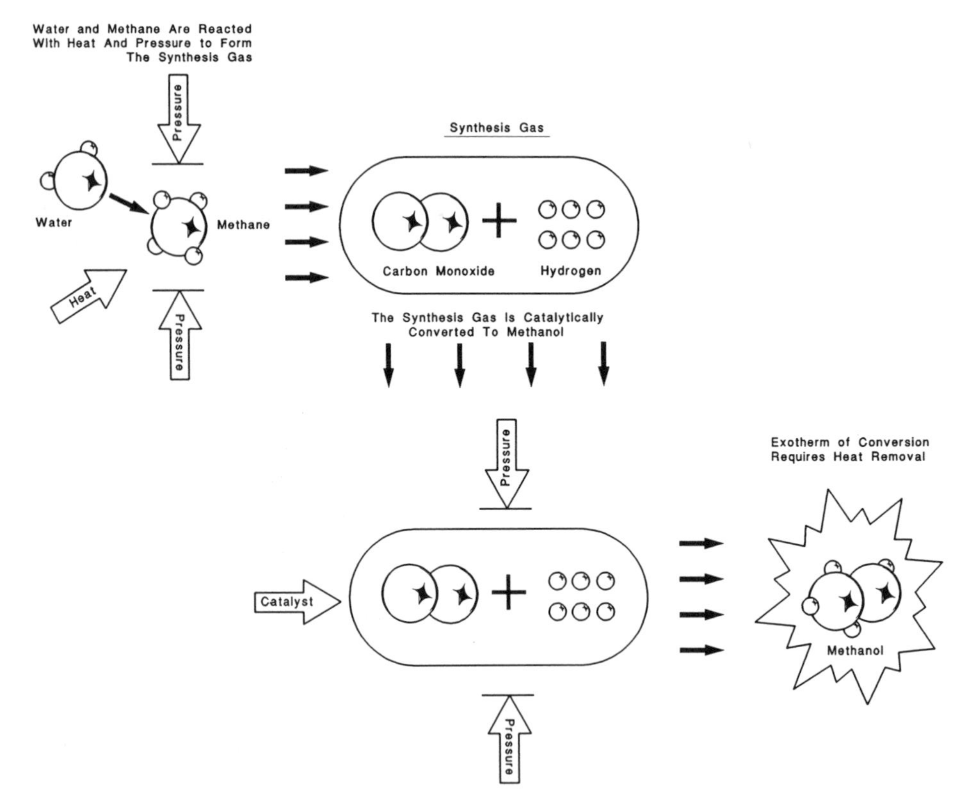

Fig. 4.1. Methanol from Natural Gas.

Methanol can also be produced from coal. A method of economically converting coal to methanol could make a number of regions energy sufficient throughout the twenty-first century. In the absence of such technology, many of the world's industrialized countries will continue to import a significant portion of their energy needs. Methanol would then be shipped into ports in the same manner that petroleum is now imported.

Environmental and Safety Factors

Environmental and safety issues occur throughout the chain of production, delivery, and consumption. The most visible impact, and the most popular subject of discussion, occurs at the end of the chain where the fuel is dispensed and subsequently converted into gaseous and particulate exhaust emissions. After nearly a century of experience with gasoline engines, the level of emissions between different vehicles is still widely divergent. Consequently, it is easy to imagine the conflicting data that can result from exhaust emission tests of different experimental methanol-fueled vehicles.

In general, a methanol engine produces lower harmful exhaust emissions than its gasoline-burning counterpart. Unburned fuel emerges from the combustion chamber as methanol, instead of hydrocarbons. Emissions of carbon monoxide are almost identical to a typical gasoline engine at an equivalent fuel/air mixture. However, methanol is happy at much leaner mixtures than gasoline, and lean mixtures produce lower CO emissions. In the final analysis, emission levels depend largely on engine tuning and management strategy. Table 4.5 shows the results of a number of studies on methanol emissions.

Methanol contains virtually no sulfur and, as a result, sulfurous emissions are nonexistent. Also, an over-rich mixture, although bad for fuel economy, does not produce the black smoke typical of an over-rich gasoline charge. However, a significant liability of methanol-burning engines is the emission of aldehydes which give the exhaust its characteristic odor. Experiments with gasoline/methanol blends demonstrate that aldehyde emissions increase in a linear relationship to the amount of methanol in the blend.[3] Formaldehyde accounts for up to 98 percent of the total aldehyde emissions from a methanol-burning engine. By comparison, only 31-54 percent of the total aldehydes produced by a gasoline-burning engine is in the form of formaldehyde. Tests conducted under U.S. Federal Test Procedure conditions demonstrated that vehicles using neat methanol (M100) had aldehyde emissions ten times higher, and unburned fuel emissions equal to those of a gasoline-fueled vehicle.[4] Formaldehyde is a major constituent of smog and also responsible for much of the associated eye

TABLE 4.5. METHANOL EMISSIONS COMPARED TO ETHANOL AND GASOLINE VEHICLES

Study	Emission Change (% relative to gasoline vehicle emissions) NMHC[1]	CO	NO_x
Methanol FFVs:			
DeLuchi, *et al.* (1988)	-33 to -4	-43 to +142	-40 to -80
Alson, *et al.* (1989)	-40 to -30[2]	equal	equal
U.S. EPA (1989)	-30	na	na
M100 Dedicated Vehicle:			
U.S. EPA (1989)	-80[2]	na	na
Chang, *et al.* (1991)[3]	lower	equal	equal
M85 Dedicated Vehicles:			
DeLuchi, *et al.* (1988)	-50 to +62	-70 to +187	-73 to +74
OTA (1990)	-40 to +20	equal	equal
Chang, *et al.* (1991)[3]	lower	equal	equal
Sperling, *et al.* (1991)	-50	0	0
Ethanol Vehicles:			
OTA (1990)	Similar to MVs	lower	equal
U.S. EPA (1990a)	Similar to MVs	equal	equal
Chang, *et al.* (1991)[3]	lower	equal	equal

[1] NMHC = Non-Methane Hydrocarbons.
[2] Adjusted for ozone-forming potentials of HC.
[3] Qualitative summary of various studies.
Source: Institute of Transportation Studies.

irritation. Exhaust aftertreatment can remove much of the formaldehyde and up to 90 percent of the unburned methanol.[5]

Fuel evaporation during processing, transportation and dispensing is also a significant contributor to the environmental impact of transportation fuels. Although methanol will produce evaporative emissions, the fuel's photoreactive emissions are less than those of gasoline, which give methanol high marks in this area.

Methanol's toxicity has been the subject of much debate regarding the relative safety of introducing the fuel into widespread use. Both methanol and gasoline are harmful if they are ingested or absorbed through the skin. However, methanol is much more readily absorbed and considerably more poisonous than gasoline. Ingestion of 60-120 cm^3 (2-4 oz) of methanol is usually fatal to an adult. A fatal dosage for a three-year-old is slightly more than 15 cm^3 (1/2 oz)—an amount that can be easily ingested in a single swallow. The fact that methanol is tasteless exacerbates the potential hazard to young children who might otherwise be repelled by the taste of gasoline.

Self-service methanol stations would also present an increased risk to consumers. Spilled methanol is absorbed through the skin much more rapidly than gasoline. Ingestion or absorption of even small amounts can result in blindness. And the familiar act of rescuing a stranded motorist with fuel siphoned from another vehicle could turn out to be a deadly act of altruism.

In general, hazards can be alleviated by appropriate product design and legislation. Anti-siphon screens for fuel tanks, service station shut-off valves set to prevent "topping-off" and the prohibition of methanol-fueled lawnmowers and trimmers, along with the attendant small containers of fuel that may be accessible to children, would reduce the hazards associated with methanol fuel. But regardless of the precautionary measures, accidental poisonings will undoubtedly occur with large-scale use of methanol.

Compatibility with Current IC Engine Design

Methanol is an excellent fuel for the Otto cycle engine. Some changes in engine design are required for M100. However, M85 (containing 15 percent gasoline) alleviates many of the difficulties associated with a methanol motor fuel. The undesirable characteristics of methanol fuel include cold-starting difficulties, lubricant contamination problems, increased engine wear, and materials incompatibilities due to methanol's corrosiveness.

Cold Starting

Unlike gasoline, methanol does not vaporize well at lower temperatures. The fuel's low vapor pressure and high latent heat of vaporization result in increasingly difficult starting as the ambient temperature drops. At a temperature of approximately 12°C, an unmodified, carbureted, SI methanol engine will refuse to start. The cold-starting temperature limit can be extended into the range of 0-5°C by improved fuel atomization, applying heat for vaporization, and by mixing methanol with fuels of greater volatility. Higher cranking speeds also help

(>110 rpm). Toyota has reported cold starts in the −20° range with M85 in their lean-burn system.[6]

Lubricant Contamination

The effects of engine lubricant contamination have been an ongoing problem with methanol fuel. During cold starts, methanol quickly enters the engine lubricating oil and can reach significant levels very rapidly. Although gasoline also enters engine oil, modern oil formulations can tolerate gasoline contamination without harm. However, methanol is not compatible with engine oil and, as a result, contamination rapidly changes the oil's characteristics. Within 15 minutes of initial cranking the lubricating oil has been transformed into an oil-methanol-water emulsion that has a markedly reduced ability to lubricate the engine. The oil remains in this condition until the bulk temperature reaches approximately 70°C, at which temperature contaminants boil off.[7]

This contamination/evaporation cycle takes place over a period of 20 minutes or longer, depending on how rapidly the oil reaches its methanol boil-off temperature. Unfortunately, short trips on the order of five to ten minutes result in continuous and progressive degradation of lubricating oil.

Increased Engine Wear

Increased engine wear has been a problem with methanol-fueled vehicles. A number of studies have confirmed that methanol fuel significantly increases the rate at which engine components breakdown, especially along the upper portion of the cylinder wall and at the top ring. Initially it was believed that methanol was washing away lubricant in these areas and leaving the cylinder walls unprotected. However, studies at Southwest Research Institute (SRI) have indicated that the corrosive effects of acids formed during combustion may also play a significant role.[8]

SRI used a water-cooled cylindrical combustion apparatus which was designed to cause condensates to form on the cylinder wall. The condensate was then collected and analyzed. It consisted primarily of water and methanol, along with small amounts of formaldehyde, formic acid, and rust-colored precipitate which turned out to be iron formate rather than iron oxide. In the presence of hydrogen peroxide (produced by methanol combustion) formic acid is transformed into performic acid which is highly reactive to iron.

Performic acid is so reactive that SRI was unable to obtain a sample that actually contained the acid. However, all the reactants were present during combustion, and the by-product of performic acid reaction with iron was also present. Findings therefore strongly suggest that performic acid is a causative element of accelerated engine degradation in methanol-fueled engines. Special formulations of lubricating oil designed to protect engines from performic acid will probably resolve the problem.

Materials Compatibility

Many of the materials used in the fuel system of a gasoline-fueled automobile are not compatible with methanol. Methanol (and to a lesser degree, ethanol) is very corrosive to lead, magnesium, aluminum, and to the plastics and elastomers that are typically found in an automobile fuel system. As a result, components such as the fuel tank, fuel pump diaphragm, carburetor, fuel gauge float, fuel pump, carburetor housing, and many seals, rings and washers rapidly degrade in the presence of methanol. Incompatibility problems rarely occur with blends containing small amounts of methanol (such as M15), but the more methanol-rich blends such as M85 or M100 require fuel systems made of methanol-compatible materials.

Natural Gas as a Transportation Fuel

Methane, the primary constituent of natural gas, is the lightest of all the hydrocarbon fuels. The methane molecule consists of one carbon atom and four hydrogen atoms. This simple and clean-burning gas is already one of the world's most successful alternative transportation fuels. Worldwide, over 700,000 motor vehicles run on natural gas and the momentum is growing for much broader application. Approximately 50,000 natural gas vehicles exist in the U.S. and another 34,000 such vehicles are in use in Canada.

Industry efforts are underway to secure a significantly greater portion of the transportation energy market, primarily through fleet vehicles that can be easily serviced by dedicated on-site refueling stations. Existing and near-term applications are based on well-proven compressed natural gas (CNG) technology. Longer-term plans include designs for automobiles that can be refueled at home utilizing a new low-pressure storage technology. The low-pressure storage system is being developed by the Atlanta Gas Light Adsorbent Research group (AGLAR). Over the past decade, the AGLAR group has served as a conduit for human resources and funds from a variety of participating utilities companies whose objective is to broaden the market for natural gas by resolving the on-board storage difficulties of CNG technology.

Fig. 4.2. The simple methane molecule is made of one carbon atom and four hydrogen atoms (top). Potential transportation applications are immense. A Russian helicopter uses LNG fuel (center). U.S. fleet applications include this Orion bus that stores CNG in rooftop cylinders (bottom). (Courtesy: American Gas Association)

The prospect of using this abundant and low-cost gas as an automobile fuel is an appealing concept, once the idea of a gaseous fuel has been accepted. This first threshold is where the average consumer is likely to become stalled with preconceived ideas about the safety and handling difficulties of natural gas. However, natural gas may be the fuel that paves the way for the ultimate transition to hydrogen as the predominant fuel of future generations. Compressed hydrogen/natural-gas blends can be delivered through the existing natural gas pipeline distribution network, and blends containing as little as 10 percent hydrogen produce a much cleaner exhaust.

Source and Supply

Natural gas is one of the most plentiful fuels in the world. The U.S. is virtually energy sufficient in natural gas. According to the American Gas Association, proven domestic reserves within the lower 48 states alone can meet current U.S. consumption levels for at least another 50 years. Improved recovery methods could increase reserves to a 200-year supply. These estimates do not include Canadian reserves or the vast U.S. natural gas deposits in Alaska. On a worldwide scale, natural gas reserves are spread between nations and continents much more evenly than those of petroleum (see Figure 4.3). As a result, political pressures and market control by individual nations is far less likely.

A large infrastructure for delivering natural gas to consumers is already in place throughout most of the industrialized world. The pipeline delivery network that supplies the U.S. is currently 25 percent under-utilized. This excess capacity alone can deliver enough fuel to supply half of that nation's transportation energy needs if vehicles were converted to run on natural gas.

The cost of natural gas, delivered to the consumer, is nearly half that of an energy-equivalent of gasoline. On a per liter equivalence basis, the cost is $0.11-$0.21 per liter ($0.40-$0.80 per gal), depending on regional pricing. Wellhead prices are expected to remain more stable than the price of petroleum. The increased predictability of supplies and prices, and the extra dollars available to consumers as a result of a switch to natural gas as a transportation fuel, would have a positive effect on the economy.

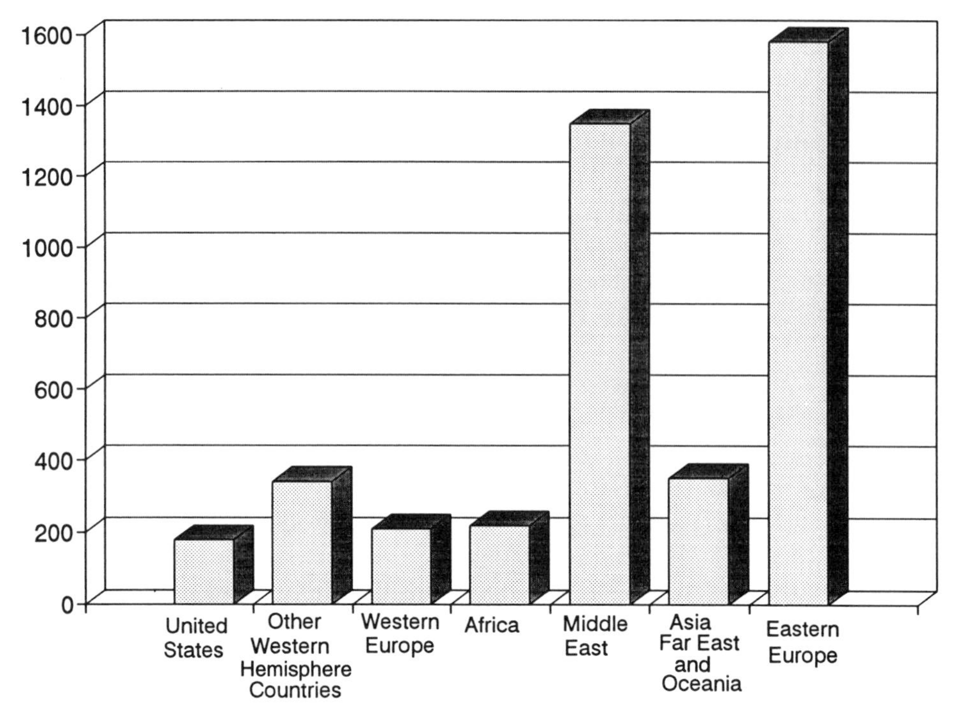

Fig. 4.3. World Natural Gas Reserves (trillions of cubic feet).

Environmental and Safety Factors

Natural gas is a very clean-burning fuel and it may be one of the safest vehicle fuels available. Technically unsound movie scenes, however, have given natural gas a bad image with dramatic portrayals of suicides and explosions. Although both of the foregoing have occurred, they are not representative of the characteristics of the fuel. The hazard of breathing air that is heavily laden with natural gas is primarily a result of the displaced oxygen, not the toxicity of the gas. As for an accidental explosion, the gas-to-oxygen ratio must be within a fairly narrow range at the time the ignition source is introduced or the gas will not ignite. This will of course occur at some point around the edges of a gas cloud, or in some areas at some moments as a room fills with escaping gas.

The safety record of natural gas vehicles (NGVs) has been well documented. Vehicles have been using natural gas as a fuel for nearly half a century in some areas of the world. Today, there are approximately 50,000 NGVs in use in the U.S. and approximately 700,000 worldwide. According to the Natural Gas Vehicle Coalition, a fire or explosion caused by a collision involving a natural gas vehicle has never been documented.

TABLE 4.6. FUEL/ENERGY PRICE COMPARISON IN U.S. DOLLARS

Energy Source	$/Common Unit	$/Gasoline Equivalent	$/Energy
Gasoline			
Unleaded (whlsale)	0.70/gal - 0.19/L	na	6.06/mmBTU - 5.74/GJ
Premium (whlsale)	0.77/gal - 0.20/L		
Methanol (whlsale)	0.42/gal - 0.11/L	0.86/gal - 0.23/L	7.42/mmBTU - 7.03/GJ
Ethanol (whlsale)	1.25/gal - 0.33/L	1.88/gal - 0.50/L	16.30/mmBTU - 15.45/GJ
Natural Gas			
Core	0.46/therm (1)	0.54/gal - 0.14/L	4.64/mmBTU - 4.40/GJ
Non-Core	0.39/therm (1)	0.45/gal - 0.12/L	3.89/mmBTU - 3.69/GJ
LPG	0.32/gal - 0.08/L	0.42/gal - 0.11/L	3.63/mmBTU - 3.44/GJ
Electricity			
Off-Peak	0.05/kW-h	1.56/gal - 0.41/L	13.54/mmBTU - 12.83/GJ
Residential	0.11/kW-h	3.57/gal - 0.94/L	30.94/mmBTU - 29.33/GJ
Hydrogen			
Steam Reformation	3.43/kSCF - 0.12/m³	1.43/gal - 0.38/L	12.40/mmBTU - 11.75/GJ
Electrolysis	12.98/kSCF - 0.46/m³	5.40/gal - 1.42/L	46.72/mmBTU - 44.29/GJ
H_2/CNG Blends (2) % by Energy Content			
5% H_2		0.50/gal - 0.13/	4.32/mmBTU - 4.09/GJ
10% H_2		0.55/gal - 0.15/	4.74/mmBTU - 4.49/GJ
15% H_2		0.60/gal - 0.18/	5.17/mmBTU - 4.90/GJ
25% H_2		0.70/gal - 0.18/L	6.02/mmBTU - 7.73/GJ
50% H_2		0.94/gal - 0.25/L	8.15/mmBTU - 9.73/GJ
75% H_2		1.19/gal - 0.31/L	10.27/mmBTU - 9.73/GJ

Based on estimates by Hydrogen Consultants, Inc.

(1) Therm = 100 ft³ (2.83 m³) of atmospheric natural gas and approximately 100,000 BTU (105.5 MJ).

(2) Prices estimated by Hydrogen Consultants, Inc.

The remarkable safety record of NGVs is probably due in large part to the strength of the high-pressure storage vessels and the stringent certification and testing standards imposed on the components of the fuel system. The development and proliferation of low-pressure technology (adsorbent storage) will likely result in thinner fuel vessels, less-stringent standards and consequently an increased risk of a ruptured fuel system. However, such an event is in many ways less hazardous than a ruptured gasoline fuel system. Unlike a liquid fuel, natural gas will not remain at the site in a highly flammable state. Instead, the escaping gas rapidly disperses into the air where it can no longer ignite.

Like gasoline and methanol, natural gas exhaust emissions vary depending on the tradeoffs made between performance, fuel efficiency, and other factors. However, natural gas is basically a low-emissions fuel. In fact, exhaust emissions are among the lowest of all the alternative fuels, with the exception of NO_x emissions. Emissions of NO_x can actually exceed those of gasoline if the natural gas engine is not appropriately tuned.

Natural gas does not contain sulfur and it does not mix with oil. Consequently, there are no sulfuric emissions, and the lubricating oil, combustion chamber, and spark plugs remain cleaner in a natural gas engine. Its gaseous nature eliminates the cold-starting and warm-up problems associated with poor vaporization of liquid fuels and thereby avoids the characteristic "rich" mixture typical of liquid-fueled engines.

The largest percentage of exhaust hydrocarbon emissions is in the form of methane which is not photoreactive. Reactive hydrocarbon emissions are extremely low. A NGV will therefore contribute very little to ozone formation. However, methane is a powerful greenhouse gas and introducing additional methane into the atmosphere may contribute to global warming. Emissions of NO_x can be controlled with a reduction catalyst as long as the mixture is stoichiometric.

TABLE 4.7. EXHAUST EMISSIONS COMPARISON
g/km* - g/mi

Fuel	Reactive Hydrocarbons Exhaust	Reactive Hydrocarbons Evaporative	CO	NO_x
Gasoline	0.22 - 0.35	0.02 - 0.04	0.87 - 1.4	0.41 - 0.66
Methanol (M85)	0.16 - 0.25	0.056 - 0.09	0.62 - 1.0	0.28 - 0.45
Ethanol (ABBE)	0.63 - 1.02	0.03 - 0.05	1.12 - 1.8	0.38 - 0.61
Natural Gas	0.12 - 0.19	0	0.06 - 0.1	0.27 - 0.44

* CARB figures are compiled in g/mi. Equivalent g/km values inserted by the author.
Source: State of California Air Resources Board (CARB)

Total emissions throughout the cycle of production and use should be moderately lower with a dedicated natural gas vehicle. Also, studies at GRI indicate that blends of natural gas and small amounts of hydrogen (up to 10 percent) can reduce already low emissions by as much as 40 percent.[9] Other studies indicate that virtually all natural gas/hydrogen blends exhibit lower NO_x and CO emissions, and that the rate of emissions reduction is greater than the rate at which the relative proportion of hydrogen is increased.[10] Chrysler Corp. completed a study in which the cost to purchase and operate natural gas vehicles in the 1998-2003 period was projected to be 16 percent lower than electric vehicles, and emissions of oxides of nitrogen would also be significantly lower. Actual effects would depend on the mix of source fuels used by electric-generating plants, which the Chrysler study projected at 55 percent coal, 13 percent natural gas, 5 percent oil, and 27 percent other sources such as nuclear and hydroelectric.[11] Based on the projected mix of fuels, the study concluded, however, that natural gas vehicles have slightly higher hydrocarbon and carbon dioxide emissions when compared to electric vehicles.

Some researchers are concerned about the uncertainty of methane's potency as a greenhouse gas, and the magnitude of increased atmospheric methane resulting from system leakages in an expanded infrastructure. At present, not enough is known to make an accurate forecast of the environmental effects of large-scale use of natural gas as a transportation fuel. On balance, expanded use of natural gas should have an overall beneficial effect on the environment.

Compatibility with Current Motor Vehicle Design

Converting a gasoline engine to run on natural gas is relatively simple. A gas induction system is installed upstream of the carburetor and timing is slightly advanced. Such a conversion will produce an engine that runs smoothly, but at slightly reduced power. These minimal modifications allow for flexible fuel operation, which means that the vehicle can be switched over to run on gasoline at the discretion of the operator.

If flexible fuel operation is not necessary, the gasoline fuel system is removed and the engine is then optimized for natural gas. Optimization consists primarily of increasing the compression ratio to a value as high as 15:1, and advancing the timing several degrees past the tolerable setting for gasoline. These modifications let the engine take advantage of the 130 octane rating of natural gas. A dedicated natural gas engine runs with increased power and efficiency, comparable to or surpassing that of a dedicated gasoline engine.

Fuel Delivery

Fuel may be introduced and metered by a mixer, which operates similar to a carburetor, or by injectors which operate similar to the injectors of a liquid-fueled engine. Until recently, the most common approach to fuel delivery has been to use a venturi or orifice system in which the incoming air pushes a mechanical element to operate a gas control valve, as shown in Figure 4.4. Since a gaseous fuel does not require vaporization, it would appear that the demands on the fuel delivery system are reduced. Quite the opposite is true: The fuel's gaseous nature requires injectors of greater volumetric capacity, and precise and rapid metering is more difficult to achieve. The process is made more demanding by the variable composition of natural gas itself and by variations in air and gas temperatures, all of which affect fuel/air ratio.

Recently, fuel injection systems have been developed that provide more precise control over fuel metering. These systems can be interfaced with traditional gasoline-style electronic fuel injection systems and thereby provide precisely controlled timing and metering. Modern electronic engine control technology is transferable to natural gas application, but may require additional inputs for fuel temperature and pressure in order to provide complete analysis and control of the variables of the natural gas fuel system.

Much remains to be learned about optimizing a natural gas engine for maximum efficiency and power, and for minimum exhaust emissions. Electronic fuel injection systems have recently been developed and still require refinement in order to match the precise fuel metering capabilities of their gasoline counterparts.

On-board Storage of Natural Gas

Natural gas for vehicular application is traditionally stored in a compressed state (CNG) at pressures up to 20.7 MPa above ambient (3000 psig). The gas may also be stored cryogenically as a liquid (LNG), or by utilizing a promising new low-pressure technology known as adsorbent storage (ANG). The effect on the engine is unchanged, regardless of the storage system.

Compressed Natural Gas

Of the nearly 190 million motor vehicles in the U.S., only 50,000 use natural gas as a fuel, and these vehicles rely almost exclusively on CNG technology. A

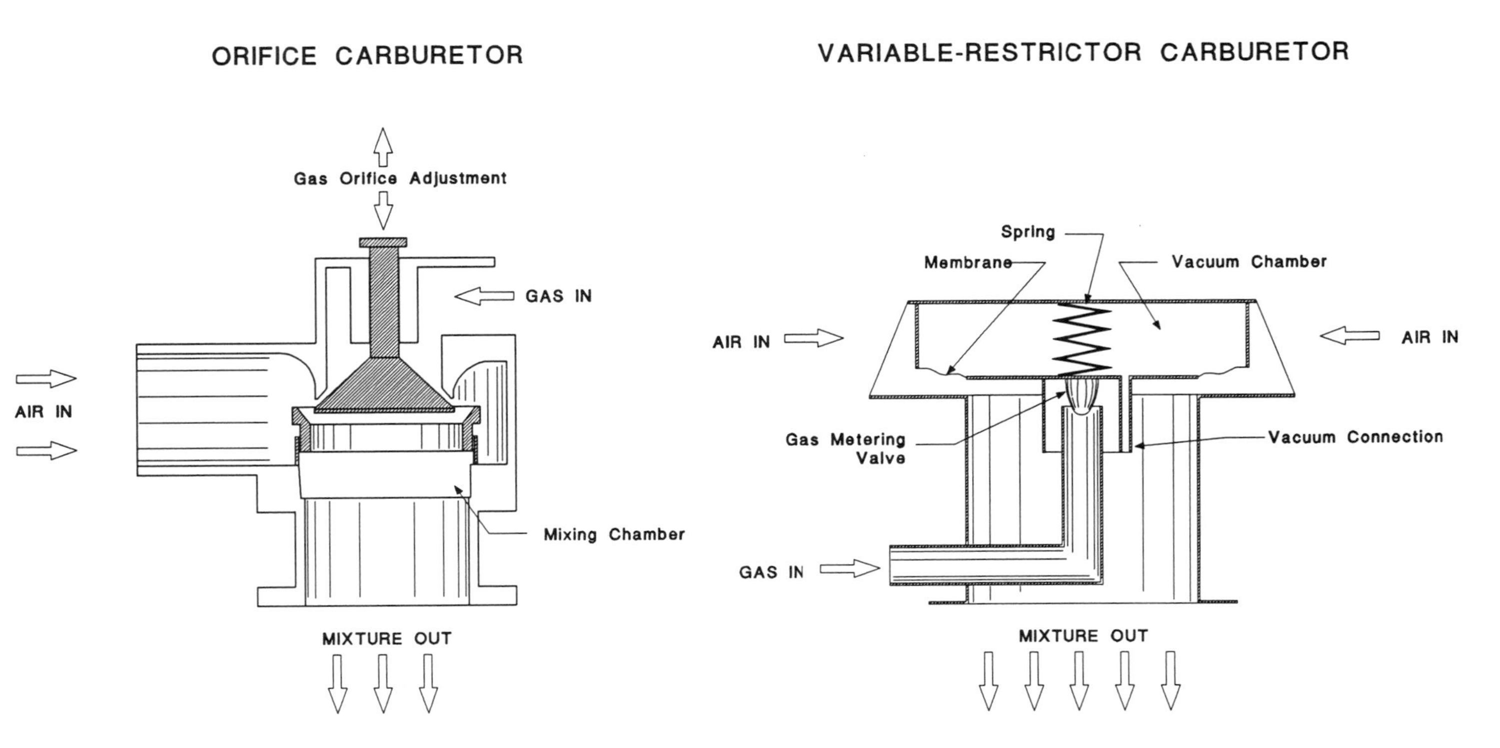

Fig. 4.4. Natural Gas Carburetors.

CNG system stores the on-board gas in heavy cylinders at pressures of up to 20.7 MPa above ambient (3000 psig). The vehicle is then refueled with high-pressure gas from a specially equipped refueling station.

The volume/range relationship of natural gas as compared to other fuels can be estimated on the basis of its energy equivalence. It takes about 0.93 m^3 of atmospheric natural gas to equal the energy available in 1 liter of gasoline (125 ft^3/gal). To equal the energy in a liter of diesel fuel requires slightly more than 1 m^3 of natural gas (135-140 ft^3/gal). Consequently, 70 cubic meters of atmospheric natural gas is required to equal the range of a 75-liter (20-gal) tank of gasoline. In order to contain this volume of CNG at 20.7 MPa above ambient (3000 psig), the storage vessel must have a volume of approximately 0.28 m^3 (10 ft^3). By comparison, an energy-equivalent gasoline tank occupies only about one-fourth the space, and a tank of diesel fuel occupies only about one-fifth the space of a CNG container.

Weight of the fuel container per given energy content is also an important consideration. Gasoline and its steel container weigh approximately 0.96 kg/L (8 lb/gal). Empty, a high-pressure steel CNG vessel weighs approximately 0.96 kg per volumetric liter (60 lb/ft^3), which calculates to about 3.77 kg per liter equivalent (31.5 lb/gal) of gasoline when filled with CNG. A fiber-reinforced aluminum CNG vessel will weigh about half that on a volume equivalence basis. A similar fiber-reinforced stainless steel vessel can weigh as little as 0.32 kg/L (20 lb/ft^3). However, reduced weight has a price penalty as can be seen in Table 4.8.

The volume, weight, and cost of energy stores defines limitations that engineers must seek to minimize with the design of the vehicle. For example, when vehicle energy efficiency is improved, allocations to expensive and heavy vessels can be reduced. Additional economies can be gained by utilizing storage vessels for structural reinforcement. CNG vessels may be positioned to provide collision protection or to lend strength to an otherwise underrated chassis. However, the periodic recertification required for high-pressure vessels complicates the challenges associated with integral fuel tanks. To obtain recertification, vessels must be removed from the vehicle and hydrostatically tested at 2.5 times their rated pressure. Consequently, quick and easy tank removal must be part of any CNG vehicle design, or regulations must be changed to accommodate new designs.

A CNG vehicle is normally refueled from a station set up to provide rapid delivery of high-pressure natural gas. A typical CNG refueling station consists of a compressor and storage vessels from which the fuel is delivered to the vehicle.

TABLE 4.8. COST AND WEIGHT COMPARISON OF CNG STORAGE VESSELS

Vessel Type	Weight Empty Vessel lb/ft³ - kg/L	Charged Weight Per Energy Equivalent of Gasoline* lb/gal - kg/L	Cost of Vessel According to Vessel Volume \$/ft³ - \$/L	Cost of Vessel According to Gasoline Energy Equivalent** \$/gal - \$/L
Steel	60 - 0.962	31.50 - 3.77	90.00 - 3.18	43.27 - 11.42
Aluminum	48 - 0.770	25.75 - 3.08	200.00 - 7.06	96.15 - 25.37
Fiber-Reinf. Stl	56 - 0.890	29.60 - 3.54	175.00 - 6.18	84.13 - 22.20
Fiber w/ Alum.	37 - 0.593	20.50 - 2.45	155.00 - 5.47	74.50 - 19.66
Fiber w/ Stainless Stl	20 - 0.320	12.30 - 1.47	325.00 - 11.48	156.25 - 41.23

* Includes approximately 90 g/L (5.6 lb/ft³) for weight of CNG at 20.7 MPa above ambient (3000 psig). Assumes a natural-gas/gasoline equivalence ratio of 0.93 m³/L (125 ft³/gal).

** Cost on a gasoline-energy-equivalence basis for high-pressure CNG vessel.

Rapid refueling is normally accomplished by a cascade delivery system. Cascade delivery utilizes the compressor to recharge a multiple of storage vessels. Fuel is then delivered to the vehicle by drawing in order from the lowest- to the highest-pressure vessel. A cascade system can refuel a typical CNG passenger car in as little as 10 minutes. The size of the compressor and the number of storage vessels are optimized for the number of daily refuelings, and the cost of the system varies accordingly. Using the Canadian experience with public natural gas refueling stations as a guideline, the average cost to retrofit a traditional service station with CNG compression and dispensing equipment is expected to be in the range of \$320,000.[12]

Regardless of the technical feasibility of operating vehicles on compressed natural gas, an undeveloped distribution infrastructure for high-pressure gas will continue to limit the fuel to fleet operations that can amortize the expense of an on-site refueling station. In a chicken-or-the-egg dilemma, a consumer-oriented distribution system for high-pressure gas is unlikely to develop in the absence of vehicles to refuel, and natural-gas-fueled vehicles will remain an unattractive option for consumers until adequate distribution is in place. The proponents of a new low-pressure storage system believe that "adsorbent storage" will effectively circumvent the dilemma.

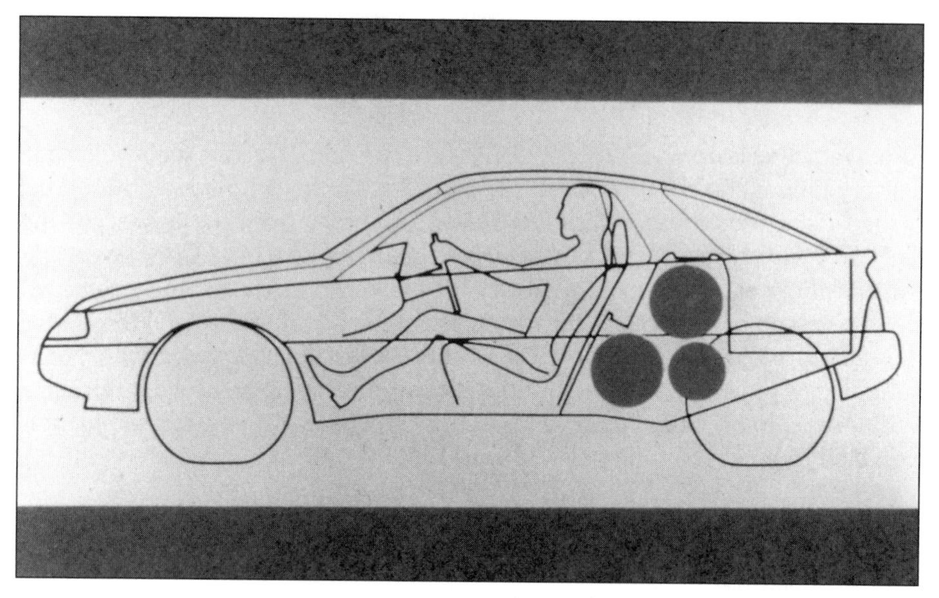

Fig. 4.5. Ford's two-seater alternative fuel vehicle made its debut at the 1982 World's Fair in Knoxville, Tenn. The two-seater vehicle runs on compressed natural gas which is stored in three cylinders behind the seats. (Courtesy: American Gas Association)

Fig. 4.6. Refueling with compressed natural gas takes only slightly longer than refueling a conventional gasoline vehicle. (Courtesy: American Gas Association)

Adsorbent (Low-Pressure) Storage

The technology of adsorbent storage has the potential to open vast new markets for natural gas in the transportation and consumer appliance markets. If development goals are realized, this new technology of low-pressure natural gas storage will make it possible to refuel boats, recreational vehicles, lawn mowers and even automobiles from a device connected to a residential gas line.

Adsorbent storage is based on the ability of some materials to assimilate methane gas (the primary constituent of natural gas). The most efficient adsorbents to date are made of activated carbon. When a standard cylinder is filled with activated carbon the amount of gas that can be stored at relatively low pressure is greatly enhanced. This principle increases storage capacity of the cylinder up to a gauge pressure of about 12.4 MPa (1800 psig) (see Figure 4.7). At 12.4 MPa above ambient (1800 psig) the presence of carbon actually degrades the storage capacity of the container by occupying room that would otherwise be available for compressed gas. Carbon adsorbency takes place at relatively low pressures. At higher pressures, an adsorbent natural gas (ANG) vessel operates as a simple compression vessel.

The relatively low operating pressures of an ANG storage system makes it possible to utilize relatively inexpensive support components. Reduced operating pressure (normally 3.5-6.2 MPa above ambient) reduces system complexity, lowers hardware costs, and eliminates hazards that might otherwise be associated with a high-pressure system. For example, an ANG compressor may be configured as a simple two-stage unit, as long as system pressure is limited to 3.5 MPa above ambient (500 psig). High-pressure CNG systems require a significantly more sophisticated, and therefore more costly, four-stage compressor. In addition, the lower pressure typical of ANG allows for less-expensive components throughout the system, including the substitution of thinner cylinders which do not require periodic recertification. This reduces manufacturing and maintenance costs, and provides the designer with additional options, including the freedom to design the storage tank as an integral and load-bearing part of the chassis.

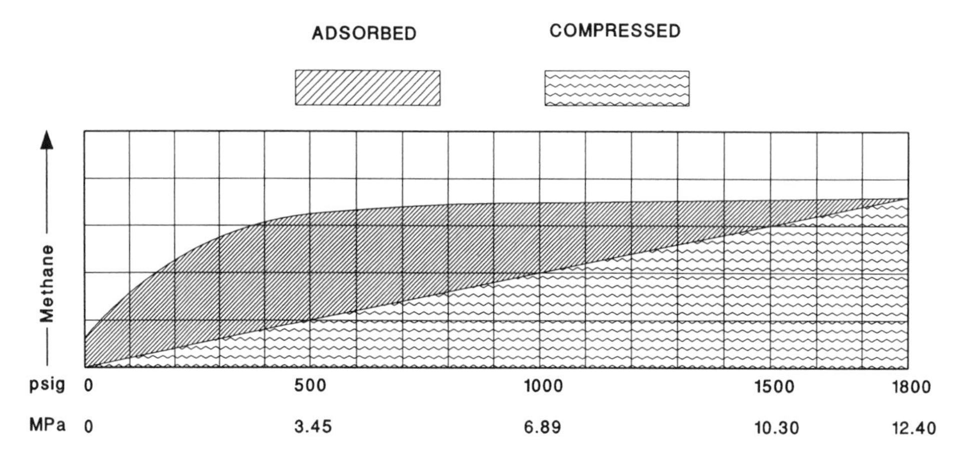

Fig. 4.7. Adsorbed/Compressed Ratio Typical of Methane Adsorbed on Activated Carbon.

Difficulties and Costs

The benefits of ANG are not without costs and engineering challenges. Methane adsorption capacity is a function of the surface area of the adsorbent and the specific number of adsorption sites on the carbon. Controlling the characteristics of the carbon to maximize its performance has been one of the most challenging problems of ANG technology. According to the Gas Research Institute (GRI), the carbon AX-21, previously manufactured by Anderson Development Company, has proven to be one of the most efficient carbons up to this time.[13]

Other problems have included contamination of the carbon bed with the normal impurities of natural gas, and managing the effects of temperature variations that result from rapid charge/discharge cycles. During rapid charge from atmospheric to 3.5 MPa above ambient (500 psig), the heat of adsorption can reduce the storage capacity by approximately 20 percent unless the sorbent bed is cooled. Later, heat may have to be reintroduced to induce complete desorption. Contamination problems have been successfully eliminated by filtering the incoming gas. A proprietary Thermal Energy Storage (TES) system developed by GRI successfully deals with exothermic and endothermic reactions to adsorption/desorption.

The final difficulty is the high cost of producing the carbon. The cost of carbon varies widely from a low of $3.30/kg ($1.50/lb) to a high of $44/kg ($20/lb). The cost of the carbon has no relationship to the material's effectiveness as an adsorbent. Until recently, however, the most costly carbons have also been the most efficient methane adsorbents, and therefore ANG could not economically compete with CNG technology. According to AGLAR's Doug Horne, the economic threshold is approximately $6.60/kg ($3/lb) for the carbon adsorbent, and that goal has been recently achieved.

Enhanced Capability ANG Storage

Recent studies at the Institute of Gas Technology (IGT) have indicated that the capacity of an adsorbent storage system is markedly enhanced by chilling the gas before introducing it into the carbon bed. According to the IGT experiments, at 3.5 MPa above ambient (500 psig) a cylinder packed with AX-21 carbon to a density of 0.4 g/cm^3 had a storage capacity of 205 Vs/V at 227 K (–46°C), and 256 Vs/V at 200 K (–73°C).* By comparison, at 293 K (20°C) and 20.7 MPa above ambient (3000 psig), a standard CNG cylinder stores about 260 Vs/V. Pre-chilling the gas also eliminated the need for the GRI's TES system. Low-temperature ANG introduces the need, however, to apply additional heat (more

---

* Vs/V = Volumes of atmospheric gas per volume of tank capacity at pressure.

than is normally required) for desorption. In general, heat to promote desorption is not viewed as a problem because both electrical and waste engine heat are readily available on-board the vehicle.

Researchers at IGT believe their low-temperature concept will reduce carbon requirements as well as eliminate the need for a system to manage the heat of adsorption. They estimate that the total reduction in required storage volume will be 40 percent with a 227 K (–46°C) pre-chill, and 52 percent with a pre-chill to 200 K (–73°C). Carbon requirements would drop by 27 and 41 percent, respectively, and storage system weight will be reduced by 59 and 67 percent, respectively. These substantial technical improvements are somewhat offset by greater costs associated with the fuel delivery system.

Creative Tank Designs for ANG Systems

The low operating pressure of an ANG system opens the possibility for tank designs that would otherwise be infeasible with high-pressure systems. High-pressure CNG containers must be cylindrical or spherical in order to contain the extremely high pressure of the gas. However, neither shape utilizes space as well as a square or rectangular vessel. The relatively low pressure of an ANG system opens the way for creative tank designs that are much more compact.

For example, a square or flat tank can be designed by stacking a series of essentially triangular cells. The efficiency gains of such a design are significant. A square tank will store 27 percent more fuel than a cylindrical tank of equal outside dimensions, and a flat tank presents even more possibilities. Using this simple approach, a flat tank might form a load-bearing floor and assume the role of the vehicle's frame. An integral frame/tank design could result in greatly increased fuel storage capability with little or no compromise in passenger or cargo area. Such tank designs are under development at Rolls Royce in England.

Home Refueling with Natural Gas

One of the most interesting aspects of adsorption storage technology is the ability to refuel at home from a standard residential gas service. A slow-fill residential system designed for overnight refueling will place significantly reduced demands on the compressor and the storage system. Such a system would have no need for heat management because excess heat would dissipate over the relatively long recharge cycle. Pre-chill capability could be provided by a small refrigeration unit. A residential refueling appliance for an ANG system could utilize a relatively inexpensive compressor designed to charge directly into the vehicle's storage vessel. Vehicle fuel costs would then be included in the consumer's monthly utilities statement.

Fig. 4.8. High-pressure residential refueling station for CNG vehicles is currently manufactured by FuelMaker Corporation of Salt Lake City, Utah. A low-pressure delivery system would be slightly smaller and about half the cost. (Courtesy: FuelMaker Corporation)

Hydrogen/Methane Blends

Hydrogen is sometimes referred to as the ultimate alternative fuel. Cryogenic or hydride storage can provide adequate on-board energy stores, but both are plagued by significant handling difficulties, high mass (especially with hydrides), high costs, and the absence of a fuel-supply infrastructure. Gaseous hydrogen utilizes essentially the same technology as CNG, but suffers from inadequate vehicle range and lack of a fuel supply infrastructure. In contrast, the infrastructure for methane (natural gas) is already in place requiring only that it be adapted to the transportation sector. Both hydrogen and methane gas can be delivered over the same pipeline distribution network. As a precursor to widespread use of hydrogen, the fuel could be introduced to the transportation sector initially as an enhancer or additive to natural gas. Gaseous hydrogen/methane blends could provide a relatively low-cost, low-risk means for a gradual transition to the so-called "hydrogen economy."

Adding hydrogen to natural gas results in significantly reduced hydrocarbon and carbon-monoxide exhaust emissions. Early results indicate that harmful emissions are significantly reduced with blends containing as little as 10 percent hydrogen. Hydrogen is also adsorbed on activated carbon. The effect of hydrogen on adsorbent efficiency with a predominantly methane blend is shown in Figure 4.9.[14]

Hydrogen as a Fuel

Hydrogen is the lightest and most simple of all the elements and yet it is capable of releasing tremendous energy when combined with oxygen and burned. Hydrogen may be used as a gaseous fuel in an IC engine, similar to the way in which natural gas is used. A more efficient way to utilize hydrogen's energy is in a fuel cell to produce electricity, which may then be converted to tractive power using an electric motor. Either way, when hydrogen is converted into energy the by-product is water. Consequently, hydrogen is normally considered a zero pollution energy source even though pollutants may be created at the primary generation site. Hydrogen is technically not a fuel but, like electricity, it is an energy carrier. Both electricity and hydrogen can be generated by utilizing virtually any primary form of energy.

Hydrogen Source and Abundance

Hydrogen is the most abundant element in the universe. Although hydrogen barely exists on the earth's surface as a free element, it has combined with a variety of other elements and may be extracted in a number of ways. The methods of producing hydrogen can be grouped into two primary categories: the chemical technologies that break down fossil fuels into more simple molecules, and electrolysis which can use any source of electrical power to produce hydrogen from water. The most economical method is to separate hydrogen from natural gas through a process called *steam reformation*. With improvements in photovoltaic and other solar-to-electricity technologies, electrolysis may ultimately provide an endless supply of cheap hydrogen for future genera-

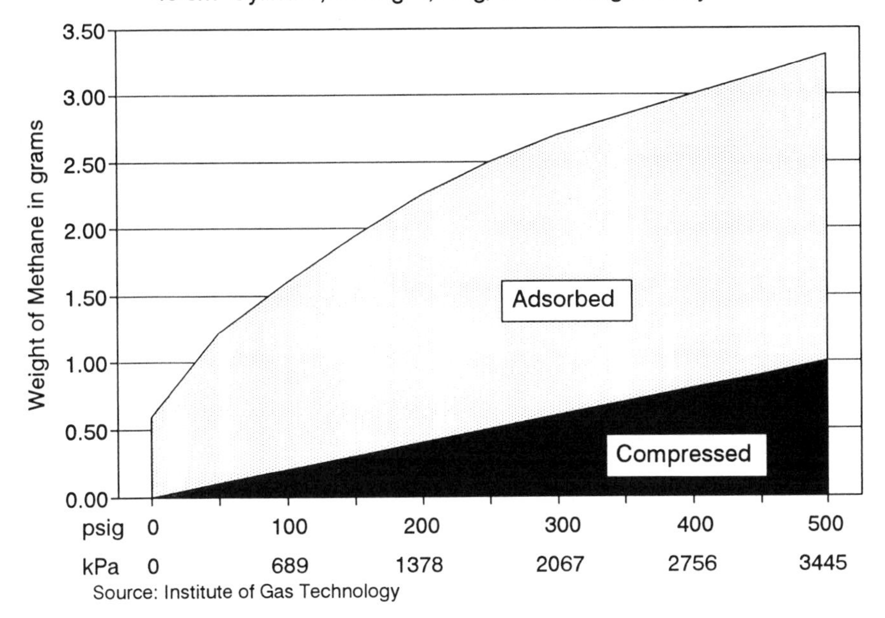

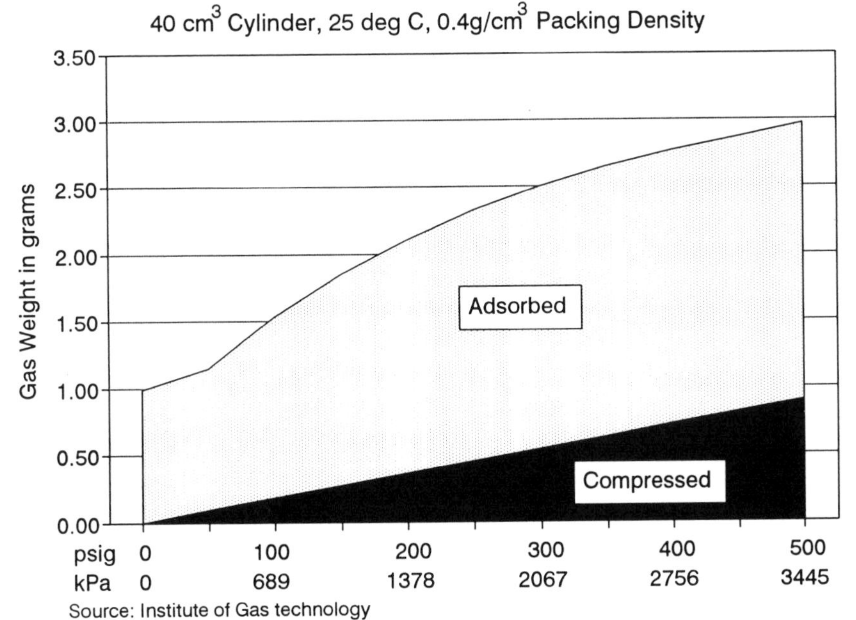

Fig. 4.9. Effects of Hydrogen on Adsorption.
(Source: Institute of Gas Technology)

tions. But today, hydrogen produced by electrolysis is significantly more expensive than hydrogen reformed from natural gas.

Worldwide production of hydrogen currently amounts to 20,860,000 t (23,000,000 U.S. tons) annually. The most extensive use of hydrogen is in the manufacture of ammonia, which consumes two-thirds of the world's production. Oil refineries use one-fifth of the hydrogen, and the remainder goes to fertilizer, dye and rocket fuel production.

Hydrogen Supply Infrastructure and Cost

The key to the acceptance of any alternative to gasoline is the availability and cost of the fuel. Today, hydrogen is both expensive and unavailable to consumers. The cost of hydrogen varies widely depending on how it is produced and on how it is delivered to the consumer. At-the-pump price estimates range from slightly more than the cost of gasoline to $6.60 or more per liter equivalent of gasoline ($25/gal). Table 4.9 outlines the production costs of hydrogen in 1989 dollars.[15]

TABLE 4.9. HYDROGEN PRODUCTION COST

Energy Source	Source Energy $ Cost ($/GJ)*	Energy Conv. Effic. (%)	Hydrogen Cost $/mmBTU ($/GJ)* Energy Cost	Capital Cost	Production Cost $/mmBTU ($/GJ)*
Nat. Gas	2.00/mmBTU (1.90)	65 - 75	2.80 (2.65)	2.80 (2.65)	5.60 (5.31)
Coal	2.00/mmBTU (1.90)	<55	4.00-5.00 (3.79-4.74)	7.00-8.00 (6.63-7.58)	12.00 (11.37)
Grid Elect.	0.03/kW-h**	63 - 81	11.00-14.00 (10.43-13.27)	8.80 (8.34)	20.00-23.00 (18.96-21.80)

* Metric units in parentheses calculated by the author
** 1 kW-h = 3.6 MJ
Source: California Energy Commission (based on 1989 estimates)

Whether hydrogen is in a gaseous or liquid state determines much about the cost and means by which it is delivered. Gaseous hydrogen may be transported through pipelines, similar to the way in which natural gas is transported. Liquid hydrogen can be economically transported by truck. Either way, transportation costs are expected to be significantly greater than those of gasoline, with trucking costs estimated at 1-1/2 times the cost of gaseous pipeline delivery. The cost to dispense hydrogen at public refueling stations is estimated to be roughly equal to the cost of dispensing natural gas through similar facilities.[15]

TABLE 4.10. HYDROGEN DELIVERY AND DISPENSING COSTS

Fuel Resource	Estimated Delivery Cost to Refueling Station $/mmBTU ($/GJ)*	Estimated Delivery Cost to Refueling Station $/kW-h**	Estimated Vehicle Dispensing Cost $/mmBTU ($/GJ)*
Gasoline	6.00-8.00 (5.70-7.60)	0.02-0.03	0.38-1.15 (0.36-1.09)
Nat. Gas by Pipeline	3.00-6.00 (2.70-5.70)	0.01-0.02	1.00-1.50 (0.95-1.42)
Hydrogen Gas by Pipeline	10.00-15.00 (9.48-14.22)	0.034-0.05	1.25-1.75 (1.18-1.66)
Liquid Hydrogen by Tanker	15.00-25.00 (14.22-23.69)	0.50-0.085	

* Units in parentheses calculated by the author
** 1 kW-h = 3.6 MJ
Source: California Energy Commission (based on 1989 estimates)

Some advocates suggest that hydrogen might be economically produced by on-site, residential units. Such systems would avoid the costs of delivering and dispensing the fuel through a conventional infrastructure. The cost of on-site production must then include amortization of generating and dispensing equipment, as well as the cost of electrical power.

Another interesting concept focuses on the existing pipeline distribution network and the finite reserves of natural gas. The idea of delivering gaseous hydrogen through the existing natural gas pipeline network is a proven concept. In the late nineteenth century, gaseous hydrogen was delivered through pipelines in the U.S. that were later converted to natural gas delivery. In Europe, some consumers were still receiving hydrogen through pipelines as recently as the 1950s. Converting the natural gas delivery network to hydrogen would create some difficulty in controlling system leaks, and it would necessitate different seals and metering equipment. However, the cost to adapt the existing system would be far less than developing a new one. An entire infrastructure for supplying energy in a gaseous form is in place, and as natural gas reserves dwindle, distributors may turn to hydrogen. The near-term development of natural gas transportation applications, perhaps utilizing adsorption storage technology and home refueling, may pave the way for the ultimate transition to pipeline-supplied hydrogen.

Hydrogen Storage Technologies

Hydrogen can be stored on-board as a compressed gas, as a super-cooled liquid (cryogenic storage), as a solid hydride, or by adsorption on activated carbon. The most common and least costly method of storing hydrogen is as a compressed gas in cylinders similar to CNG technology. For cryogenic storage, hydrogen must be liquefied by reducing the temperature of the gas to 20.4 K. A properly designed cryogenic container can limit the boil-off rate to between 2 and 5 percent per day at an internal pressure of 0.6 MPa above ambient (approximately 100 psig).

Hydride and adsorption storage are both relatively new technologies. The principle of hydride storage is based on the propensity of certain materials to assimilate hydrogen under pressure to form a hydrogen compound (metal hydride). The hydrogen is later released by applying heat to the hydride. The hydride can store more hydrogen for a given volume than is possible with cryogenic storage. However, hydrides are extremely heavy and expensive and tend to react sluggishly to demands.

Adsorption on activated carbon is essentially the same technology used for natural gas adsorption storage. However, hydrogen adsorption takes place at significantly lower temperatures.

Table 4.11 provides a comparison of different hydrogen storage technologies to a conventional tank of gasoline. Even if the hydrogen were stored as a liquid, a hydrogen-fueled IC vehicle would have to carry four to five times the volume of fuel to equal the range of a gasoline-fueled IC vehicle.

Compatibility with the Internal Combustion Engine

With very little modification, a standard internal combustion engine will run well on hydrogen. The three primary problems associated with hydrogen fuel include reduced power in comparison to gasoline, a tendency to flashback into the intake manifold, and embrittlement of iron combustion chamber components.

TABLE 4.11. GASOLINE AND HYDROGEN SYSTEM ATTRIBUTE COMPARISON

Storage Technology	Fuel Weight and Volume Fraction* (WF/VF)	Mass Energy Density (kW-h/kg)**	Volumetric Energy Density (kW-h/L)**	Infrastructure	Propulsion System Interface***	Notes 1) Advantages 2) Disadvantages
Gasoline Tank	WF = 70% VF = 80%	9.0	8.1	Low-pressure tanks and pumps	No	1) Simple system; low cost 2) Crash safety
Compressed Gas Cylinder	WF = 1-5% VF = 60%	0.4 to 2.0	0.4 to 0.6	High-pressure tanks and compressor	No	1) Simple system; low cost 2) High pressure; safety; low capacity
Liquid/Slush Tank	WF = 10-25% VF = 55-72%	3.9 to 9.9	1.5 to 2	Medium-pressure tanks and compressor; cryogenic refrigeration	Yes	1) Mass capacity; volumetric capacity 2) Evaporative loss; low temperature; safety
Hydride Tank	WF = 1.5-4%	0.6 to 1.6	0.6 to 1.6	Low-pressure tanks and compressor	Yes	1) Crash safety; volumetric capacity 2) Poisoning; loss of capacity; heavy
Activated Carbon Adsorption Tank	WF = 4-8%	1.6 to 3.2	0.6 to 0.95	Medium-pressure tanks and compressor; refrigeration equipment	NA	1) Lightweight; crash safety 2) Refrigeration temperature; evaporative loss; high volume

* Fuel Weight Fraction = Fuel Weight/(Fuel Weight + Storage System). Fuel Volume Fraction = Fuel Volume/(Loaded Storage System Volume).

** Hydrogen higher heating value = 39.4 kW-h/kg. Assuming a 2.5 to 1 efficiency ratio in favor of a fuel cell vehicle, to achieve the same energy density as a gasoline tank the hydrogen energy storage system must have a mass energy density of 3.6 kW-h/kg and a volumetric energy density of 3.24 kW-h/L. (1 kW-h = 3.6 MJ)

*** Propulsion System Interface refers to the need for the storage system to be connected to the fuel cell for thermal or other reasons.

Source: California Energy Commission

Reduced power results from the reduced volumetric energy of a stoichiometric charge of hydrogen. Several causes have been suggested to explain the phenomenon of flashbacks. Flashbacks occur when hydrogen is mixed with air and then carried into the intake port as a combustible charge. For some time, experimenters were uncertain of the cause of flashbacks. However, the effectiveness of water injection and cooled exhaust gas recirculation in reducing the tendency to flashback indicates that the primary cause is the presence of hot spots at the cylinder intake port. Although mechanically difficult, Japanese researchers have successfully demonstrated a design for a direct injection system that does not deliver the hydrogen until ignition is desired.[16] Additional work in Japan also suggests that two-cycle engine characteristics might be especially complementary to the flame characteristics of hydrogen.[17]

No aftertreatment is required for a hydrogen-fueled IC engine. Emissions consist primarily of steam with a trace amount of hydrocarbons from engine lubricating oil. The thermal efficiency of a hydrogen IC engine is slightly greater than that of the same engine optimized for gasoline.

Fuel Cell

ADVANTAGES

Any discussion of hydrogen fuel would be incomplete without the mention of fuel cells. The most efficient way to convert hydrogen into energy is by using it as fuel in a fuel cell. A fuel cell converts hydrogen directly into electricity at an efficiency of 50-60 percent. Depending on the design, a fuel cell can operate up to 2-1/2 times more efficiently than an internal combustion engine. A fuel cell can be quickly refueled, whereas a battery must be recharged over a much longer period. Table 4.12 compares emissions and efficiency of a hypothetical vehicle configured for various power systems.

TABLE 4.12. VEHICLE EFFICIENCY BY TYPE

Vehicle Type	Emissions	% Energy Efficiency
Gasoline IC Engine	Low	12-15
Hydrogen IC Engine	Ultra-Low	12-16
Fuel Cell Electric Vehicle	Zero	30-40
Fuel Cell/Battery Electric Vehicle	Zero	25-35

The primary technical disadvantages of the fuel cell include low specific power, slow start-up and an inability to respond to peak demands. A hybrid fuel cell/battery system, however, can resolve the problem of peak demands. Such a system would rely on the battery for high transient loads and utilize the fuel cell to provide a continuous output to recharge the battery. Fuel cell costs are also prohibitively high. Table 4.13 provides a comparison of existing fuel cell technology, along with improvements expected by the year 2002.[15]

TABLE 4.13. COMPARISON OF POWER DENSITY

Propulsion System	1991 Power Density W/kg	Power Density Expected by 2002 W/kg
IC Engine	200-500	200-600
Phosphoric Acid Fuel Cell	35-80	50-100
Solid Polymer Fuel Cell	50-150	100-250
Alkaline Fuel Cell	50-200	100-300

Source: California Energy Commission

In its present state of development the fuel cell is still too large and costly for automotive application. With further development, power densities could approach those of the IC engine. If costs can be reduced, the compact, rapid-starting solid polymer cell may ultimately be the most adaptable to transportation application.

Electric Power

Like hydrogen, electricity is not a fuel. However, its potential for replacing the primary transportation fuel, gasoline, is significant. In the U.S., the transportation sector is responsible for 63 percent of the nation's petroleum consumption. But less than 5 percent of the nation's electrical energy comes from petroleum fuels. Electrical energy is produced from a variety of alternative fuels and energy sources. Consequently, electric vehicles are vehicles that run on the source fuels used by utilities companies to generate electrical energy. Table 4.14 lists the U.S. Energy Information Administration breakdown of electrical energy sources, along with projections through 2010.[18]

TABLE 4.14. U.S. SOURCE FUEL FOR ELECTRIC GENERATION WITH PROJECTION THROUGH THE YEAR 2010

ExaJoules (10^{18})* - Quadrillion (10^{15}) BTUs (Quads) Per Year

Electrical Energy Source	1990 EJ - Quads	2000 EJ - Quads	2010 EJ - Quads
Distillate Fuel	0.105 - 0.10	0.20 - 0.19	0.16 - 0.15
Residual Fuel	1.29 - 1.22	1.96 - 1.86	1.64 - 1.55
Natural Gas	3.09 - 2.93	5.08 - 4.82	6.03 - 5.72
Steam Coal[1]	16.94 - 16.06	18.93 - 17.94	23.84 - 22.60
Nuclear Power	6.48 - 6.14	6.83 - 6.47	7.04 - 6.67
Renewable Energy/Other[2]	3.91 - 3.71	5.94 - 5.25	6.59 - 6.25
Total (ExaJoules - Quads)**	31.80 - 30.14	38.54 - 36.53	45.29 - 42.93

* EAI figures are given in quadrillion BTUs. ExaJoules calculated by the author.
** ExaJoule figures may not result in the correct totals because of rounding.
1 Coal-fueled steam generation includes consumption by independent power producers.
2 Includes electricity generated to serve the grid from hydroelectric, geothermal, wood waste, municipal solid waste, other biomasses, wind, photovoltaic and solar thermal sources, plus waste heat, and net electricity imports.
Source: U.S. Energy Information Administration.

In 1989, total U.S. energy consumption amounted to 85.7 EJ (81.2 quads) and oil provided 42 percent of the energy, or about 36 EJ (34.1 quads). The transportation sector consumed 27 percent of the nation's energy, but that sector relies almost totally on petroleum motor fuel. Consequently, 63 percent of the nation's total oil supply was consumed by all forms of U.S. transportation in 1989. That amounted to 10.85 million barrels of oil per day, or over 23 EJ (22 quads) of annual energy consumption. Sixty-two percent of the energy consumed by the transportation sector is consumed as fuel by passenger cars and light-duty trucks (1989).

Although the electric vehicle is itself an efficient consumer of energy, when the entire energy cycle is considered, the advantage tends to diminish. Considering the overall energy chain, electric vehicles consume 10-50 percent less primary energy than their gasoline-fueled counterparts, depending on the particular energy chain. Additionally, electric cars can reduce petroleum consumption by approximately 95 percent because a switch to electric cars amounts to a switch to the source fuel used to generate electricity. If all U.S. gasoline vehicles were electric powered in 1989, U.S. oil consumption would have been reduced by 6 million barrels per day, even if the overall energy consumption of vehicles remained the same. In 1989 the U.S. imported oil at the rate of 7.12 million barrels per day.[19]

When the production of energy is assigned to a centralized facility, benefits of scale and consolidation also emerge. Producing energy becomes far more simplified and efficient. System attributes such as weight and size are no longer critical. The cost of a fixed generating system can be evaluated over the long term, rather than the short-term life of an automobile energy system. Long-term amortization can justify investment in efficiency and emission controls that perhaps could not be justified with the relatively short service life of a vehicle.

Storage technologies, often discussed as major hurdles for natural gas and hydrogen vehicles, are not especially relevant to fixed-site application. Natural gas, for example, is delivered to a fixed site through pipelines as it is consumed. Hydroelectric, solar, and wind generation are also easily utilized renewable resources for the fixed site. Conversely, these energy sources are virtually impossible to adapt to a mobile system. Fixed sites can be more easily converted to different fuels. System changeovers, upgrades and routine maintenance are implemented by professional technicians, and the sheer size of operations provides much greater economic resources. Such is not the case with individual vehicles, each with its own self-contained and somewhat unique energy system.

Electric power might therefore be viewed as the ultimate alternative fuel. The environmental and economic profile of electrical power reverts to the characteristics of the fuels used for its generation. The industry's heavy reliance on coal will undoubtedly result in an increase in emissions of sulfur oxides as more coal is burned to produce more electrical energy for automobiles. Even so, studies have shown that electric vehicles are much cleaner than gasoline-fueled IC-engine cars. The actual reduction in harmful emissions varies according to the emissions profile of the generating plant (see Chapter Five).

Today, electric vehicles suffer from the same limitations that stifled their development nearly a century ago. The battery has remained a comparatively poor transportation energy source. Ultimately, these limitations will be resolved. Solutions could come in the form of a storage device, or an on-board generation system such as the fuel cell. Or, the electric vehicle of the future could assume a different role in transportation. Urban cars and neighborhood cars based on existing technology could improve air quality in the inner cities and reduce dependence on petroleum fuels. Should a breakthrough in technology occur, the alternative transportation energy source of predominance will surely be the carrier—electricity.

Alternative Fuels and Alternative Cars in the 21st Century

Today's automobile, as well as our attitudes and expectations of it, are the natural by-products of the low cost and great abundance associated with its energy system. Transportation energy in the form of gasoline has been abundant and inexpensive throughout the twentieth century, and especially so in the U.S. American driving habits and tastes in automobiles have been nurtured by an unending supply of fuel at prices that are typically less than half the cost of the same fuel in Europe and Japan. For $1.25, U.S. consumers can purchase a gallon of gasoline ($0.33/L) that will propel 1600 kg (3500 lb) of hardware and passengers over a distance of 32 km (20 mi) and at speeds of up to 160 km/h (99 mph). The same $1.25 will buy a loaf of bread or two soft drinks. In the U.S., bottled water is more expensive than gasoline.

Petroleum has been the energy bargain of the twentieth century, but it probably will not be on sale throughout the twenty-first century. Unless vast new reserves are discovered, OPEC will soon own the last remaining reserves of easily recoverable oil, and the price is not likely to remain as low as it is today. Petroleum will ultimately be priced according to the limited commodity that it is. The increasing scarcity and expense of motor fuel will then begin to reshape our attitudes towards personal transportation vehicles. An alternative fuel or energy source must eventually be adopted for transportation. Alternative vehicle types such as urban cars, commuter cars, and sub-cars, also are likely candidates for the new paradigm in personal transportation.

Although modest energy supplies can be provided at affordable costs, duplicating the convenience/energy profile of gasoline is technically challenging and economically infeasible with existing technology. There is an ancient maxim to the effect that: "If the mountain will not come to Mohammed, then Mohammed must go to the mountain." In a sort of going-to-the-mountain approach, the vehicle can be reconfigured to reduce its appetite for energy and thereby move closer to the more modest price/energy/storage capabilities of many alternative fuels. By reducing vehicle energy requirements, many of the technical and economical challenges of alternative energy sources can be more easily managed. In this regard, alternative fuels and alternative cars are complementary components of tomorrow's world of plentiful, clean, and affordable energy for personal transportation.

References

1. Vernon P. Roan and Thomas A. Barber, "The New Breed of Hybrid Vehicles," SAE Paper No. 810270, Society of Automotive Engineers, Warrendale, PA, 1981.

2. Quanlu Wang, *et al.*, "Emission Impacts, Life-Cycle Cost Changes, and Emission Control Cost-Effectiveness of Methanol-, Ethanol-, Liquid Petroleum Gas-, Compressed Natural Gas-, and Electricity-fueled Vehicles," Institute of Transportation Studies of the University of California, Davis, CA, May 1993.

3. Coinage, A., H. Menrad and W. Bernhardt, "Alcohol Fuels in Automobiles," Alcohol Fuels Conference, Inst. Chem. Eng., Sydney, 9-11 August 1978.

4. Frank Lipari and Fayetta L. Colden, "Aldehyde and Unburned Fuel Emissions from Developmental Methanol-Fueled 2.5L Vehicles," SAE Paper No. 872051, Society of Automotive Engineers, Warrendale, PA, 1987.

5. H.G. Adelman, D.G. Andrews and R.S. Devoto, "Exhaust Emissions for a Methanol-Fueled Automobile," SAE Paper No. 720693, Society of Automotive Engineers, Warrendale, PA, 1972.

6. Robert M. Siewart and Edward G. Groff, "Unassisted Cold Starts to -29°C and Steady-State Tests of a Direct-Injection Stratified-Charge (DISC) Engine Operated on Neat Alcohols," SAE Paper No. 872066, Society of Automotive Engineers, Warrendale, PA, 1987.

7. Shirley E. Schwartz, Donald J. Smolenski, and Sidney L. Clark, "Entry and Retention of Methanol Fuel in Engine Oil," SAE Paper No. 880040, Society of Automotive Engineers, Warrendale, PA, 1988.

8. Thomas W. Ryan, Thomas J. Bond, and Richard D. Schieman, "Understanding the Mechanism of Cylinder Bore and Ring Wear in Methanol Fueled SI Engines," SAE Paper No. 861591, Society of Automotive Engineers, Warrendale, PA, 1986.

9. Personal conversation with Chris Blazak, Gas Research Institute.

10. Gregory J. Egan, "Near Term Introduction of Clean Hydrogen Vehicles Via H_2-CNG Blends," paper presented at the Fourth Canadian Hydrogen Workshop, Toronto, Canada, November 1 & 2, 1989.

11. "Alternate Fuels Update," editorial in *Automotive Engineering*, May 1993, p. 8.

12. Margaret K. Singh, "A Comparative Analysis of Alternative Fuel Infrastructure Requirements," SAE Paper No. 892065, Society of Automotive Engineers, Warrendale, PA, 1989.

13. "Economic Analysis of Low-Pressure Natural Gas Vehicle Storage Technology," Task 3 Topical Report, March 1989-April 1990, Gas Research Institute.

14. "Gaseous Fueled Vehicles: A Role For Hydrogen," paper presented at the National Hydrogen Association's 2nd annual U.S. Hydrogen Meeting, Washington, D.C., March 13-15, 1991.

15. "Hydrogen-Fueled Vehicles Technology Assessment Report," draft, California Energy Commission, June 29, 1991.

16. S. Furuhama and Y. Kobayashi, "Development of a Hot-Surface-Ignition Hydrogen Injection Two-Stroke Engine," *International Journal of Hydrogen Energy*, 9-33, pp. 205-213, 1984.

17. S. Furuhama and T. Fukuhama, "High Output Power Hydrogen Engine with High Pressure Fuel Injection, Hot Surface Ignition and Turbocharging," *International Journal of Hydrogen Energy*, 11-6, pp. 399-407, 1986.

18. "1991 Annual Energy Outlook," U.S. Energy Information Administration.

19. "International Petroleum Statistics Report," U.S. Department of Energy, Energy Information Administration, February 1991.

Chapter Five

Electric and Hybrid Vehicles

Courtesy: Syd Mead, Inc.

The energy produced by the breaking down of the atom is a very poor kind of thing. Anyone who expects a source of power from the transformation of these atoms is talking moonshine.

Ernest Rutherford, Physicist (1933)

Energy is the most abundant resource in the universe. Unfortunately, it is the resource in shortest supply on-board an electric vehicle (EV). As a direct replacement for gasoline, the modern traction battery is a relatively poor energy source. A lead/acid EV battery pack typically weighs 450 kg (990 lb) or more, takes up to eight hours to recharge, yet delivers less than one percent of the energy available from an equal mass of gasoline. To make up for this extreme technical disadvantage, greater emphasis on energy conservation and vehicle efficiency in EV design is essential. An electric car is more than just a standard car powered by a different energy source. It is a new type of vehicle that requires a new level of emphasis on energy efficiency. Newer batteries of greater specific energy are on the horizon. But technical solutions alone may not close the gap between EV capabilities and consumer expectations. In the final analysis, the vehicle itself, as well as its place in the transportation system and the marketplace, must be approached with a different mind-set.

By positioning the consumer-oriented EV in its proper role as an adjunct to, rather than a replacement for conventional vehicles, the gap between technical capabilities and consumer expectations can be narrowed. Electric cars, because of their obvious technical distinction, offer an ideal opportunity to create vehicle use distinctions in the minds of consumers. Correctly positioned and marketed, electric urban cars can make cities cleaner and quieter, and nations more energy sufficient. Without such a paradigm shift away from the idea of multi-purpose, high-performance automobiles and toward specialized vehicle types of more limited capabilities, the present opportunity for a switch to EVs could easily fall victim to a backlash of frustrated and disillusioned consumers.

Aside from the penalty in available energy, the electric car is a superb transportation device. EVs have no exhaust emissions, they are virtually maintenance-free, and they can be "refueled" at home utilizing the existing infrastructure. With electric vehicles, fuel is converted into usable energy at a remote facility instead of on-board the automobile. This shift in concept simplifies the tasks of controlling emissions, utilizing alternative fuels, and producing maximum energy from minimum resources. Just to control emissions, for example, each conventional vehicle (CV) requires its own mobile environmental protection system, along with the individual inspection and maintenance required to keep the system in good operating condition. With over 400 million automobiles in the world (a figure that could double in 20 years), the cost and complication of installing and maintaining individual vehicle systems is enormous. If emissions were controlled at the comparatively small number of centralized generating facilities, the job would become many times more manageable and much more cost-effective.

The inherent fuel efficiency of large, stationary generating sites is well established. In addition, stationary sites can easily utilize a variety of alternative fuels. Natural gas, considered an excellent alternative transportation fuel, already produces about 10 percent of electrical energy in the U.S. Over the next ten years one third of all new U.S. generating capacity will be fueled with natural gas. Coal is one of the most abundant sources of inexpensive energy. Today, 55 percent of electrical energy used in the U.S. comes from domestic supplies of coal. Nuclear power produces 19 percent of U.S. electrical energy and 20 percent comes from a variety of alternative sources including hydroelectric, solar energy, geothermal energy and wind generation. Less than 6 percent of the nation's electrical energy comes from petroleum. Consequently, a switch to EVs is also a switch to alternative energy sources. And according to industry estimates, enough excess capacity already exists during nighttime off-peak hours in the U.S alone to recharge 20 to 40 million electric vehicles.

Today, the limitations of the battery seem to prevent this otherwise ideal vehicle from becoming a serious contender in the personal transportation arena. Regardless of its limited specific energy, the lead/acid battery is still the only viable choice as a commercial EV energy storage device. With renewed development efforts in the U.S., Europe, and Japan, better batteries may evolve by the end of the decade. But at this stage of development, most researchers do not expect to match the performance of liquid-fueled conventional vehicles with new battery technology. Just to double or triple the performance of the lead/acid battery at an acceptable cost presents enormous technical challenges. Alternative battery systems are burdened in some way and to varying degrees by high costs, difficult serviceability, low service life, poor power or energy characteristics, or some other combination of attributes that make them presently infeasible as a consumer product.

Electric cars already have a legitimate role in transportation, even with the existing limitations of electrical energy storage technologies. Using available technology, mission-specific EVs have the potential to markedly reduce harmful emissions and lower the world's dependence on petroleum motor fuels. Over the near term, integrating EVs into the transportation system is primarily a marketing challenge, rather than one of technology. Consumers should therefore not be oversold on technological solutions. Over the longer term, however, when long-life, high-output batteries become technically and commercially viable, the electric car will likely experience a renaissance that will forever change the shape of personal transportation and leave liquid-fueled vehicles for specialized applications.

Electric Cars—Their Energy Source and Emissions

A number of studies have indicated that a switch to electric cars can significantly lower harmful emissions and, to a lesser degree, reduce the total energy consumed by the world's fleet of private automobiles. Electric cars also have the potential to substantially reduce the world's consumption of petroleum. About 95 percent of transportation's energy now comes from petroleum fuels. An electric car typically consumes about 95 percent less petroleum than an equivalent CV. This is because EVs are powered by the source fuels used to generate electricity. Electric cars are the ultimate alternative-fuel vehicles.

France, Belgium, Hungary, and Sweden all derive a large portion of their electric power from nuclear generating plants. In these countries electric cars are largely nuclear powered. In the U.S., approximately 55 percent of electrical energy comes from coal, which makes U.S. EVs predominantly coal-powered cars. Worldwide, fossil fuels, primarily coal and natural gas, still account for nearly 64 percent of all electrical generating capacity. However, only a very small percentage of electrical energy comes from petroleum. Regardless of the regional mix of source fuels, a switch to electric cars manifests as a dramatic move away from petroleum.

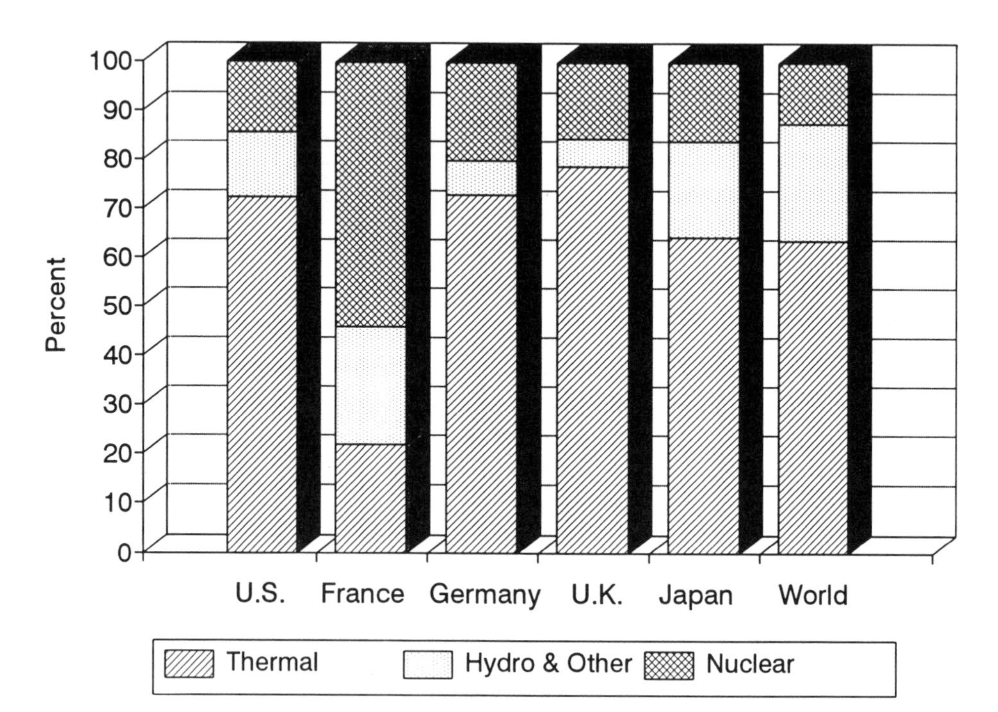

Fig. 5.1. Source of Electric Power.

In order to estimate the potential impact of EVs on primary energy consumption, the entire energy chain, as well as the differences between the two energy systems, must be considered. With conventional cars, feedstocks, primarily petroleum, are refined into motor fuels and delivered to local service stations. Fuel is then dispensed into vehicles where it is converted into power while the vehicle is underway. With electric cars the conversion of fuel into power takes place at a centralized generating facility. Power is then delivered by electrical transmission lines and ends up at individual wall outlets ready to charge EV batteries. Later, batteries are discharged as the vehicle is driven. Various losses occur and emissions are produced throughout both energy chains before power is ultimately delivered to the drive wheels. Consequently, the impact of EVs on primary energy consumption, as well as on the resulting emissions, must account for efficiencies and emissions throughout the entire chain or comparisons will not be accurate.

When energy is produced, the cycle begins with the process of recovering petroleum, mining coal, extracting natural gas, or producing biomass feedstocks for fuel. Efficiency throughout the chain, from feedstock recovery to conversion into usable energy, varies according to the particular technology. Extraction, mining and recovery efficiencies are already high and future improvements are unlikely. However, generating plant efficiency (conversion efficiency) is expected to improve over time as older, less-efficient plants are retired and replaced with newer designs. According to estimates by the Institute of Transportation Studies (ITS), natural-gas-fueled generating plants may be as much as 20 percent more efficient and coal-fired plants may be 10 percent more efficient by the year 2011.[1]

When petroleum is the source fuel for electrical power, about 26.7 percent of the energy contained in petroleum arrives at the wall-plug as electrical energy. Nationwide, only 5.9 percent of U.S. generating capacity comes from petroleum. Trends point to increased use of coal and natural gas, both in the U.S. and in the rest of the world. The overall energy efficiency of these two source fuels is 29.9 and 26.8 percent, respectively. Southern California Edison, the largest producer of electrical energy in the world, has projected that the increased load of EVs will be supported primarily by electrical energy produced from natural gas. By the year 2010, 60.4 percent of U.S. generating capacity will come from coal, 14.1 percent will come from natural gas, 14 percent will come from nuclear plants, and 8 percent will be derived from renewable sources (geothermal, wind, solar, hydro). Only 3.5 percent will be produced from petroleum. Because of increased conversion efficiencies and the trend toward more efficient source fuels, electric cars will become more energy efficient, even if vehicle technology does not improve. In a report for the Institute of Transportation Studies (ITS) at Berkeley, California, Wang and DeLuchi estimated the overall energy efficiencies for seven energy-producing chains in the year 1995, as shown in Figure 5.2.[2]

Compared to wall-plug electricity, approximately double the amount of primary energy arrives at the service station as liquid motor fuel. However, wall-plug electricity has already been converted into usable power while liquid motor fuel has not. In order to compare the consumption of primary energy between the two vehicle types, vehicle efficiency must be included. Automobile energy efficiency is highly variable, and improvements that might be expected from advances in technology are difficult to project. Even though internal combustion (IC) engine technology is relatively mature, fuel consumption still varies widely between different CVs. EV technology is embryonic, and energy consumption also varies significantly between different experimental EVs. As a result, estimates of future EV energy efficiency tend to be even more divergent than similar projections of CV efficiency. In general, however, researchers agree that EVs will consume less primary energy. Using their estimated 1995 U.S. source fuels mix and EV technology as a basis, Wang and DeLuchi projected that EVs in the U.S. will consume 13-29 percent more primary energy than IC vehicles if petroleum, natural gas, or biomass is the source fuel, and 30-35 percent less energy if coal is the primary energy source.[2] In a later report on the impact of EVs in the Southern California Edison district, Wang and Sperling estimated

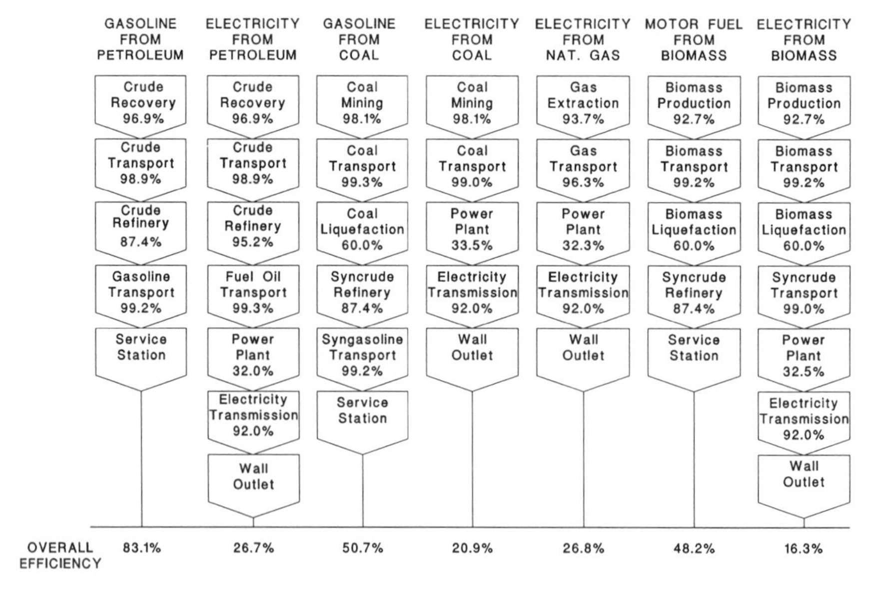

Fig. 5.2. Overall Efficiency for Seven Energy-Production Chains Estimated for 1995. (Source: Ref. [2])

that, with year 1996 technology and source-fuels mix, EVs will consume 10.3 percent less energy than an equal mix of IC vehicles.[1]

Although methods vary in sophistication, simple comparisons between EV and CV energy consumption can be made by assuming a charger/battery efficiency for the EV, and converting the CV's fuel consumption into its energy equivalent. If charger efficiency is assumed as 80 percent, and battery efficiency as 75 percent, then wall-plug energy consumption will be 1.66 times the EV's on-the-road energy consumption. Table 5.1 shows results calculated for a number of different vehicles.

TABLE 5.1. ELECTRIC AND CONVENTIONAL VEHICLE ENERGY CONSUMPTION COMPARISON

Vehicle	Weight (kg - lb)	Driving Cycle	kW-h/mi (from Battery)	kW-h/mi (from wall-plug)	km/L - mpg	kJ/km - Btu/mi
Impact	1321 - 2910	City (3)	0.180	0.298	-	669 - 1020
TEVan	2633 - 5800	City (1)	0.410	0.680	-	1508 - 2322
4-seat BMW	1634 - 3600	ECE (2)	0.268	0.446	-	998 - 1523
ETV-1	1780 - 3920	FUDS (3)	0.293	0.486	-	1088 - 1660
U.S. CAFE (5)		FUDS (3)	-	-	12 - 28	2916 - 4448
'90 Geo Storm	908 - 2000	FUDS (3)	-	-	22.5 - 53	1555 - 2372
Hypothetical Commuter (6)	450 - 990	City (4)	-	-	32 - 75	1093 - 1667

(1) Figures were obtained over a random urban driving cycle.
(2) The European ECE cycle is a composite of the U.S. Federal Urban Driving Schedule and the Federal Highway Schedule.
(3) FUDS is the Federal Urban Driving Schedule which is used by EPA to test urban fuel economy.
(4) Estimated for urban driving cycle.
(5) Average fuel economy of new cars registered in 1990.
(6) See Chapter Three for energy consumption of hypothetical 450 kg (990 lb) commuter car.

Results must then be adjusted to account for the upstream losses that occur when electrical energy and motor fuel are produced. A comparison of vehicle primary energy consumption for the most common source fuels is shown in Table 5.2. It is important to note that the CVs included in the comparison do not represent the average fuel economy of the existing fleet. Fleet-average fuel economy is about 8.5 km/L (20 mpg) in the U.S., compared to 9 km/L (21.4 mpg) in Japan, 11.5 km/L (27 mpg) in France, 10.8 km/L (25.5 mpg) in the U.K., and 9.5 km/L (22.5 mpg) in Germany, although new-car fuel economy is higher in all countries.[3] The Geo Storm and the hypothetical commuter car were included to illustrate the effect of improved CV fuel economy. Conversion/production efficiencies in Table 5.2 are taken from Fig. 5.2. In their studies, Wang and DeLuchi estimated sub-compact car petroleum-to-gasoline primary energy consumption at 5440 Btu/mi (3567 kJ/km).

TABLE 5.2. PRIMARY ENERGY CONSUMPTION BY VEHICLE TYPE IN kJ/km*

Vehicle	Gas. from Oil (83.1%)	Elect. from Oil (26.7%)	Gas. from Coal (50.7%)	Elect. from Coal (29.9)	Elect. from NG (26.8%)	Gas. from Biomass (48.2%)	Elect. from Biomass (16.3%)
GM Impact	-	2506	-	2237	2496	-	4104
TEVan	-	5648	-	5043	5627	-	9252
4-seat BMW	-	3738	-	3338	3724	-	6123
ETV-1	-	4075	-	3639	4060	-	6675
'90 Geo Storm	1871	-	3067	-	-	3226	-
U.S. CAFE	3509	-	5751	-	-	6050	-
Hypothetical Commuter	1315	-	2156	-	-	2268	-

* Btu/mi = kJ/km × 1.609 × 0.9478.
(Parentheses indicate percentage of primary energy delivered either to wall-plug as electricity, or to service station as motor fuel.)

A number of variables will impact the relationship between future EV and CV energy consumption. The degree to which vehicle technology and source-fuel mixes affect primary energy consumption can be seen by selectively comparing different vehicles and source fuel chains. When the energy consumption of GM's Impact using electrical energy produced from coal is compared to the U.S. CAFE vehicle using petroleum fuel, the electric-powered Impact is approximately 36 percent more energy efficient.[4] When Impact is compared to the hypothetical commuter car, the commuter car is about 41 percent more energy efficient. Comparing the experimental ETV-1 to U.S. CAFE results in nearly 4 percent more energy used by the EV. EVs are not expected to make significant market penetration until early in the twenty-first century when generating plant conversion efficiencies and EV technology will be much improved. Concurrently, petroleum recovery efficiencies are likely to fall as petroleum supplies diminish and the remaining deposits become more difficult to recover. However, CV fuel economy will also improve over time, and non-petroleum motor fuels will tend to unlink CVs from the effects of diminishing petroleum resources. The table therefore serves more as a basis for comparisons rather than a projection of future vehicle energy efficiencies.

Regardless of the diversity in estimations of primary energy savings, research indicates that EVs will produce significantly less harmful emissions and markedly reduce the world's dependence on petroleum fuels. Wang and DeLuchi concluded that in many U.S. regions, electric cars will reduce transportation petroleum consumption by more than 90 percent under their 1995 technological scenario. Even in worst-case areas, such as cities like New York that rely heavily on petroleum as a source fuel, EVs will reduce petroleum use by 63-65 percent. In Chicago, Houston, and Los Angeles, sub-compact car petroleum consumption, when replaced by comparable EVs, will drop by 98.0, 98.2 and 91.9 percent, respectively.[2] Most of the reductions come from replacing petroleum fuel with the source fuels used to generate electricity, rather than reducing primary energy consumption. Electric cars also excel in their ability to reduce atmospheric pollution.

Several studies have compared air pollution between electric cars and conventional IC vehicles. Initially, some had speculated that large-scale use of electric cars may do little more than replace tailpipe emissions with the emissions from large generating plants. If energy consumption is similar between the two vehicle types, then generating electrical energy for EVs might simply shift emissions to a different source with little overall effect on air pollution. However, a number of studies in the U.S. and Europe have verified that large reductions in hydrocarbon (HC) and carbon monoxide (CO) emissions can be expected if conventional vehicles are replaced by electric vehicles. A study at Electric Power Research Institute (EPRI) estimated that in U.S. urban areas, substituting EVs for CVs will reduce non-methane organic gases (NMOG) by 98 percent, nitrogen oxides (NO_x) by 92 percent, and CO by 99 percent. Additionally, EPRI estimates that EVs will produce only half the carbon dioxide (CO_2) of a conventional vehicle.[5] The International Energy Agency (IEA) estimates similar reductions in CO_2 emissions in most OECD countries, with variations depending on local source-fuel mixes.[6]

In another study, Wang and Santini surveyed six driving cycles in four U.S. cities and found that HC and CO emissions are consistently reduced by approximately 97 percent, regardless of the driving cycle or the local source fuels. This is because conventional vehicles produce large amounts of HC and CO emissions per unit of energy, while generating plants produce limited amounts. High HC and CO emissions result largely from cold starts and short trips that do not allow individual vehicles to become fully warmed up. They also concluded that NO_x

emissions could either be reduced or increased, depending on the local mix of source fuels. Sulfur oxides (SO_x) and particulate matter (PM) may increase with expanded use of EVs due to the much smaller SO_x and PM emissions of conventional IC vehicles.[7] Greater SO_x and PM emissions are due primarily to the use of coal for generating electricity. Emissions in this category are substantially lower in regions that rely more heavily on natural gas, and nearly eliminated in regions that rely primarily on nuclear power.

Studying the effect on greenhouse gases, DeLuchi concludes that with the existing source-fuels mix, EVs are likely to have a mixed effect on CO_2 emissions, one of the primary gases responsible for global warming. Compared to CVs using reformulated gasoline, electric vehicles using coal as a primary energy source produce significantly more CO_2 emissions. Natural gas results in slightly less CO_2 emissions, and nuclear power reduces emissions by 98 percent, when comparing EVs to CVs. Because most (but not all) greenhouse gas emissions are proportional to energy consumption, DeLuchi concludes that increased energy efficiency is the key technical variable in reducing greenhouse gases produced by transportation. According to DeLuchi, the largest long-term reduction in greenhouse gases will come from substituting non-fossil energy sources such as biomass or solar power.[8]

Electric Vehicle On-Board Energy Flow

Electric cars operate with a closed energy system. All the energy is self-contained and in general the supply cannot be extended by adding some type of fuel. Within this closed system the energy/work loop can operate with nearly equal efficiency in both directions, either converting electrical energy into work, or converting vehicle kinetic energy back into electrical energy. By comparison, conventional IC vehicles operate with an open energy system. When additional energy is needed, more fuel is added. Vehicles in this category process energy in one direction only, by transforming fuel into work. They cannot do the reverse and transform work (kinetic energy) into fuel. The inherent economy of the closed EV energy system, along with its equally inherent scarcity, encourages conservation through on-board energy management. Table 5.3 provides a comparison of the on-board energy allocation between the two vehicle types.

Mechanical and electrical losses occur throughout the EV power system. And just as with the IC vehicle, losses vary in magnitude according to system design and operating conditions. Also, batteries have different charge/discharge efficiencies, and efficiencies vary according to load. Jet Propulsion Laboratory (JPL) tested a number of state-of-the-art electric cars under sponsorship of the

TABLE 5.3. VEHICLE AVERAGE OPERATING EFFICIENCY

	Electric	Spark Ignition
Motor or Engine	80%	23%
Drivetrain	82%	78%
Battery and Charger	70%	—
Overall Vehicle	46%	18%

U.S. Department of Energy (DOE) to determine the energy flow on-board an electric car. Dynamometer tests of the ETV-1 Electric Test Vehicle equipped with a chopper-controlled, separately excited DC motor indicated losses at various steady-state speeds, as shown in Figure 5.3.

Regenerative Braking

During deceleration, the brakes of a conventional vehicle convert kinetic energy into heat and discharge the energy into the air. Regenerative braking is designed to reclaim this otherwise wasted energy by generating a charge to the battery during deceleration. Under the most ideal conditions the efficiency of the loop, from battery to kinetic energy and from kinetic energy back to the battery, is no greater than 50 percent. Theoretically, up to half of the energy required to overcome the inertia of acceleration can be recovered by using regenerative braking on deceleration. An early study by McConachie indicated that a 50-percent recovery efficiency would result in slightly more than 50-percent increase in range over a city driving cycle.[9] The 50-percent increase in range is based on the assumption that regeneration will be effective in all instances and that it completely replaces the vehicle's conventional braking system (Table 5.4).

Different driving schedules indicate different recovery potentials. Over a particular driving schedule, the actual increase in range depends on the practical limitations of an optimized regenerative braking system. A number of studies with experimental regeneration systems have shown widely divergent results. During tests at JPL using the General Electric/Chrysler-built ETV-1 Electric Test Vehicle, approximately 42 percent of the kinetic energy stored in the vehicle actually made it back to the battery terminals.[10] Energy accepted by the battery is then reduced by charging efficiency, another characteristic that varies between battery types, temperature, state of charge, and many other variables. In actual practice, regenerative braking normally increases vehicle range on the order of 5 to 15 percent. However, GM reports that in tests with Impact, regeneration produces a 25-percent increase in range.

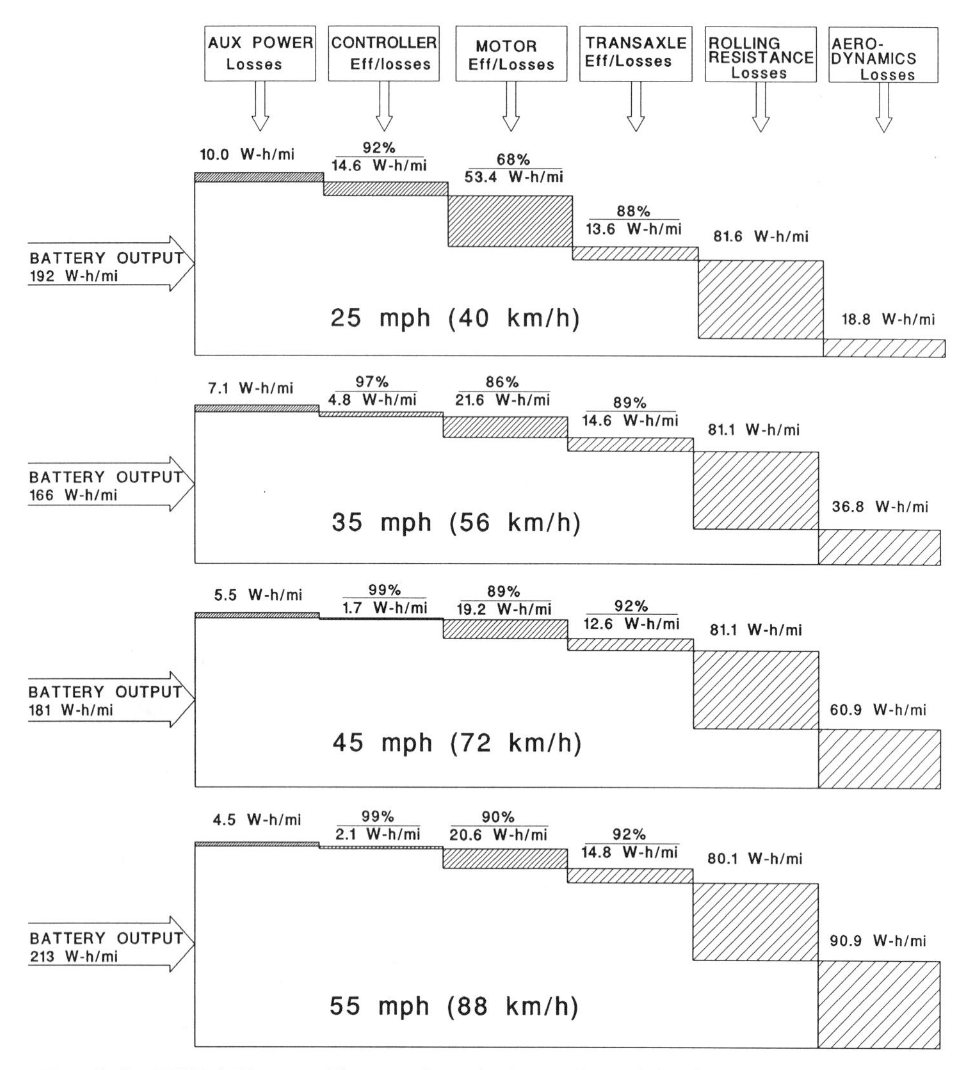

Fig. 5.3. ETV-1 Energy Flow at Steady-State Speed (W-h/mi = 2.237 kJ/km).

Typically, regeneration systems are not designed to reclaim all the energy available on deceleration. Designers tend to configure systems with emphasis on the upper speed ranges. Both mechanically and electronically it is difficult to maintain regeneration at very low vehicle speeds. Also, batteries vary in their ability to accept high transient inputs, and they become less efficient at higher loads. As charge rate climbs, internal resistance goes up and energy is increasingly converted into heat. At the upper extremes, batteries may be damaged.

TABLE 5.4. VEHICLE OPERATING SCHEDULE OVER A CITY DRIVING CYCLE 1500-kg VEHICLE WITH 4.1-LITER ENGINE BRISBANE METROPOLITAN AREA

TEST VEHICLE	C1	C2	C3	C4
Trip Duration (minutes)	63	78	67	59
Average Speed (km/h)	18.2	14.9	17.2	19.6
Maximum Speed (km/h)	66	62	65	62
No. of Idle Periods	60	98	81	57
Percent of Time Idling	33.9	35.8	33.6	32.3
No. of Brake Applications	142	166	168	141
Percent Time Braking	22.8	16.9	20.9	21.4
Overall Energy Used (kJ/km)	670	756	745	720
Rolling Resistance (kJ/km)	242	232	237	242
Energy Dissipated (mJ)	8.26	10.1	9.8	9.23
% Range Increase with 50% Regeneration Efficiency	47	53	52	50

Note: Test covered a 19.33-km city route which included moderate grade variations, a speed limit of 60 km/h and no freeway driving.

Human factors engineering can improve energy recovery. Initially, regenerative braking was controlled by the accelerator pedal. Suddenly releasing the pedal caused maximum regenerative braking which also created high current transients, as well as maximum demand on drive-wheel traction. Driveability was a problem, especially if drive-wheel traction was lost. With modern systems, regeneration that exceeds the equivalent of IC engine compression braking is controlled by the brake pedal. Increased brake pedal pressure also increases regeneration current, as well as vehicle braking effect. Hydraulic brakes are modulated to balance braking forces between front and rear wheels and to prevent loss of traction.

Mechanical Overview

Electric cars are much more simplified mechanically, and significantly more complex electronically. The electric motor has few moving components (essentially, the rotor/armature and bearings). In general, the transmission is simplified as well. EV transmissions typically have only two discrete ratios and no reverse. Typically, the vehicle is powered in the reverse direction by electronically reversing the rotation of the motor. With a few exceptions, batteries are passive systems with no moving parts. Processes take place on an

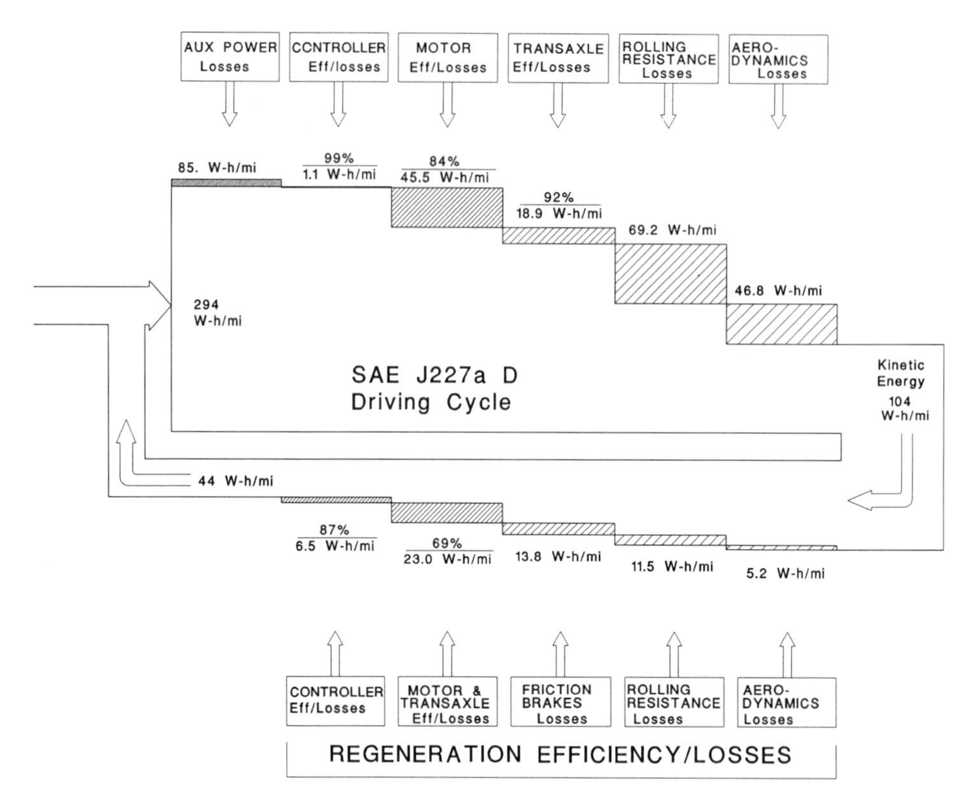

Fig. 5.4. ETV-1 Energy Flow with Regenerative Braking (W-h/mi = 2.237 kJ/km).

electrochemical level rather than mechanically. It is therefore tempting to regard the electric vehicle as a relatively simple machine, but complexity that may be lacking on a mechanical level is more than offset by increased complexity on an electrical level. Controlling the motor, regulating vehicle systems, interfacing operator inputs, and managing the flow of energy within the system are complex and demanding tasks. To properly manage the system requires continuous feedback from a variety of subsystems and a multitude of real-time computations, which can be accomplished only with sophisticated hardware/software architecture.

The Electric Motor

The electric motor is the heart of the electric vehicle's tractive power system. It functions as a sort of bidirectional conduit between mechanical work and electrical energy, converting electrical energy into mechanical work and converting mechanical work into electrical energy. In one form or another, the machine utilizes a rotating set of windings that interact electromagnetically with a stationary set of windings. When electrical potential is applied, the windings establish magnetic fields which cause the armature or rotor to rotate and thereby develop power. An electric motor will run indefinitely with little or no maintenance, and it produces no harmful emissions. It develops power smoothly and quietly and can achieve an efficiency of 85-95 percent.

Electric motors exhibit different operating characteristics depending on their design. Also, motors are designed to run on either direct current or alternating current. Traditionally, DC motors are used in applications requiring variable speed and frequent acceleration and deceleration of large-inertia loads. Normally, a DC machine is heavier, smaller, produces more power for its size, and is more expensive than an AC motor. In the past, DC motors were used in most electric vehicle applications because of their high starting torque and their ability to run on direct current from the battery. Because of improved inverters—devices that convert DC into AC power—AC induction motors have gained favor in recent years. Induction motors are less costly to manufacture, they are lighter, and they use no commutator brushes so less maintenance is required.

Direct-current motors are designed as either *series*, *shunt*, or *permanent magnet* machines. Classifications refer to the way in which the field is energized. Traditionally, the most common EV motor has been the series machine. In a series machine the field is connected in series with the armature (Figure 5.5). The series motor develops high stall torque. At a given voltage, torque and speed are inversely related and power is constant. Consequently, the series motor is often referred to as a "constant horsepower" machine. It is used in applications requiring high pull-away torque and a broad torque band. The series motor has no inherent speed limitation and in an unloaded condition it will continue to accelerate until bearing drag equals torque, or until it destroys itself.

A shunt machine has field windings that are connected in parallel with the armature (Figure 5.6). The motor is essentially a constant-speed, variable-output machine. When load is applied, it delivers power at slightly reduced rpm and current increases in proportion to load. Unlike a series motor, the shunt motor develops very little stall torque. In applications that require greater pullout torque, a shunt motor may be equipped with series windings to improve low-rpm torque characteristics. Such a motor is then referred to as a *compound machine*.

With a shunt or compound machine, the field may be separately excited to provide independent control of the field and armature circuits. The relationship

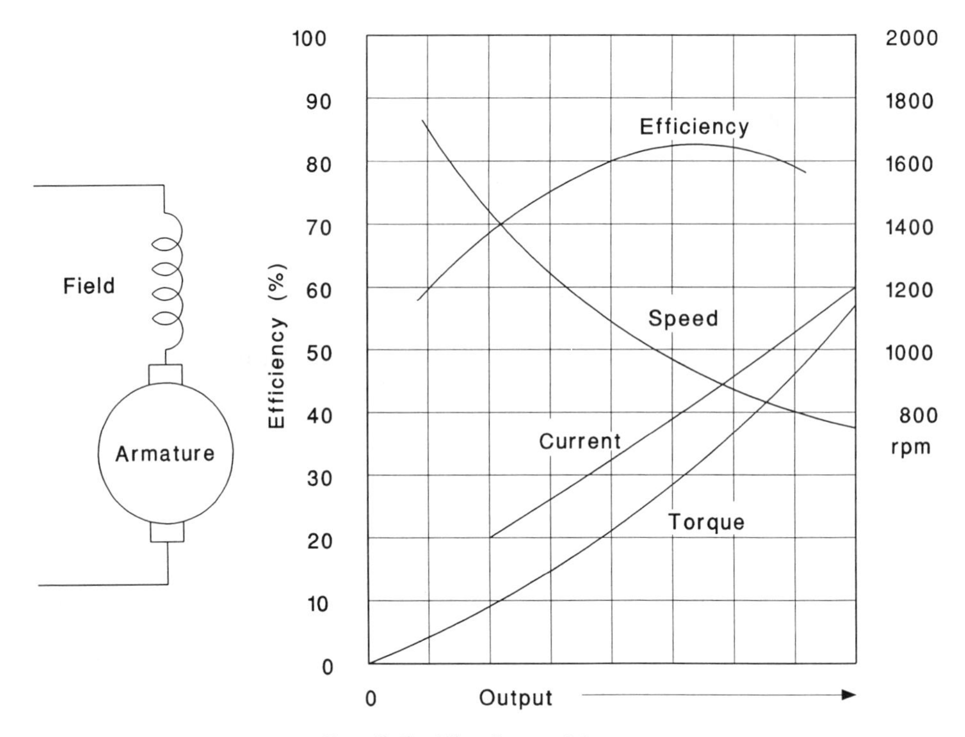

Fig. 5.5. The Series Motor.

between field and armature current determines the base speed of the motor. Increasing field voltage and/or reducing armature voltage reduces the motor's base speed. Decreasing field voltage and/or increasing armature voltage results in a higher base speed. When the motor is running at less than base speed, both torque and current increase and the motor will accelerate until it reaches full rpm. When motor rpm is significantly greater than base speed, counter emf (electromotive force) reverses the flow of current until motor rpm is reduced to base speed.

Controlling the relatively low-current field is a simple method of controlling the speed of the motor. Depending on the design of the machine, a 3:1 speed ratio can be achieved with field control alone; however, motoring efficiency suffers as the field is weakened. In EV applications, a shunt motor is often equipped with independent control of both the armature and field circuits. This provides maximum control of the machine's motoring/generating characteristics throughout a broad rpm range.

The permanent magnet motor utilizes permanent magnets to establish the field. This slightly improves motoring efficiency, but it limits the control options available with an electrically excited field.

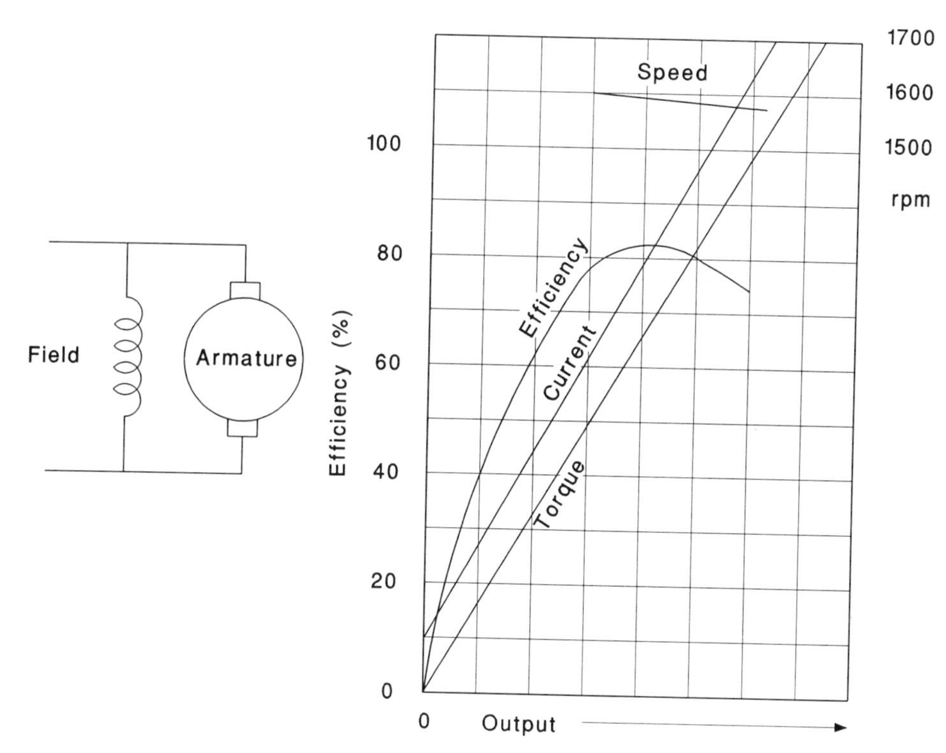

Fig. 5.6. The Shunt and Compound Motor.

Alternating Current for Electric Cars

The advantages of alternating current for vehicle motive power have been recognized for many years. Motors designed for alternating current are lighter, less expensive, they can be more efficient, and they require less maintenance because they lack the commutator brushes common to DC motors. However, the battery supplies direct current which must be converted into alternating current in order to run an AC machine. Until recently, DC-to-AC inverters were large, complicated, inefficient, and expensive devices, which placed AC systems at a disadvantage. However, recent advances in high-power transistors and microprocessors have made AC power conditioning practical. As a result, AC systems appear to be the wave of the future, although proponents still exist in both camps.

The advantage of an AC system centers around the use of a three-phase induction motor which has higher specific power and is less costly to manufacture than an equivalent DC motor. Higher specific power results primarily from the higher operating speeds possible with an AC machine. A direct-current motor is limited in rpm because of the limitations of commutation. With no commutator, an induction motor can easily operate at speeds in excess of 10,000 rpm. Alternating current motors can be air-cooled or liquid-cooled and they offer good flexibility in packaging. They can vary considerably in the ratio of length to diameter without undue penalties in performance. At the extremes, very long rotors tend to become unstable at lower speeds, and very short rotors are less efficient. Two-pole motors tend to be favored over four-pole designs because they operate at twice the rpm per given frequency. When current is chopped at twice the frequency, switching losses increase.

TABLE 5.5. ELECTRIC MOTOR/SYSTEM ATTRIBUTES

Motor Developer/Type	Motor Power Density kg/kW (1)	L/kW	Voltage V	Electronics Power Density kg/kW (1)	L/kW	Trans. Y or N	System Peak Power kW
General Electric DC Sep. Exc. (ETV-1)	3.0	0.97	108	1.5	1.7	N	30
General Electric AC Induction (ETX-1)	1.2	0.13	192	1.3	1.4	Y	40
General Electric AC PM Sychr. (ETX-II)	0.98	0.15	192	0.69	0.91	Y	52
General Electric AC Induction (MEVP)	0.60	0.14	340	0.55	0.45	N	56
Cocconi Eng. AC Induction (Impact)	0.5	0.17	320	0.35	0.66 (2)	N	90-100
Pentastar DC Sep Exc. (TEVan)	1.5	0.45	176	0.80	1.8 (2)	Y	50

(1) All specific power and volume values based on maximum motor power.
(2) Includes DC-DC converter and battery charger.
Source: Ref. [16]

Compared to DC systems, the power conditioning and control required for alternating current results in a much more complicated system. But inverters normally contain components that can be interlaced to serve as wall-plug chargers, which economizes on hardware and somewhat offsets the increased costs. Another consideration has to do with battery cell voltage. Battery packs for AC systems are normally configured for higher voltages (~200 V). Higher operating voltages enable the system to operate at reduced current levels, which reduces inverter switching losses. However, conventional battery systems require a greater number of cells to produce higher voltages, which tends to increase costs, reduce battery performance, and decrease system reliability. Today, either alternating or direct current is an appropriate choice for an electric car. With further development, AC systems are likely to become more prevalent, and DC systems may become outmoded. Table 5.5 compares power system attributes between a number of AC and DC systems.

The Controller

The controller is the device that controls vehicle speed. In a sense, an EV controller "throttles" the electric motor by modulating and conditioning the current. Controllers for DC machines are relatively simple compared to the complex devices necessary to control AC machines. Techniques of controlling DC motoring speed include resistance control, voltage stepping (selectively connecting batteries in series or parallel), and converting the current into high-frequency pulses using a pulse width modulator. Controllers for AC machines are more complicated. Power conditioning units for AC systems normally include a three-phase inverter, a pulse width modulator, and microprocessor waveform conditioning and real-time feedback and signal processing.

Inexpensive DC speed controllers, such as the devices used in golf carts, are typically based on resistance control or voltage stepping. However, neither system is adequate for an electric automobile. Today, electronic control is considered the only practical means of speed control for electric cars. Moreover, the role of modern controllers has been expanded to include a number of functions having to do with on-board energy management, regeneration, and battery charging.

Pulse Width Modulator

Pulse width modulators are the most efficient and reliable electronic controllers for DC machines. These devices operate by rapidly turning the motor on and off. Current is literally "chopped" into segments of either full terminal voltage or zero voltage. This high-frequency chopped current results in an average voltage that is less than battery terminal voltage. The actual value of applied voltage depends on the length of the on and off cycles and the number of cycles per second. Motoring speed is thereby controlled by varying the duration and frequency of the on/off bands. Control is smooth and stepless and usually accompanied by an audible hum. Pulse width modulators are commonly referred to as "chopper controllers" or simply "choppers" because of the chopped nature of the current.

There are primarily two types of choppers, the SCR chopper and the transistor chopper. SCR choppers are larger, more expensive, and emit a louder hum than transistor-based devices. Operating frequency is a maximum of 1000 pulses per second for SCR units, and up to 20,000 pulses per second for transistor choppers. Before high-power transistors were developed, SCR choppers were used in applications requiring high output current. Today, the transistor chopper has largely replaced the older SCR units. Depending on the output, operating efficiency of a transistor chopper extends from a low of about 85 percent to a maximum of 99 percent (see Figure 5.7). Often, a relay is installed to bypass the chopper at full voltage to eliminate internal losses.

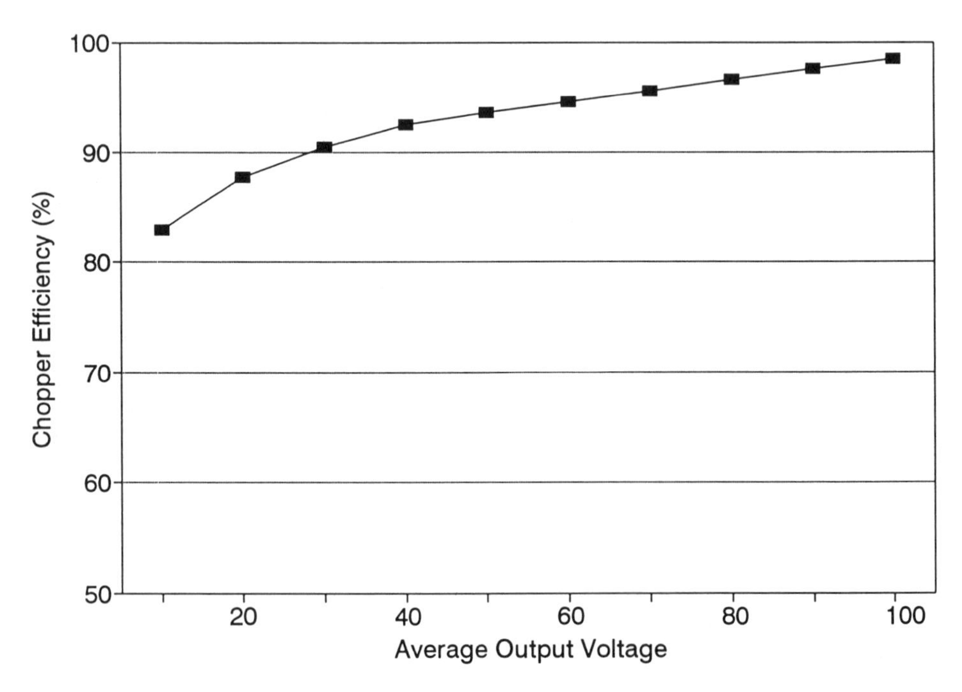

Fig. 5.7. Chopper Controller Efficiency by Output Voltage.

Controllers for Alternating Current

Alternating current reverses direction many times per second to produce a sinusoidal waveform. A three-phase system is essentially the same as three ordinary single-phase systems with the three single-phase waveforms shifted out of phase by one third of the cycle. Power conditioners/controllers for AC motors change DC current from the battery into quasi-sinusoidal AC current having the appropriate frequency and magnitude for proper system performance. Inverters vary in type and control strategy as well as in the details of implementation. In general, systems are based on a chopped current in conjunction with microprocessor signal processing as needed to control the magnitude, frequency, and slip-angle of the waveform. Greater chopping speeds and internal processing result in greater losses to power conditioning, and larger, heavier, and more costly control units. AC controllers also tend to be more noisy than their DC counterparts. Only in the last few years have components become available that make AC controllers power-efficient and cost-effective.

CVT Speed Control

A continuously variable transmission (CVT) is usually not considered a speed control device for an electric vehicle. However, experiments at Quincy-Lynn in the late '70s demonstrated that such a transmission can control vehicle speed and thereby eliminate the electronic speed controller and provide torque multiplication in the process. In the CVT-equipped Quincy-Lynn electric car (see Urba Electric in Appendix), the compound-wound DC motor was set to run continuously at full voltage, and vehicle speed was controlled by varying the transmission shift position. To accomplish this, the transmission was configured to shift according to a signal supplied by a floor-mounted pedal which replaced the normal accelerator pedal. Transmission shift position tracked the depression of the pedal and thereby controlled vehicle speed.

Quincy-Lynn's prototype system contained relatively little logic. When the pedal was at half-position, the transmission would also assume a 50-percent upshift position which would accelerate the vehicle to 50 percent of its maximum speed. When the pedal was fully depressed, the transmission would fully upshift and accelerate the vehicle to maximum speed. The ratio spread was adequate to smoothly control vehicle speed from 16 to 97 km/h (10 to 60 mph) while the motor remained at a steady-state rpm. Below 16 km/h (10 mph), the transmission was allowed to slip, just as the clutch of a manual transmission would be slipped during start-up. This was also controlled by the accelerator pedal. Releasing the pedal below a pre-established speed disengaged the transmission. The system's general layout is shown in Figure 5.8.

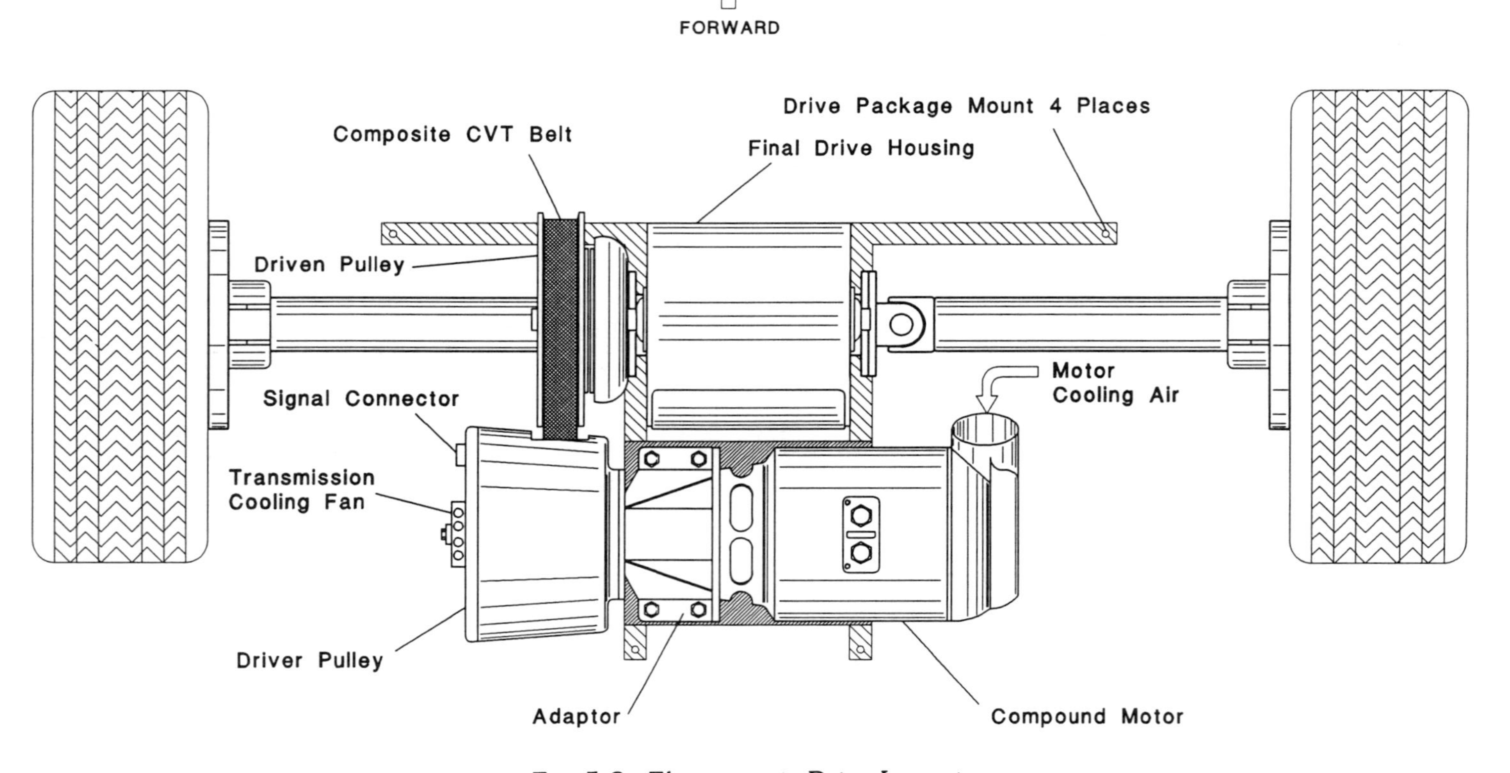

Fig. 5.8. Electromatic Drive Layout.

Regenerative braking was a natural by-product. Since transmission shift position tracked the pedal position, reducing pedal pressure caused the transmission to downshift. This slowed the vehicle and accelerated the motor which increased counter emf and reversed the flow of current. The amount of braking force and the value of regeneration current was controlled by the pedal position in relation to vehicle speed. The driveability of the system was excellent. Speed control was natural and there was no sense that anything unusual was taking place in the drivetrain.

The Quincy-Lynn car demonstrated the feasibility of CVT speed control. However, the prototype did not explore the system's potential beyond this basic function. A number of modifications were outlined for a second-generation system, but further development was not funded. For example, belt-squeeze in the proof-of-concept unit was controlled by springs in the driven pulley, which made belt-squeeze dependent on shift inputs rather than torque requirements. Belt-squeeze must be controlled at both the drive and driven pulleys and regulated according to torque in order to achieve maximum transmission efficiency. A field control loop would have broadened the vehicle's speed range and provided significantly better control of regeneration. By relying on counter emf, regeneration was limited to the upper speed ranges and the system's full potential for reclaiming energy was not realized. In the prototype, the motor ran continuously even when the vehicle was at rest. Efficiency would have been improved if the motor was shut off when the vehicle stopped. The free-running load of the motor was approximately 850 W.

Reduced transmission efficiency during shift and at high reduction ratios is the primary engineering challenge of a CVT speed control system. The Electromatic Drive transmission was dynamometer tested at 96-percent efficiency at full upshift, but data was not obtained during shift and at high reduction ratios. Literature contains widely divergent reports of CVT efficiencies during shift and at high ratios. In general, efficiency suffers during both conditions. However, it is difficult to estimate the degree to which improperly tuned units may have impacted test results. A non-lubricated V-belt CVT can be quite efficient throughout the range of operating modes, provided that belt-squeeze is appropriately matched to torque. Electronically controlled toroidal and lubricated steel belt drives also can be used as EV speed controllers. Additional information on CVT drives is provided in Chapter Three.

Integrated Electronic Controls

Like its IC engine counterpart, the electric tractive system has reached a level of sophistication that demands microprocessor logic for appropriate management.

Once microprocessors are introduced, multitasking and systems integration become possible. A state-of-the-art Electronic Control Unit (ECU) can process feedback from various vehicle systems, sense and analyze operating conditions, then execute driver inputs in ways that balance system response and optimize performance. The ECU can manage the regenerative braking system, control the transmission shift schedule, and integrate a variety of electrical and mechanical functions that may be required to keep the vehicle precisely tuned to operating conditions and driver inputs. A fluid and integrated shift schedule can provide torque multiplication in response to demand, keep the motor in its region of maximum efficiency and manage the flow of current into and out of the battery. Improved driveability also is a primary goal. The electric vehicle should emulate the feel and response of the familiar IC vehicle. A modern EV might also incorporate drive-by-wire, brake-by-wire, and steer-by-wire, and thereby eliminate mechanical connections between driver command modules and vehicle systems.

Powertrain control tasks must be done in real time and involve simultaneous and complex calculations. As events and status are sampled, data may be deposited in memory and shared with other subsystems. Some tasks must perform unique calculations in order to determine the appropriate torque, braking, and shift schedules. Other tasks can then use the information from the schedules to make appropriate outputs to solenoids, clutches and operator interface systems. Battery state-of-charge will affect shift schedule biases and regeneration response to brake pedal pressures. Control parameters for shift schedule, motor current and frequency in AC systems, or field/armature current and ratio in DC applications, depend on the complex relationship between vehicle speed, transmission ratio, brake input, vehicle acceleration/deceleration responses and battery state-of-charge. During regeneration, the system must sense loss of traction and modulate the hydraulic brakes, balancing hydraulic braking with regenerative forces, and balancing both against vehicle conditions. In addition to powertrain control tasks, the ECU may also provide status, warning and diagnostic information to the driver. It might also delay responses to certain driver inputs such as force downshift or reversing at higher speeds, and provide default detection, systems override, or even systems shutdown when appropriate conditions exist.

In the process of achieving maximum energy efficiency and improved driveability, the mechanically simple electric automobile necessarily becomes an inherently complex machine. Hybrid vehicles place an even greater demand on system computers and software engineering. It is only through appropriate hardware and software architecture that electric and hybrid/electric vehicles can achieve superior driveability and performance.

On-Board Energy Storage

Since the automobile's inception, the absence of an appropriate energy storage device has been the Achilles' heel of electric cars. There are a number of ways in which to store energy. Kinetic energy can be mechanically stored in a spinning flywheel. But flywheels of high specific energy require very high rotor speeds, which introduces a number challenges having to do with materials capabilities, bearing life, and system costs. Fuel cells can generate electrical energy as long as they are supplied with fuel. But slow start-up, low power density, poor reactivity to peak demands, and high costs tend to eliminate them as a viable mid-term power source for electric cars. Expectations for near- and mid-term electric energy systems have therefore centered on development efforts with the secondary (rechargeable) battery.

One obstacle that has delayed battery technology has been the high cost of research and development efforts, which has been compounded by the absence of a significant market for electric cars. Today, the environment is more favorable on both counts. Governments and consumers are more interested in environmentally friendly alternatives to gasoline, and as a result, far more resources are being committed to development efforts. In September 1991, the U.S. Department of Energy, along with Ford, General Motors and Chrysler, joined forces to develop advanced battery technologies for future electric cars. Together under the United States Advanced Battery Consortium (USABC), vast new resources, up to $100 million per year, have been committed to development efforts. Similar projects are underway in Europe and Japan.

An energy storage system for automobiles must meet many criteria and often requirements are at odds with each other. The energy source must have high specific energy and high specific power. But in general, a battery's specific energy and its specific power are inversely related: When one is increased the other decreases. Battery materials must be plentiful, inexpensive, non-toxic and either recyclable or easily disposable without harm to the environment. Recharge time and maintenance must be low, and life must be high, preferably equal to the life of the vehicle. Unfortunately, the chemical changes that produce current are also changes that decompose battery materials. Self-discharge is another attribute typical of battery couples. A vehicle battery might rest unused for a period of days or even weeks, and power must be instantly available on start-up. Yet all batteries experience some degree of self-discharge. Many of the presumed requirements for battery performance were largely dictated by the characteristics of the liquid-fueled heat engine, and it may be unrealistic to expect batteries to match them. In order to establish realistic development goals, the USABC has identified primary criteria for advanced electric vehicle batteries, shown in Table 5.6.

TABLE 5.6. USABC ADVANCED BATTERY PRIMARY GOALS

	Mid-Term	Long-Term
Power Density (W/L)	250	600
Specific Power (W/kg) (80% DOD*/30 s)	150 (200 desired)	400
Energy Density (W-h/L) (C/3 discharge rate)	135	300
Specific Energy (W-h/kg) (C/3 discharge rate)	80 (100 desired)	200
Life (years)	5	10
Cycle Life (cycles) (80% DOD*)	600	1000
Power and Capacity Degradation (% of rate spec.)	20	20
Ultimate Price ($/kW-h) (10,000 units @ 40 kW-h)	<$150	<$100
Operating Environment	30 to 65°C	40 to 85°C
Recharge Time (h)	<6	3-6
Continuous Discharge in 1 h (% of rated energy capacity) (no failure)	75	75

* Depth of Discharge

In general, battery systems fall into three classifications: the *aqueous systems*, the *ambient-temperature lithium systems*, and the *high-temperature systems*. Lead/acid batteries, various nickel systems, and flow batteries are aqueous systems. Zinc/bromine is the aqueous flow system receiving the most attention; however, the mechanical complications of pumps and the safety hazards of bromine are two major concerns. Another interesting new system is the vanadium redox flow battery under development at New South Wales University in Australia. Low energy density may limit EV application, but much remains to be learned. Ambient-temperature lithium batteries are also in the early stages of development. These batteries are plagued by high costs and rapid deterioration of the positive electrode. In their favor is their characteristically high cell voltage and low weight. High-temperature systems include lithium/metal sulfide, sodium/metal chloride, and sodium/sulfur batteries. Batteries in this class have high energy density but their high operating temperature and corrosive solutions create difficult technical and safety challenges. The lead/acid battery is still the mainstay of electrical storage technology and it is the keystone by which other systems are compared. Table 5.7 provides a comparison of a number of battery couples.

When comparing battery performance, it is important to consider the state of existing technology and the nature of available information. Technology is rapidly improving and what is true today may be much different in a few years or even a few months. Also, much of the information is established under controlled and often ideal conditions and therefore may not be strictly applicable

TABLE 5.7. BATTERY SYSTEM COMPARISON

Battery Type	Energy Density W-h/kg	Power Density W/kg	Cycle Life	Recycle % of Materials	Energy Efficiency
Lead/Acid	40	130	750	97%	65%
Aluminum/Air	200	150		75%	35%
Lithium/Iron Disulfide	>130	>120	1000	50%	
Lithium/Polymer	100	100	400	50%	
Nickel/Cadmium	56	200	2000	99%	65%
Nickel/Iron	55	130	1500	99%	60%
Nickel/Metal-Hydride	80	200	1000	100%	90%
Nickel/Zinc	80	150	200		65%
Sodium/Sulfur	100	120	500	50%	85%
Vanadium Redox	50	110	400	100%	89%
Zinc/Air	120	120	135	75%	60%
Zinc/Bromine	70	100	500		65%

in a real-world environment. Even within a battery category, much about its performance depends on the tradeoffs that designers have made between specific energy, specific power, and cycle life. A battery that demonstrates exceptional energy density may still be infeasible as a consumer product because of other negative attributes. Performance data is affected by discharge cycle with slower and more steady discharges yielding greater total energy and longer life. Manufacturing costs are often difficult to project with an experimental system. In summary, a number of factors affect performance data, and many factors other than performance are critical to a battery's viability as a consumer product. The filters on the way from the laboratory to the marketplace are multi-layered.

In the field, the characteristics of a vehicle's battery system are much more inconsistent than any of its other components or subsystems. Battery performance depends on age, charging procedures, previous discharge history, discharge rates and even controller characteristics. For example, the energy density of a lead/acid battery may be degraded by 30 percent or more when subjected to periods of high discharge. The pulsed discharge of chopper controllers also have a detrimental effect on specific energy. Battery performance is also subject to the complex and sometimes cantankerous nature of the chemical processes involved.

Projecting the cost of an advanced battery system is complicated by the lack of experience in producing the volumes typical of automotive tractive applications.

Experimental batteries may appear advantaged or disadvantaged by early cost estimates. Cost projections are necessarily based on highly variable estimates of materials and processes. Projections often must rely on estimates based on undeveloped technologies and production processes, which implies a high degree of uncertainty. When pilot production begins on a new EV battery design, such as is now the case with Ni/Cd and Ni/Fe batteries, the high prices typical of low volumes may have a negative psychological effect by having established a price, albeit one that is not based on appropriate volumes.

Lead/Acid (Pb/Acid)

Lead/acid batteries are still the battery of choice for electric cars. They have been around since 1854 and essentially have matured along with the automobile industry. Like an incumbent politician, the lead/acid battery has the advantage of being an established entity. Familiarity in product technologies breeds a level of confidence that can come only with experience. For much the same reasons, the lead/acid battery is also relatively inexpensive. Having been in high volume production for most of the twentieth century, development, tools and facilities costs have been amortized and manufacturing techniques and economies are well established. The product, as well as ongoing refinements of the technology, are therefore less costly, and the results of improved battery design tend to be more predictable and reliable. A new battery technology must fight an uphill battle on all fronts, first to demonstrate its technical superiority over the lead/acid battery, then to instill confidence that unforeseen problems will not develop in the field, and finally to compete with an established product during start-up operations.

Technical advantages of the lead/acid battery include its simple manufacturing process and a relatively high power density. Lead/acid batteries are also easily recycled and the infrastructure is in place. Disadvantages include the relative scarcity of lead as a raw material, the battery's vulnerability to damage due to sulfation and deep discharge, and its low specific energy.

Although lead/acid batteries are maligned because of their limited capacity to store energy, maintenance problems are the characteristic that has most frustrated EV fleet owners. This often surprises people because electric cars are supposed to be maintenance-free. Theoretically, EVs are low-maintenance vehicles, and with appropriate engineering they should require little maintenance in actual practice. Early vehicles have been largely experimental and often have not received the investment in engineering typical of most automotive products. For maximum life, batteries should be properly broken in by discharging to approximately 80 percent DOD (depth of discharge) for the first 10 to 15 cycles. If they are not properly "formed," cycle life as well as energy density may suffer. Batteries must be periodically "equalized" by a very low charge rate maintained for a period after the batteries are fully charged. If batteries are not charged to gassing voltage, energy density can drop due to electrolyte stratification. Overcharging can damage batteries as well.

Batteries self-discharge and they must be regularly supplied with a maintenance charge to prevent damage. This problem is not limited to the lead/acid battery. All battery types self-discharge at varying rates. At Quincy-Lynn we would often store prototype electric cars for extended periods, unattended. Batteries were usually dead when a vehicle emerged from a long hibernation. Moreover, they could not be revived by recharging. The problem was sulfation, which begins to occur when a lead/acid battery is unused for approximately 30 days. In general, maintenance problems can be easily managed by technical solutions, as long as the vehicle is regularly used and recharged with a properly designed charger.

The most tenacious problem of the lead/acid battery is its low specific energy. Within limits, the specific energy of a lead/acid battery can be increased by increasing the area-to-volume ratio of plate active materials. Essentially, thinner plates increase the battery's ability to store and deliver energy. A primary limiting factor is that of service life. When the area-to-volume ratio of active material is increased in order to increase specific energy, the positive plate breaks down more rapidly and battery life suffers. Consequently, there is an inverse relationship between specific energy and service life (see Figure 5.9). One method of increasing the surface-to-volume ratio of active plate material is by utilizing the tubular or gauntlet plate design. The tubular plate is designed to increase plate integrity at higher surface-to-volume ratios and thereby allow batteries of greater specific energy while still retaining an acceptable cycle life. The tradeoff here is one of cost: Tubular plate batteries are more costly to manufacture.

One variant is the "flow-through" battery which utilizes forced electrolyte circulation to improve battery performance. Circulating the electrolyte increases active material utilization by allowing the electrolyte to access the material's micropores. Circulation also eliminates specific gravity and heat gradients and minimizes the overcharge required to keep batteries equalized. In general, however, flow-through batteries suffer from reduced service life, primarily because of the more aggressive electrochemical action promoted by electrolyte circulation. Nevertheless, specific energy and specific power are significantly improved. Experiments at Johnson Controls, Inc., have demonstrated increases of 67 percent in specific energy and 54 percent in specific power due to the flow-through design.[11] Advantages of the flow-through battery include:

1. Specific energy is significantly increased.
2. Performance is improved over the Simplified Federal Urban Driving Schedule (SFUDS).
3. The battery is less sensitive to peak discharge rates.
4. The ability for thermal management is provided.
5. Active material utilization is improved which reduces demand on natural resources.

A promising new design is the Horizon battery (formerly Electrosource). It was developed by a group of aerospace engineers who adopted technology from strategic countermeasure devices. It is made of a co-extruded proprietary lead filament. Lead wire is extruded around a thin fiberglass filament under extreme

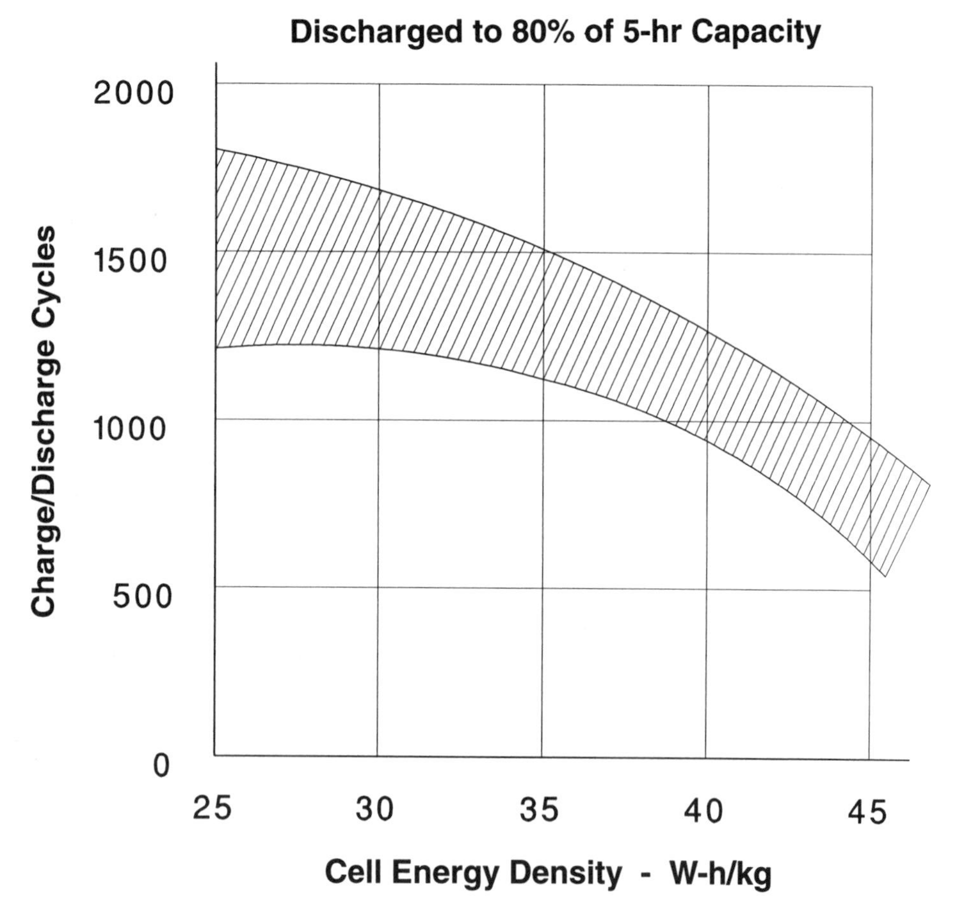

Fig: 5.9. The Relation of Energy Density to Cycle Life.

pressure. Strength comes from the glass-filament core, which eliminates the need for antimony, calcium or other alloys normally required to stabilize plate materials. The resulting high-density, fine-grain lead also results in extended plate life. Wire is woven into a grid which is coated with a proprietary paste and assembled into electrical plates. Plates are positioned horizontally rather than vertically in finished modules. The result is an advanced battery design that is expected to cost less than a conventional lead/acid battery (see Figures 5.10 and 5.11).

Cycle life is about four times greater than a conventional lead/acid traction battery. This feature alone constitutes a significant improvement over traditional designs. Additionally, the battery can tolerate an eight-minute recharge to 50-percent capacity, or a half-hour recharge to 99-percent capacity. Specific energy at the C-3 discharge rate is over 50 W-h/kg. Specific power is 500 W/kg at 100-percent charge, and 220 W/kg at 80-percent DOD. Life is approximately 900 cycles to 80 percent DOD at the C-2 rate.

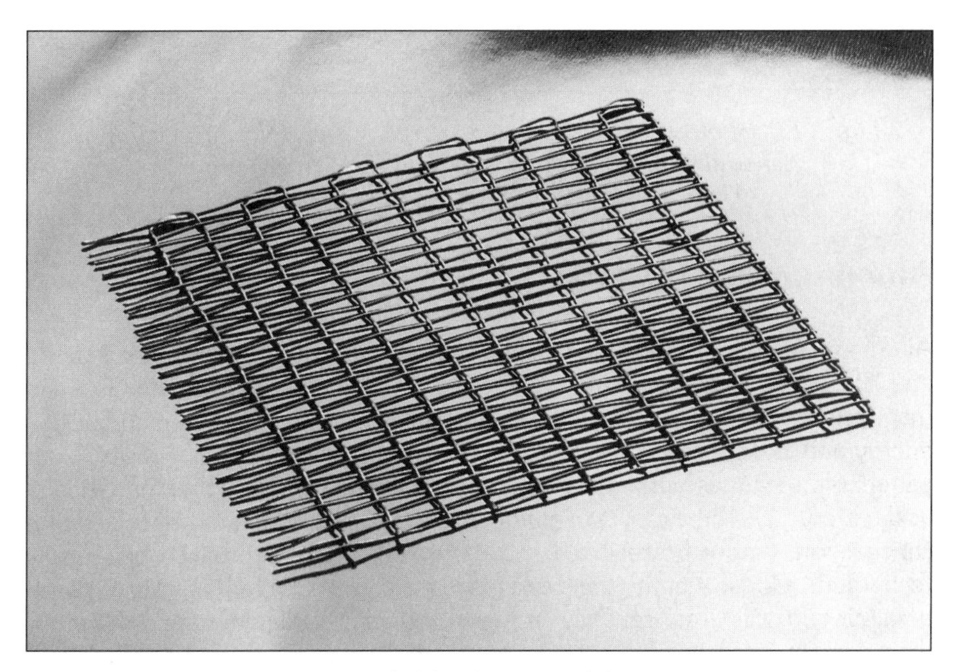

Fig. 5.10. Woven co-extruded lead wire grid design creates a stronger structure without increasing costs, and improves specific power and specific energy. (Courtesy: Horizon Battery Technologies, Inc.)

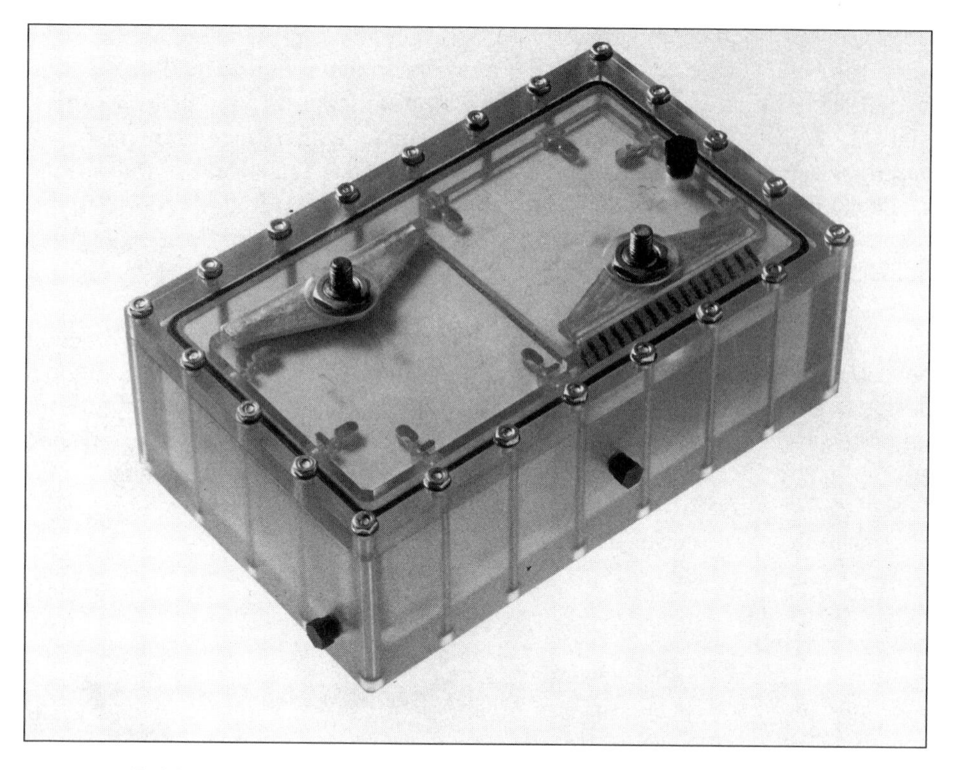

Fig. 5.11. A prototype battery in an acrylic case. Pilot production is scheduled for a new plant built by BDM International. (Courtesy: Horizon Battery Technologies, Inc.)

Aluminum/Air (Al/Air)

Aluminum/air batteries are the first in a series of systems that use a metal anode and an atmospheric oxygen cathode. Iron/air, zinc/air, lithium/air, and magnesium/air are just a few. Metal/air batteries tend to exhibit high specific energy and low specific power. Theoretically, the energy density of an Al/air battery is three times higher than Zn/air and seventeen times higher than a lead/acid battery. The breakdown of aluminum that occurs during the electrochemical reaction cannot be reversed so the Al/air battery cannot be electrically recharged. Consequently, the device is not a true secondary battery. Aluminum/air batteries are essentially "refueled" by adding aluminum and therefore they operate more as a fuel cell. To recharge the battery, the aqueous alkaline electrolyte is replenished with water and the waste aluminum hydroxide is removed. Approximately every fourth time the system is refurbished the aluminum anode must also be replaced.

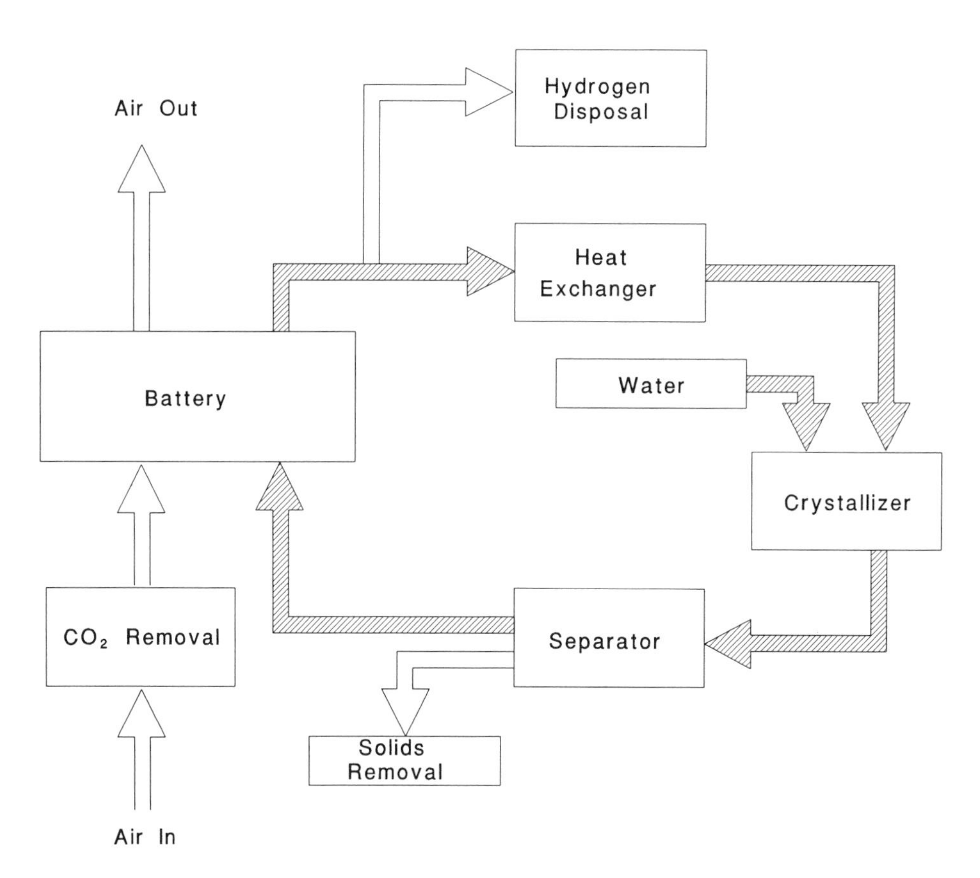

Fig. 5.12. Aluminum/Air Flow System.

An aluminum/air battery is a very mechanical device (Figure 5.12). Incoming air is scrubbed of CO_2 and dehumidified before entering the battery. Pumps circulate electrolyte between the cells and a separate reservoir. A crystallizer removes soluble by-products from the electrolyte to create an aluminum hydroxide slurry. Concentrated aluminum hydroxide is then washed into another receptacle for later removal. The system is complex and efficiency is low. Although aluminum is a plentiful metal, the mining and smelting process is very energy-intensive.

Lithium/Iron Sulfide (Li/FeS)

Lithium/iron sulfide batteries are still in the early stages of development. The lithium/iron monosulfide battery utilizes a lithium alloy anode and an iron monosulfide cathode. The electrolyte is a molten lithium chloride and potassium chloride salt solution. Operating temperature is about 450°C. Lithium/iron disulfide is a similar high-temperature battery that utilizes a positive electrode of lithium-iron disulfide. This battery has a molten electrolyte of lithium chloride, lithium bromide, and potassium bromide. It operates at a slightly reduced temperature (400°C), the electrolyte is more corrosive, and it has about twice the energy density of a lithium/iron monosulfide battery. Both technologies are expected to result in a more compact battery of significantly greater energy density. Vehicle range may be as much as eight times greater than an equivalent vehicle powered by lead/acid batteries, and twice that of a Na/S-powered vehicle. Both lithium/iron sulfide couples are highly experimental.

Nickel/Cadmium (Ni/Cd)

Nickel/cadmium batteries were developed in 1901. Ni/Cd batteries for EVs are currently available from SAFT. The SAFT EV battery has a cathode made of sintered nickel chemically impregnated with a mix of hydroxides. The anode is a plastic bonded cadmium composite. The electrolyte is a solution of potassium and lithium hydroxide. Open-circuit cell voltage is 1.35 V. SAFT's STM5-200 6-volt module is slightly smaller than a standard lead/acid traction battery, and at 25.5 kg, about the same weight. Specific energy is 52 W-h/kg and specific power 210 W/kg. Life can be as high as 2000 cycles. SAFT estimates that Ni/Cd-powered EVs will run 200,000 km before batteries must be replaced.

The Ni/Cd battery is superior to the lead/acid battery in all areas except price. SAFT's STM5-200 EV battery retails for $1,235 each. Chrysler bought them for their TEVan, which uses 35 modules (monoblocks), and paid $1,100 each for a total of $38,500 per vehicle. Over the life of the batteries, replacement cost calculates to $0.19 per kilometer ($0.31 per mile), which equates to approximately $6.19 per gallon of gasoline at 20 mpg, and electricity is not included. High costs reflect the low production volumes, and prices are expected to drop as volume increases.

Technical problems have to do with the scarcity and toxicity of cadmium, and to a lesser degree, with the battery's memory effect. If Ni/Cd batteries are not completely discharged or fully recharged, they tend to remember state-of-charge extremes and then behave as though they have less capacity. In electric car applications, batteries may take on the characteristics represented by the most common trip distance. However, they can be reconditioned by running them through a number of complete charge/discharge cycles. The relative scarcity of raw materials is not so easily dispatched. Although nickel is abundant, cadmium is not. Consequently, manufacturers are concerned about the implications of the large quantities of cadmium that might be required for Ni/Cd-powered electric cars. Finally, cadmium is very toxic and carcinogenic, and disposal is strictly regulated.

Nickel/Iron (Ni/Fe)

The nickel/iron battery was invented by Thomas Edison at the beginning of the twentieth century. These batteries have a negative electrode of iron and a positive electrode of nickel oxide. The electrolyte is potassium hydroxide. Advantages include a smaller, lighter package with greater specific energy than the lead/acid battery. Long service life is also characteristic. Nickel/iron batteries can weigh up to 20 percent less, store as much as 50 percent more energy, and last four times longer than similar lead/acid batteries. Volumetric energy density is about 113 W-h/L. Another advantage is that production capabilities already exist and batteries are packaged in 6-volt modules, similar in size to that of traditional lead/acid modules. Nickel/iron batteries are produced by Eagle-Picher Industries. Life and specific energy in comparison to lead/acid batteries are shown in Figure 5.13.

Disadvantages include higher initial costs and a relatively higher self-discharge rate. Eagle-Picher produces only 1500 nickel/iron traction batteries per year. Batteries are therefore essentially hand-made. As a result, the single-unit price of their NIF200 module is $1,300. According to Darrel Ideker, Plant Manager at E-P's Joplin, Missouri, facility, if production were increased to just 10,000 batteries per year, per-unit cost would drop to approximately $400. Accounting for increased service life, the price may ultimately be on par with that of a lead/ acid battery. As for self-discharge, the NIF200 module will lose about 13 percent of its charge in seven days. With regular use, self-discharge does not present a problem.

Nickel/Metal Hydride (NiMH)

Nickel/metal hydride batteries have been under development since the '70s, both at Ovonic Battery Co. in the U.S. and at SAFT in France. SAFT supplies nickel/cadmium batteries for a variety of applications, as well as for EVs. Ovonic supplies conventional round cells for use in flashlights, laptop computers, cellular phones and other small appliances. A nickel/metal hydride battery can be fully recharged in about 15 minutes. Energy density is twice that of a lead/acid battery, or about 80 W-h/kg. Batteries are compact, with a volumetric energy

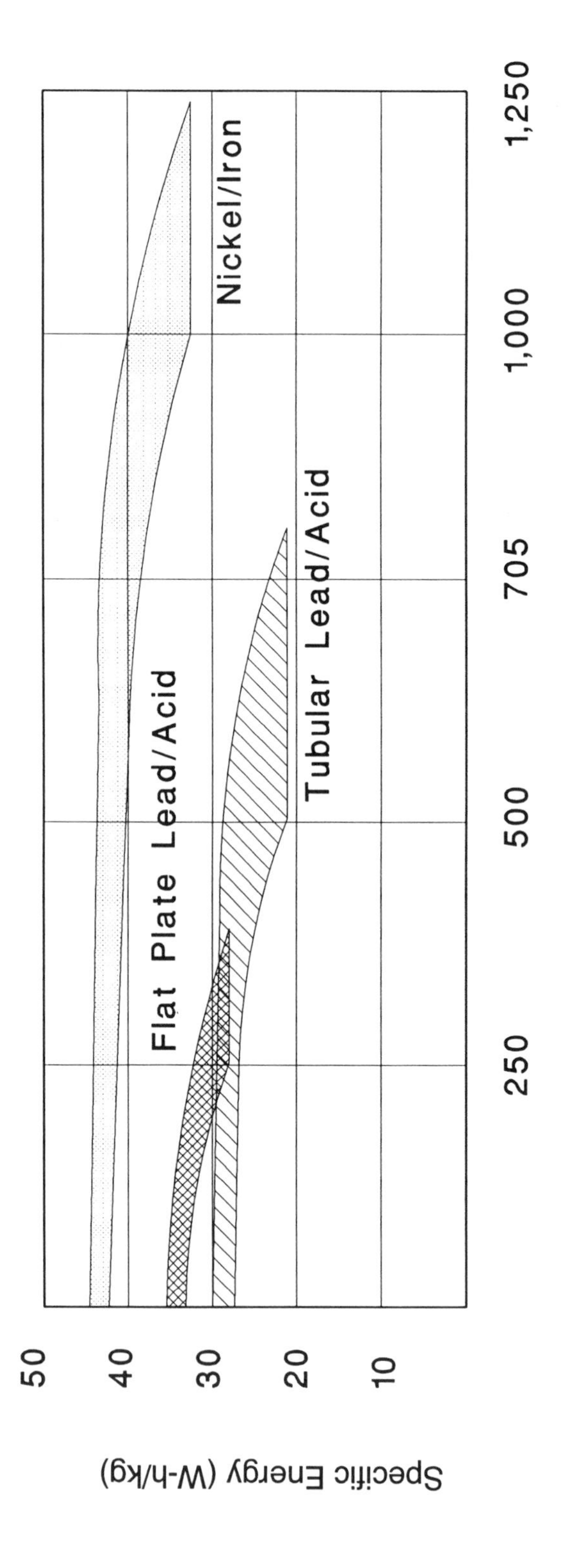

Fig. 5.13. Specific Energy Versus Cycle Life.

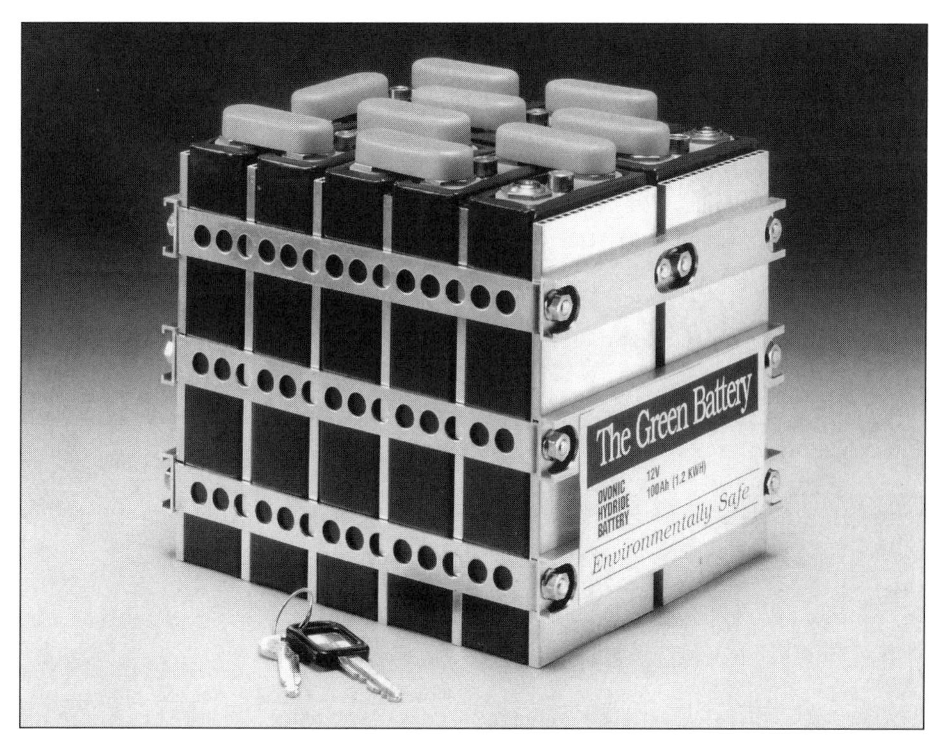

Fig. 5.14. Ovonic Prototype 12-V EV Battery.
(Courtesy: Ovonic Battery Company)

density as high as 215 W-h/L. In May 1992, the USABC awarded an $18.5 million contract to Ovonic to develop a mid-term EV battery.

Early on, it was generally believed that NiMH batteries lacked high specific power. What was not understood is that existing batteries had been developed for specific applications where energy density was more important than power. The relationship of cell materials, primarily vanadium, nickel, titanium, zirconium and additional transition metals, can be altered to tailor battery characteristics to the application. Cells developed specifically for EV application can achieve a peak power of 175 W/kg at 80-percent DOD. Excellent performance at continuously high discharge rates are also characteristic. The battery is tolerant of extremes in temperature with over 85-percent capacity available at –20°C. Cycle life is extremely high: Batteries have been tested up to 1000 cycles at 100-percent DOD; cells discharged to 30-percent DOD have lasted up to 10,000 cycles.

A high rate of self-discharge has been a problem with NiMH batteries. Early versions typically lost about 50 percent of their charge when left at room temperature for 30 days. Tests on extended C-size cells indicated a 15-20 percent self-discharge in seven days.[12] With improved materials and cell design, batteries can retain 65-80 percent of their charge over a 30-day period. Advanced separator materials can increase charge retention to about 93 percent over the same period.[13] Batteries are totally sealed, abuse-resistant and maintenance-free. Based on existing EPA protocols, NiMH batteries could be discarded in landfills as non-hazardous and non-polluting waste.

Nickel/Zinc (Ni/Zn)

Nickel/zinc batteries have existed for nearly 100 years. This battery couple has a theoretical energy density of about twice that of a lead/acid battery. It has excellent power density and it can deliver power even at extremely low temperatures. A lead/acid battery is virtually dead at –40°C, while the Ni/Zn battery can still deliver 40-60 percent of its 20°C capacity. A typical Ni/Zn cell has a positive electrode of nickel oxide, a negative electrode of zinc, and a potassium hydroxide electrolyte. Materials are abundant and inexpensive, and manufacturing processes are relatively straightforward.

The primary difficulty with the Ni/Zn couple is plate instability and its propensity to grow dendrites as it recharges. During recharge, zinc is redeposited unevenly or in the wrong places. As a result, plates begin to change shape and composition. In addition, dendrites grow between the positive and negative plates and ultimately cause the cells to develop shorts. Life is typically limited to about 100-150 cycles.

Attempts to resolve the problem of plate instability have centered on reformulated and flowing electrolytes and improved separator materials. Recent work at Electrochimica Corp. with a reformulated chemistry for the zinc electrode-electrolyte system shows much improved life with little compromise in the performance of the nickel electrode. Experiments at Electrochimica with prismatic cells incorporating their new stabilized chemistry have demonstrated cell life of 600-800 cycles with a 60-percent retention of original capacity.[14]

Sodium/Sulfur (Na/S)

Sodium/sulfur batteries were developed in the '60s by Ford Motor Company. More recent development has been done by Chloride Silent Power and ABB in Europe. Several automobile manufacturers have built experimental vehicles powered by Na/S batteries. Na/S batteries have high specific energy as well as high specific power. Volumetric energy density is about 128 W-h/L. On the basis of weight, Na/S batteries have about three times the energy density of a lead/acid battery.

Na/S batteries have two liquid electrodes separated by an ionically conductive ceramic electrolyte. Molten sodium is the anode and molten sulfur-sodium polysulfide functions as the cathode. Individual cells are cylindrical and consist of the tubular ceramic electrolyte which is filled with molten sodium on the inside and surrounded by molten sulfur on the outside. The cell's aluminum case is hermetically sealed and unvented. On the order of 360 individual cells are then assembled into strings and housed in a thermal enclosure designed to keep temperature within the operating range. Temperature must be maintained above 300°C for the battery to operate. Operating temperature is typically on the order of 350°C. Load generates heat and increases internal temperature. Batteries must therefore be heated for start-up, insulated to avoid heat loss under very light or zero-load conditions, and cooled at higher loads to avoid excessive heat build-up. The cell stack is normally cooled by circulating air through the enclosure. Open circuit cell voltage at 40-100 percent charge is 2.072 V. Below 40-percent state of charge, cell voltage slopes downward to a 1.72 V cutoff which is considered the fully discharged voltage.[15]

High manufacturing costs, low service life, and safety issues have been the primary problems of the Na/S battery. Although both sodium and sulfur are low-cost elements, high costs result from the attendant demands of sealed cells and their thermally controlled environment, the necessary multiple cell interconnects, bypasses, and fuses, in addition to the safety measures associated with high energy density and corrosive, high-temperature materials. Safety issues appear to be manageable through adequate fusing, improved electrical insulation, and crush-resistant cells and enclosures. Adequate life is also in sight: In 1977, cell life was approximately 70 cycles; today, mean time to cell failure is approximately 1000 cycles. When cells are assembled into batteries, life declines because of the combined failure rate of individual cells and subsystem components. Battery architecture and failure management strategies are pushing battery life closer to aggregate cell life. Manufacturing costs are the final hurdle.

Vanadium Redox Flow Battery

Development of the vanadium redox battery began in 1985 at the University of New South Wales in Australia. Like aluminum/air and zinc/air systems, a vanadium redox battery can be recharged by adding an active material. In a sense, the battery can be refueled. Unlike the other systems, however, the electrical charge is actually carried by the liquid electrolyte. Positive and negative electrolyte solutions are kept separate by a thin membrane as they flow through the battery's half-cells. Electrolyte can be recharged and stored, fully charged, outside the vehicle then dispensed as needed for an "instant recharge." Recharging can also be done conventionally without changing electrolyte. Energy efficiency is extremely high: about 90 percent, or 87-88 percent when pumping losses are considered. Pumping losses occur because the electrolyte is pumped through the cells from separate reservoirs (Figure 5.15). Another important characteristic is that discharge rate has little effect on energy density. Also, the battery can be completely discharged without damage. To increase available energy, larger electrolyte reservoirs are installed. The cell stack remains unchanged.

Because the charge/discharge cycle does not involve solid phase changes, there is no shedding or shorting, and electrolyte life is virtually infinite. The system can tolerate high charge rates, so recharging can be done in a fraction of the time required to recharge a typical lead/acid battery. On-board recharging time is about 45-60 minutes. When electrolytes are dumped into a high-capacity stationary site, they can be electrically recharged in about 5-10 minutes. The ability to refuel with charged electrolytes and recharge depleted electrolytes during off-peak hours makes the system ideal for load-leveling purposes. Load leveling improves the efficiency of electrical generating facilities and thereby reduces the cost of generating electrical power.

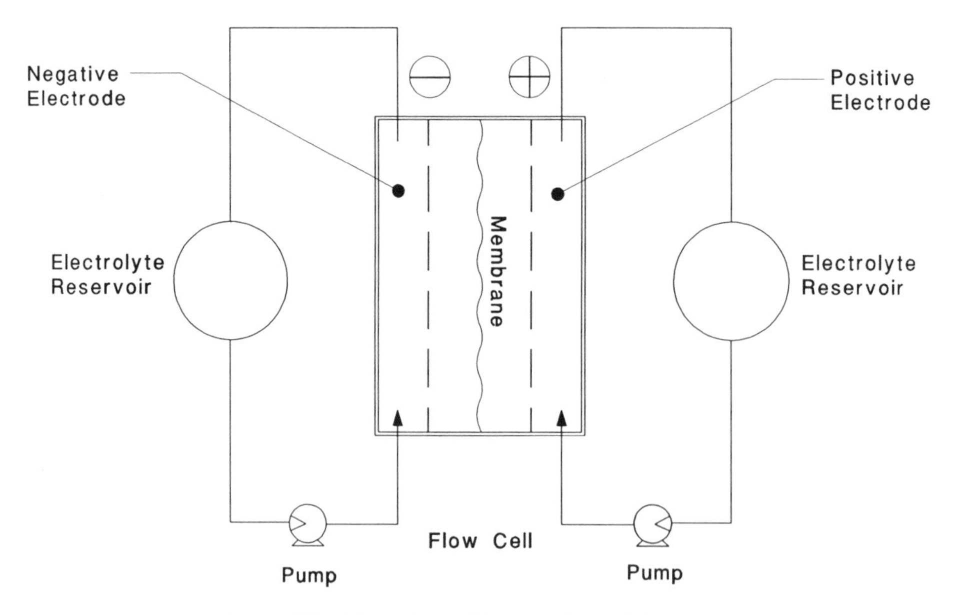

Fig. 5.15. Vanadium Redox Flow Schematic.

The disadvantage of the vanadium battery for electric car applications is its relatively low specific energy compared to Na/S and metal/air batteries. Specific energy is about 25 W-h/kg. Increasing the concentration of vanadium in solution from 2 molar to 3 molar or higher increases specific energy to about 40 W-h/kg. In-situ chemical regeneration of the positive electrolyte while the car is running allows a 90-percent volumetric reduction of the positive electrolyte and a doubling of the negative electrolyte. This increases specific energy to about 80 W-h/kg. Cost is relatively higher with smaller systems. Smaller systems appropriate for electric vehicle applications cost significantly more per kW-h than larger stationary industrial applications. Because the system is relatively new, much is still unknown about the economies of producing vanadium redox batteries in large quantities. Developers are optimistic that cost may ultimately match that of the lead/acid battery.

Zinc/Air (Zn/Air)

Zinc/air batteries have been around for some time. Their primary application has been in hearing aids and other small electronic devices where high energy density at low current is important. Until recently, zinc/air batteries were not considered rechargeable. Recharging causes the growth of dendrites that ultimately short-out cells. However, Dreisbach Automotive, Inc. (DEMI), in Santa Barbara, CA, has resolved the problems of dendrite growth and produced a rechargeable zinc/air EV battery system. The battery's high specific energy is the attribute that has sparked interest in EV application. In 1991, a zinc/air-powered Honda CRX won the Inaugural Solar and Electric 500 race held at Phoenix International Raceway by leading the pack for two hours on a single charge. Table 5.8 lists the steady-state energy consumption and range of DEMI's converted Honda CRX.

TABLE 5.8. DEMI HONDA CRX RANGE AND ENERGY CONSUMPTION AT VARIOUS SPEEDS

Nominal Speed		Energy Consumption		Range	
mph	km/h	W-h/mi	W-h/km	miles	km
30	48	124	77	322	518
35	56	145	90	275	442
45	72	186	116	215	346
55	88	228	142	175	282
65	105	270	168	148	238
75	121	311	193	128	206

Source: Ref. [16]

Fig. 5.16. Honda CRX Powered by Zinc/Air Battery. (Courtesy: Arizona Public Service Company)

Energy density of a zinc/air battery can be varied to fit the application. Early versions had an energy density of roughly 180 W-h/kg. However, specific power was marginal. As a result, power for the converted Honda had to be augmented with a Ni/Cd battery pack weighing 159 kg (350 lb). In more recent designs, some of the specific energy is sacrificed in favor of increased specific power. When specific energy is reduced to 120 W-h/kg, specific power increases to 120 W/kg, or nearly equal to that of a lead/acid battery. With increased specific power, the Ni/Cd battery pack is no longer required for most automotive applications. If greater acceleration is desired, power can still be augmented with a Ni/Cd battery pack. Full recharge of the zinc/air battery takes about six hours and a Ni/Cd pack can be continuously float-charged by the zinc/air system.

DEMI's zinc/air cells use a gas-permeable air cathode made of carbon, plastics, and proprietary catalysts. The anode is a zinc paste similar to common flashlight anodes. Electrolyte is potassium hydroxide in a water gel, also similar to flashlight batteries. Air is typically ducted in from the front of the vehicle, routed through the battery, then discharged at the rear. Airflow both cools the battery and provides the necessary reactive oxygen. Humidity and CO_2 content must be controlled within specific limits. Consequently, incoming air is electronically monitored, scrubbed of CO_2, and conditioned for humidity before it enters the battery. These requirements are met by a specially designed DEMI Air Manager/ Battery Box.

Cycle life is not especially high (135 cycles). However, when energy density is considered, zinc/air vehicles can run 40,000 km (25,000 miles) before batteries need replacing. Zinc is plentiful, environmentally friendly, and costs are low. DEMI estimates the cost of zinc/air batteries, modeled after the most recent experimental battery, at about $50 per kW-h. Charge/discharge efficiency is approximately 60 percent, figuring about 80 percent efficiency on discharge and about 75 percent on charge. Additional tests to determine zinc/air's potential in automotive tractive applications have been contracted by the USABC.

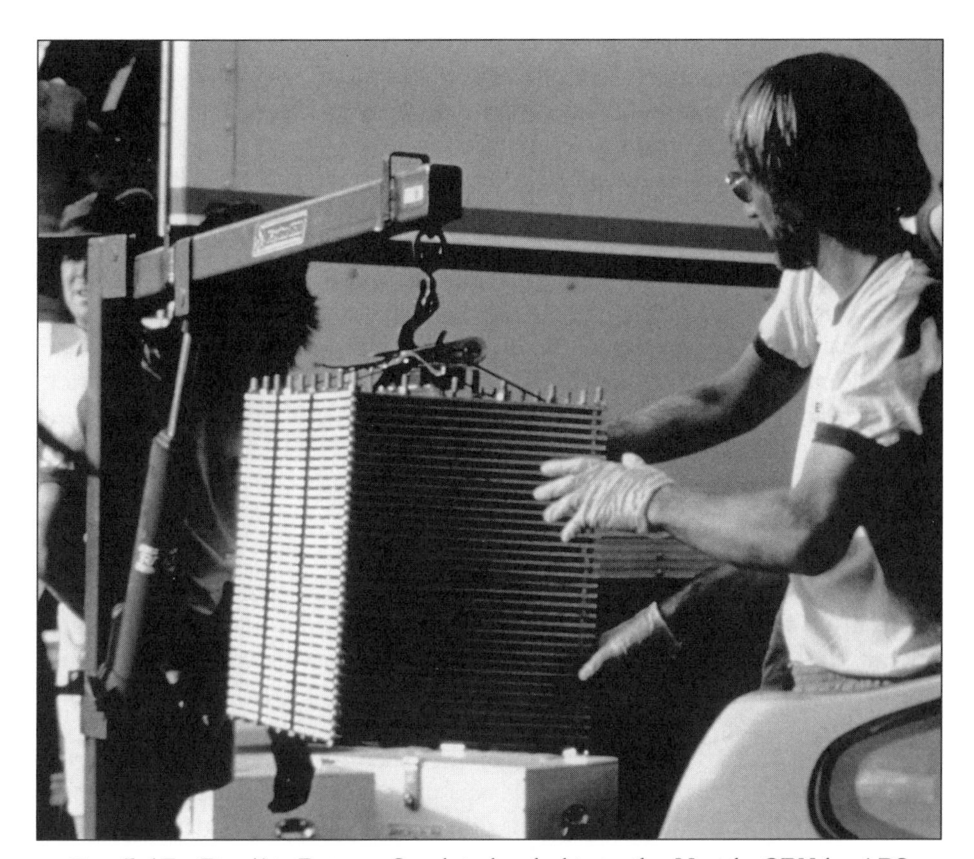

Fig. 5.17. Zinc/Air Battery Stack is loaded into the Honda CRX by APS technician. Rear seat is removed and the car becomes a two-seater. (Courtesy: Arizona Public Service Company)

Zinc/Bromine (Zn/Br)

Zinc/bromine is an aqueous flow battery system that has been under development for a number of years. Development began at Exxon Research & Engineering in the 1970s and has continued at Johnson Controls, Inc. (JCI), and at SEA in Austria. According to JCI, the advantages of their bipolar zinc/bromine battery are high specific energy and relatively low manufacturing costs. High specific energy is due primarily to the nearly all-plastic battery stack. The battery consists of a number of thin, conductive, carbon-filled plastic bipolar electrodes, thin microporous separators, and monopolar terminal electrodes. The liquid electrolyte is an aqueous solution of zinc bromide, bromine complexing agents, and optional supporting salts. Salts may be added to increase the electrolyte conductivity. Materials are abundant and inexpensive.

Deep discharge does not affect battery life, and self-discharge is minimal. One of the liabilities is the need to circulate the electrolyte, which is used to control battery temperature and prevent dendrite growth. JCI has built experimental batteries in the 65-80 W-h/kg range and projects manufacturing costs of approximately $55 per kW-h in high volumes. The battery operates at ambient temperature and service life of 1000-2000 cycles is expected with additional development.

The primary drawback of the zinc/bromine system is the highly corrosive and toxic nature of electrolyte solutions. Liquid electrolyte can cause severe burns, and inhalation of the vapor can be fatal. Consequently, there are concerns about the potential hazards of zinc/bromine batteries in the event of a collision. However, as the need requires, managing such hazards is not without precedent.

Fuel Cells

Fuel cells can convert fuel into electrical energy at an efficiency of about 50-60 percent, or about 2-1/2 times more efficiently than an internal combustion engine. They can use a variety of fuels including methanol, ethanol and natural gas, as well as gaseous or liquid hydrogen. When alternative fuels are substituted for pure hydrogen, efficiency is somewhat reduced. A fuel cell is comprised of catalytically activated electrodes (anode), the oxidant (cathode) and the electrolyte. Alkaline fuel cells, which utilize an aqueous solution of potassium hydroxide in combination with hydrogen and oxygen as fuel and oxidant, have been used to power spacecraft since the mid-'60s.

A fuel cell operates in the reverse of the electrolysis of water, during which electricity is used to split water into gaseous hydrogen and oxygen. In fuel cells, hydrogen and oxygen are recombined into water and in the process a flow of electrons from the anode to the cathode is established. In terrestrial transportation applications, oxygen can be extracted from air, and hydrogen can be extracted from a variety of hydrocarbon fuels, or carried on-board in gaseous or liquid form. When alternative fuels are used, a reformer must be employed to extract hydrogen from the fuel before it can be used in the reactive process.

Alkaline fuel cells suffer from slow start-up, low specific power, and slow response to changes in load. Consequently, they are not good candidates for land transportation applications. The technologies receiving the most attention include the phosphoric acid fuel cell (PAFC), and proton-exchange-membrane (PEM), as well as efforts to develop the reformers needed for reforming hydrogen from liquid hydrocarbon fuels. Work is also proceeding on solid oxide fuel cells which operate in the range of 1000°C. These high-temperature cells exhibit higher fuel efficiency, greater power density, and they do not require external reformers.

Fuel cells may be the transportation power source of choice over the long term. Today, the technology is embryonic and the ultimate cost and performance of future power units are difficult to project.

Kinetic Energy Storage

Flywheels are another promising energy storage device. They store energy mechanically instead of electrochemically. The idea of using flywheels to store energy for vehicles has been around since the 1800s. Forty years ago, buses in Switzerland were powered by large flywheels that were recharged at each stop. Near the end of his career, William Lear of Lear Jet fame became interested in kinetic storage for transportation application. In order to store more energy in a spinning flywheel, either the speed, the mass, or both are increased. To avoid undue mass, flywheels are typically operated at higher speeds. A lower mass spinning at higher speeds results in greater specific energy. Modern high-speed flywheels operate in a vacuum at speeds of 100,000 rpm or more. The rotor is normally suspended in magnetic bearings. Energy is exchanged electromagnetically, rather than mechanically in order to avoid the difficulties of mechanically interfacing with the high-speed rotor.

Controlling disturbances that can destroy the rotor, and developing materials and designs that can tolerate the high stresses, have been the greatest technical challenges. Because the greatest velocity is at the rim, concentrating the mass at the rim results in the greatest amount of stored energy. However, rim-type anisotropic flywheels can quickly exceed the structural limitations of materials because of unequal distribution of radial stresses. When the mass is concentrated at the rim, stress increases toward the hub and the material's cross-section declines in the reverse of the build-up of stress. For a while, it looked as though the hub-biased isotropic design might provide the answer to improved stress distribution and greater energy density. Isotropic flywheels utilize a geometry in which the mass is concentrated at the hub and decreases toward the rim. Materials are thereby evenly stressed. Unfortunately, the capacity to store energy is more limited with the isotropic architecture.

More recently, emphasis has returned to the anisotropic flywheel where work has concentrated on designs that employ fiber composite rims connected by spokes or a web to the hub. However, as the rotor expands at high speeds, maintaining even stress distribution and dynamic stability between the hub and rim has remained a problem. A promising new design is the system under development by American Flywheel Systems (AFS) in Bellview, Washington. The AFS design employs a number of axially positioned tubes spaced evenly around the hub, between the hub and the rim. As the hub and rim expand at different rates, the tubes compensate by distorting, thereby maintaining rim/hub stability and avoiding the effects of stress differentials between the two components.

AFS plans to package the unit in modules approximately the size of a conventional lead/acid traction battery. Inside each evacuated module, two carbon fiber counter-rotating rotors will spin at speeds of up to 200,000 rpm. Speeds in this range result in specific energy of 225-270 W-h/kg. As with other advanced flywheel designs, specific power is limited only by the capacity of the electrical system.

The initial cost of a flywheel tractive system may be the most difficult hurdle. However, AFS estimates that lifetime costs will be much lower than a conventional electrochemical battery system. Life of a new kinetic battery is estimated at 4000 cycles, or about 400,000 km (250,000 miles). Although manufacturing costs have not been established, AFS believes that cost per vehicle kilometer could be as little as 10-13 percent as much as a comparable lead/acid battery.

TABLE 5.9 COMPARISON OF GM's IMPACT WITH AFS KINETIC BATTERY VERSUS LEAD/ACID BATTERY

	GM's Impact with Lead/Acid Batteries	GM's Impact with AFS Flywheel
Total Energy	16.2 kW-h (propulsion pack)[1]	43.6 kW-h
Number of Modules	26 (propulsion pack)[1]	20
Weight of Storage Devices	480 kg (propulsion pack)[1]	225-270 kg
Specific Energy	30 W-h/kg[1]	193-226 W-h/kg
System Life	40,000 km	400,000 km
Vehicle Range	80-200 km	480-970 km

[1] Impact figures updated by author.
Source: American Flywheel Systems, Inc.

Hybrid/Electric Vehicles

One method of increasing EV range is to supplement the battery's reservoir of energy with a liquid-fueled heat engine. Depending on how the system is configured, the heat engine can offset demand on the battery by powering a generator to produce electrical energy, or it can deliver power to the drive wheels to either partially unload the motor or even entirely propel the vehicle. Either way, a portion of the energy required to propel the vehicle is supplied by the heat engine and range is thereby extended. Details of system interface and energy management strategy depend on the complex relationship between the vehicle mission and performance goals, and the tradeoffs made when the system is designed. Over the years, consensus regarding hybrid drives has gone through cycles of ascending and descending favor. Hybrids became popular in the late '70s, interest waned in the mid-'80s, and has picked up again in the '90s.

In theory, the hybrid/electric vehicle combines the good qualities of the electric car with those of the IC engine to create one vehicle with the best of both worlds. Heat engines are not constrained by limited energy supplies, but their efficiency suffers during part-load operation. Batteries suffer primarily from low specific energy. Combining the two energy systems might therefore produce a vehicle of greater energy efficiency and extended range. In a load-leveling approach, an underrated heat engine could operate in its region of greatest fuel efficiency, regardless of vehicle road load. The battery then provides for peak demands. For optimum efficiency, the heat engine is not used as long as range requirements are within the capability of the battery. For extended range, the heat engine is employed. Regardless of the operating schedule, power from the IC engine supplements the energy available from the battery and range is thereby extended, perhaps enough to match that of a conventional automobile.

In actual practice, a hybrid/electric vehicle can end up combining the worst of both worlds, rather than the best. Depending on the operating schedule and one's perspective, a hybrid vehicle can just as easily be criticized for resulting in an electric car that consumes gasoline and produces harmful emissions, or an IC vehicle with the extra expense, mass, and limited performance of an electric car. A hybrid vehicle contains two propulsion systems, neither of which may be adequate for the vehicle mission. And if one of the systems is adequate, that necessarily puts the value of the companion system in doubt. Further, a hybrid system is complicated and expensive, if for no other reason than redundancy. Hybrids are generally heavier than either an all-electric or a conventional IC vehicle. Although hybrid drives may be simple in concept, real benefits come only from carefully matching vehicle systems and operating schedule to the vehicle mission.

Hybrid Configurations

Hybrids are normally classified as either *series* or *parallel*. In the series hybrid the heat engine powers a generator which either charges the battery or supplies power directly to the propulsion circuit and thereby reduces demand on the battery. A parallel hybrid utilizes two tractive systems in parallel. In this configuration, both systems deliver power to the drive wheels. Either system may be utilized independently to propel the vehicle, or both systems may be utilized simultaneously for maximum power. A sub-category of the parallel system is called the split hybrid. In the split hybrid, one system powers one pair of wheels and the other powers the other pair. Tractive power is supplied by two mechanically independent systems.

Most of the development activity with hybrid designs occurred between 1978 and 1984. During this period, the Jet Propulsion Laboratory (JPL), General Electric and the Aerospace Corporation conducted a variety of studies under U.S. Department of Energy sponsored programs which were designed to determine the benefits of hybrid/electric vehicles. These studies began with in-depth computer simulations of hybrid vehicles over complex driving cycles. Later, vehicles were actually built and tested. The JPL/Aerospace studies indicated that the parallel hybrid configuration resulted in a smaller, lighter drivetrain and produced a vehicle with the best overall performance. Series hybrids were limited by the high mass of large engine/generator sets (gensets), and the difficulty of packaging high-output units in the small space available on-board a predefined Hybrid Test Vehicle (HTV). However, it was determined that the parallel hybrid produced a mechanically more complicated system, and the operating schedule and systems interface was more difficult to control. Today, with advances in genset packaging, series hybrids no longer suffer from the same limitations. Because of their simple design and control requirements, the series configuration may ultimately be the preferred design.

Series Hybrid

Series hybrids utilize a heat engine to power an on-board generator, as shown in Figure 5.18. The genset is used primarily as a range extender and the vehicle operates on battery power alone for most of the kilometers traveled. When the trip length is not within the capabilities of the battery, the genset is switched on to provide additional range. For longer trips, the genset might be automatically turned on when the battery reaches a predetermined depth of discharge. Early designs, which utilized generators in the 5-10 kW range, could have a significant effect on driving range only at relatively low urban speeds. To provide greater range at higher speeds, much more powerful gensets are required. Improved technology has resulted in greater output from smaller, lighter units, and modern series hybrids can now match the performance capability of parallel hybrids.

Quincy-Lynn's hybrid/electric, Town Car, is an early series design developed in 1980 (see Appendix). At the time, Town Car was envisioned primarily as an electric urban car with an on-board generator assist. The goal was therefore to minimize the size of the generator and thereby minimize the additional mass, as well as the degree to which the vehicle would rely on IC engine power. The relatively lightweight system employed a 3-kW IC engine to drive an on-board DC generator. When the generator was energized it started the engine which was set to run continuously at approximately 80 percent of maximum output. It produced an output of slightly more than 2 kW which was delivered either to the batteries or to the propulsion system, depending on the load. Range at a steady-state 56 km/h (35 mph) was 169 km (105 miles) and fuel consumption at the same speed was 40 km/L (93 mpg). On electric power alone, steady-state 56 km/h (35 mph) range was 105 kilometers (65 miles). Assist from the genset was conservative, even for the period in which it was designed.

Today, the trend in series hybrids is toward greater range, even equaling the range of multi-purpose conventional vehicles. This implies much larger generating systems. A state-of-the-art generating system is sized according to a predetermined cruising speed and vehicle range requirements. Consequently, the maximum-range driving cycle must be defined before the system can be designed. The battery is sized according to peak power demands, and the genset

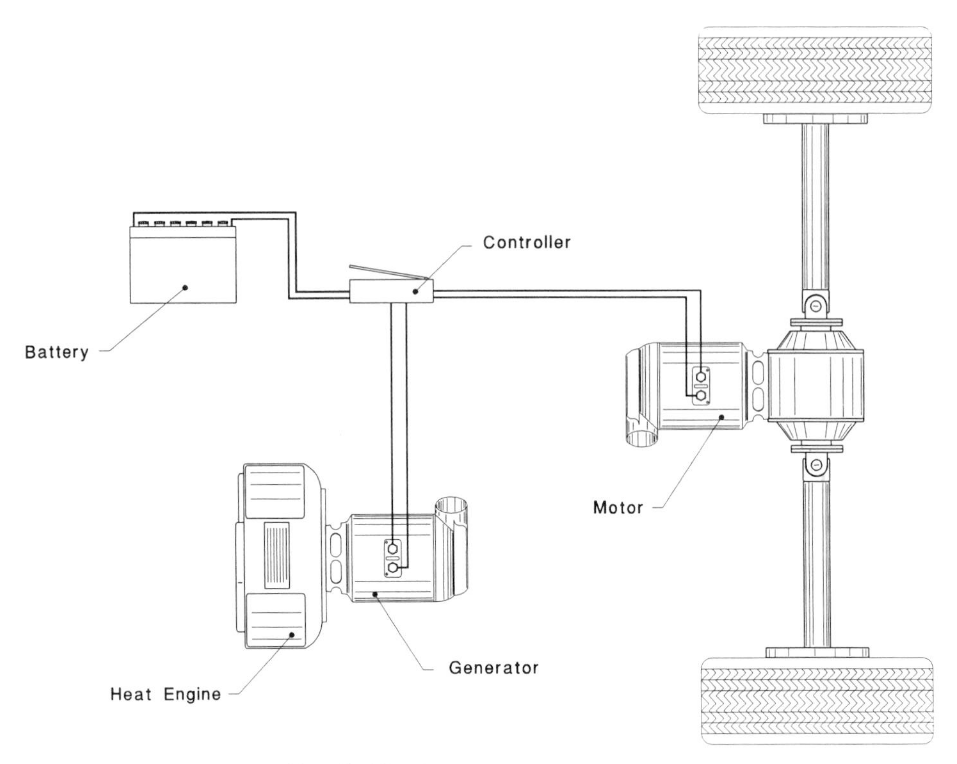

Fig. 5.18. Series Hybrid Layout.

is sized to allow for approximately 80-percent DOD at maximum vehicle range. During hybrid operation at higher speeds, energy is drawn only partially from the battery. Power for acceleration and passing comes primarily from the battery. At reduced cruising speeds, excess generating capacity is used to recharge the battery. Although generating system efficiency is high because the heat engine is loaded into its region of minimum bsfc, vehicle efficiency actually results from using wall-plug electricity for at least 80 percent of the vehicle kilometers traveled.[17] If the genset exceeds these requirements, vehicle efficiency suffers and the justification for electric power becomes more difficult to support.

The advantage of the series hybrid is its mechanically simple design in comparison to the parallel hybrid. Emissions from the IC engine can be easily controlled by preheating the catalyst with electric power from the battery before the genset is turned on. Once the system is energized, it then runs continuously at a predetermined output, which minimizes fuel consumption and further simplifies emissions control. Fuel economy for a state-of-the-art series hybrid vehicle over the Federal Urban Driving Schedule (FUDS) and Federal Highway cycles (FHWC) is greater than a conventional IC vehicle.[17] Typical series performance characteristics are shown in Table 5.10.

TABLE 5.10. SERIES HYBRID PERFORMANCE CHARACTERISTICS

	Battery Electric Mode				Series Hybrid Mode					
	FUDS		FHWC		FUDS		FHWC			
Vehicle Type	W-h/km	Range[1] (km)	W-h/km	Range (km)	mpg[2]	Eff.[3] (%)	mpg	Eff. (%)	Acceleration Times (sec)	
Minivan	185	93	188	86	26.1	0.85	26.4	0.88	4.7	12.1
Microvan	136	96	132	93	35.6	0.85	37.5	0.88	4.7	12.5
Compact Car	116	99	103	107	41.8	0.86	47.8	0.87	4.3	11.0

(1) Usable range to 80% DOD
(2) Gasoline fuel and min bsfc = 300 g/kW-h
(3) Average efficiency from engine output to inverter input
Source: Ref. [17]

Parallel Hybrid

In a parallel hybrid (Figure 5.19), power from the heat engine is delivered directly to the drivetrain rather than to a generator. Unlike a series system, which can recharge batteries when the vehicle is at rest, a parallel hybrid system works only when the vehicle is moving. The engine may be capable of propelling the vehicle by itself, or it might provide only a portion of the required tractive power. Although mechanically more complicated, a parallel hybrid avoids the losses of converting mechanical power into electrical power before it is put to work in the vehicle. A heat engine is smaller, lighter, and more power-intensive than an electric motor or generator. Consequently, designers are free to utilize greater install power and thereby improve vehicle performance and range capabilities. Another attribute has to do with the impact of supplemental IC power on battery design requirements. Since peak power is supplied by the IC engine, batteries can be designed for lower specific power and greater specific energy.

Range and performance are determined primarily by the size of the heat engine. Details of the control scheme determine the vehicle's fuel consumption and emissions. During short trips, the vehicle operates primarily on electric power. During longer trips and at highway speeds, the vehicle becomes IC-engine-dominant and fuel consumption and emissions become more like those of a conventional automobile. Because driving patterns are heavily skewed toward shorter trips, a parallel hybrid will operate most of the time as an electric car and

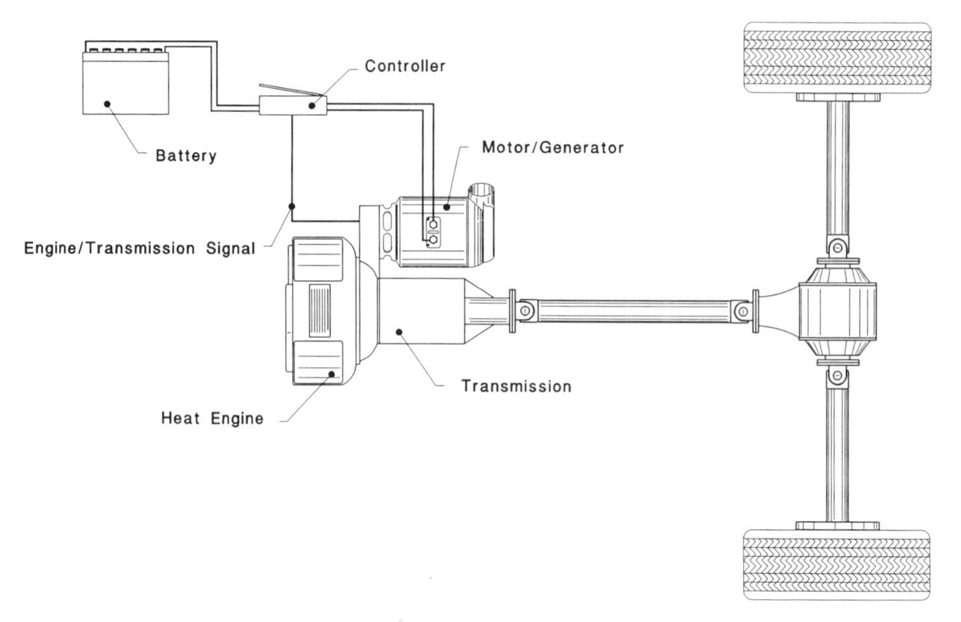

Fig.5.19. Parallel Hybrid Layout.

thereby reduce emissions and dependence on petroleum fuels. Transitions between system dominance must be smooth and fluid. A velocity threshold is normally used to determine the bias between IC engine and electric systems. Below a predetermined velocity, the vehicle is propelled only by the electric system and the IC engine turns on only if power is needed for acceleration. Above this speed the IC engine is continuously running and vehicle tractive power is augmented by the electric system. Ideally, the transition speed will vary depending on battery state of charge.

Compared to the series configuration, a parallel hybrid has advantages as well as disadvantages. In concept, the two configurations function much alike. In both cases, maximum advantage is derived from the electric drive to the degree that it is emphasized over the IC engine drive. Load sharing is an important strategy. In order to effectively load share, the relationship between engine and driveshaft speeds must be highly fluid, which in the parallel hybrid would best be accomplished with a continuously variable transmission. Systems interface and control is inherently more complex, but it is only through precise control that the parallel hybrid can excel. Because the engine in a parallel system operates over a broader power band, average fuel economy is likely to suffer. Additionally, an engine that cycles on and off typically produces more emissions. Although a parallel hybrid allows greater performance flexibility through greater reliance on

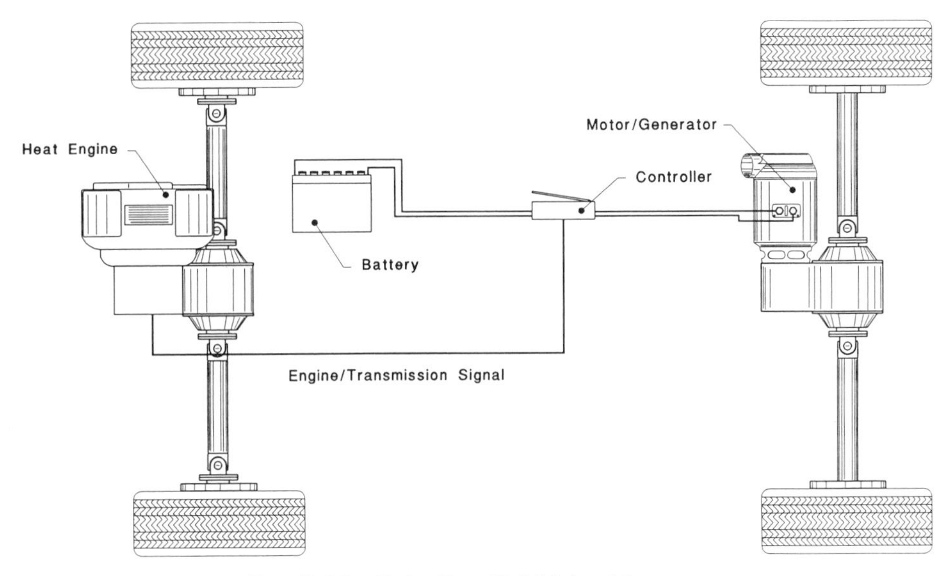

Fig. 5.20. Split Parallel Hybrid Layout.

the heat engine, for the same reasons it may produce more emissions and consume more liquid fuel. To the degree to which it de-emphasizes electric power, it also loses the benefits of electric power.

Electric Vehicles and Vehicle Downsizing

A conflict of technical attributes exists between the idea of vehicle downsizing to minimize energy consumption, and the larger battery packs required for greater on-board energy stores. Energy stores increase in proportion to the size and mass of the battery. However, the need for energy decreases when the size and mass of the vehicle is reduced. One strategy for making peace between these two opposing characteristics is to allocate a greater portion of the vehicle's mass to the battery. When more of the vehicle is allocated to the battery pack, vehicle range and performance increase. Nearly 40 percent of the mass of GM's high-performance, two-seater Impact is comprised of the battery. Although this approach can be applied to both small and large vehicles, the most cost-effective approach is to reduce vehicle size in order to reduce energy requirements and thereby allow for a smaller, less costly battery pack.

The high cost of batteries is one of the largest consumer disincentives associated with electric vehicles. Chrysler estimates the cost of replacing TEVan's 817 kg

(1800 lb) Ni/Cd battery pack at approximately $6,000.* Replacements for Impact's lead/acid batteries, which weigh 499 kg (1100 lb), are estimated at $2,000. Batteries represent a significant portion of first costs, as well as a significant outlay at various points in the life of the vehicle. When the demand for energy is reduced, the size and cost of motors, batteries, and controls are also reduced. Reducing the vehicle's demand for energy can significantly reduce first costs, as well as maintenance costs.

With the possible exception of GM's Impact, designers in the U.S. have tended to favor larger vehicles that can accommodate larger battery packs with greater stores of energy. In Europe, the trend is just the reverse, toward smaller vehicles with more modest appetites for energy. Smaller EVs require fewer batteries for the same range and performance. They also imply a greater emphasis on space-efficient packaging and batteries of greater volumetric energy density. By limiting payload capacity, the size and mass of the vehicle can also be reduced. Two-occupant vehicles are adequate for the 90th-percentile trip in the U.S. For two-thirds of U.S. drivers, a vehicle with a range of 85-150 km (53-93 miles) would meet the range requirements of the 95th-percentile trip.[18]

If the EV were no longer envisioned as a replacement for conventional cars, performance comparisons would cease to be as meaningful. If 10 percent or more of vehicle trips were left to larger IC vehicles of greater specific energy, EVs could be significantly downsized, battery packs could be smaller, and costs could be reduced. Electric vehicles present both an opportunity and an imperative to encourage the idea of mission-specific vehicle design at a time when consumers may already be more inclined to see the new vehicle type through different eyes.

Horlacher AG

Horlacher AG in Switzerland specializes in fiberglass-reinforced plastic (FRP) transportation, industrial, and consumer products. Their interest in automotive design predates the existing company when the uncle of founder Max Horlacher did much of the coachwork for Bugatti. Today, the family-owned company builds at least one prototype electric vehicle each year in their on-site model shop. In 1992, their NaS-Sport set a world record by driving from Zurich to Geneva, then cruising for four hours in Geneva traffic in order to exceed their goal of driving 500 km (310 miles) without recharging (Figure 5.21).

---

* Chrysler's estimate is based on increased quantities and cannot be compared with the $38,500 cited earlier as the cost of the first batteries for TEVan.

Fig. 5.21. The sleek two-seater NaS-Sport weighs only 450 kg (990 lb) and can run 547 km (342 miles) on a single charge. Power is supplied by a sodium/sulfur battery developed by ABB in Heidelberg, Germany. The car consumes just 5 kW-h per 100 km (62 miles), or less than half the energy of GM's Impact. (Courtesy: Horlacher AG)

Fig. 5.22. Horlacher City (left) and Horlacher Sport are both two-seater electric cars. The Sport is longer and lower and includes a rear luggage area. (Courtesy: Horlacher AG)

Horlacher is also a leader in low-mass vehicle safety research. The company has developed and crash-tested a number of low-mass cars that incorporate their own "hard shell" concept in which a crush-resistant belt improves crash survivability (see Chapter Seven). Other designs include the Horlacher City and the Horlacher Sport (Figure 5.22). Their vehicles have been tested with a variety of different battery couples.

Esoro AG

Esoro's E301 is another excellent example of an ultralight EV design (Figures 5.23 and 5.24). The E301 was developed by Esoro AG in cooperation with the aerospace company Bucher Leichtbau AG, both of Switzerland. E301 is a frameless design that is built 90 percent of fiberglass composites. Ultralight materials are an essential contributor to vehicle downweighting and energy conservation. According to Esoro's studies, reducing body weight by 1 kg ultimately allows a reduction in curb weight of as much as 1.4 kg. Drivetrain and batteries can then be downrated accordingly without affecting vehicle performance. With a Ni/Cd battery pack weighing only 260 kg (572 lb), the lightweight car can cruise for 100-150 km before recharging.

Fig. 5.23. The sleek two-seater Esoro E301 will cruise 100-150 km on its 260-kg (572-lb) Ni/Cd battery pack. Curb weight is 620 kg (1365 lb). Dimensions are 3.07 m long, 1.55 m wide, and 1.42 m high. Seating is a 2+2 layout. (Courtesy: Esoro AG)

Fig. 5.24. The Esoro E301 from the rear. Doors move out then forward to provide easy access in cramped parking spaces. Hatchback provides access to behind-the-seat storage. (Courtesy: Esoro AG)

Renault Zoom

Renault's Zoom is an electric urban car designed for multi-car city dwellers. In France, 28 percent of households have at least two cars. Average vehicle occupancy in France's cities is 1.18 passengers. Consequently, Zoom was designed to carry two people on local city trips. Parking space in Europe is also more limited than in the U.S. To reduce space requirements, the entire rear suspension assembly rotates forward and under the car to reduce overall length and let the car fit into tighter parking spaces. Renault estimates that with two million of these vehicles on the road traveling 40 km (64 miles) per day, only two percent of that nation's generating capacity will be consumed.

Zoom has an urban driving range of 150 km (93 miles). Steady-state range at 50 km/h (31 mph) is 260 km (161 miles). The car's maximum speed of 120 km/h (75 mph) is also much greater than might be needed by the average city dweller. Zoom can accelerate from 0 to 60 km/h (0-37 mph) in less than six seconds. Power comes from a 25-kW self-synchronous AC motor which is supplied with power from Ni/Cd batteries located under the bench seat. Curb weight is 800 kg (1762 lb) of which 350 kg (770 lb) is comprised of batteries.

Fig. 5.25. Renault's urban car Zoom features a tuck-under rear suspension that reduces overall length for parking. Doors move laterally away from the body, then swing up so occupants can enter and exit the vehicle in cramped parking spaces. (Courtesy: Renault)

According to Renault, a market for special-purpose urban vehicles is already emerging in Europe. Zoom is a concept car that may provide the basis for future production vehicles.

Fiat's Downtown

Downtown is a recent prototype that reflects the emphasis on urban transport at Fiat (Figure 5.26). The car's 165 kg (363 lb), 18.6 kW-h sodium/sulfur battery provides enough power to carry three occupants a distance of 190 km (118 miles) over an urban driving cycle. Range at a steady 50 km/h (31 mph) is 300 km (186 miles). Downtown accelerates 0-80 km/h (0-50 mph) in 21 seconds and maximum speed is 100 km/h (62 mph). Curb weight is 700 kg (1540 lb).

Fiat obviously subscribes to the ideal three-occupant seating as representing the best compromise between seating capacity and trip requirements (Figure 5.27). The vehicle also incorporates the experimental VENUS (VEhicular Navigation

Fig. 5.26. Fiat Downtown is a prototype urban car from Fiat. It carries three over a 190-km city driving cycle. (Courtesy: Fiat)

Utility System) navigation system which indicates the best route to the selected destination, as well as distance-to-destination, continuous distance remaining throughout the trip, and parking accommodations at the destination. Although designed for maximum energy and space efficiency, Downtown does not speak to consumers in terms of compromise. Instead, it embodies the latest in creature comforts and high-tech urban commuter accommodations.

As with GM's Impact, motors are built into the front wheels for more efficient packaging (Figure 5.28). A disadvantage of this motor design is higher unsprung weight. The battery and electronics are tightly packaged in the rear of the vehicle where every crevice is efficiently utilized (Figure 5.29).

Fig. 5.27. The one-plus-two seating arrangement of the Downtown includes rear seats that can be reconfigured to form combination utility table/restraints for children. (Courtesy: Fiat)

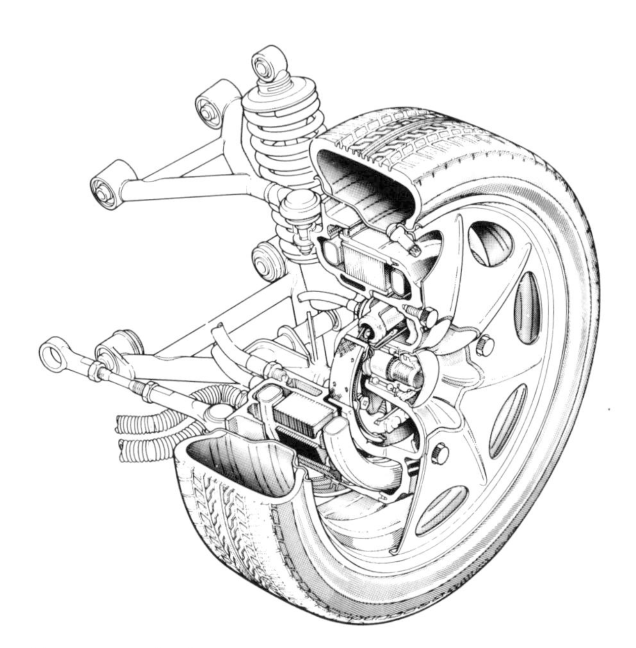

Fig. 5.28. Downtown's front motor/wheel assembly. One 5-kW motor is built into each front wheel in a manner similar to Impact. (Courtesy: Fiat)

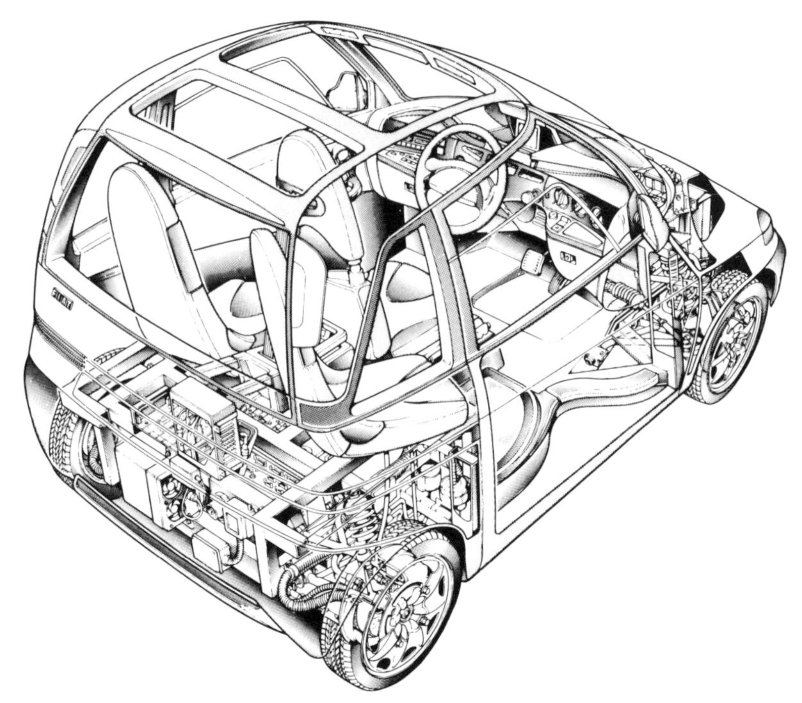

Fig. 5.29. Downtown's battery and control assembly; every crevice is efficiently utilized. (Courtesy: Fiat)

Chrysler's TEVan

A bias toward larger platforms in the U.S. is reflected by Chrysler's TEVan, which is the first U.S. electric vehicle to receive Federal Safety Certification (Figures 5.30 and 5.31). The product's orientation toward commercial applications in which higher vehicle costs are more justifiable is also in keeping with the van configuration. The first of approximately 50 production vehicles was delivered in the first quarter of 1993. Based on the company's Dodge Caravan platform, the totally electric vehicle utilizes a General Electric DC motor and controller, in conjunction with SAFT's STM5-200 Ni/Cd batteries. The 68 kg (150 lb), 52-kW motor operates up to 8000 rpm and provides a top speed of 117 km/h (73 mph). Range is up to 240 km (150 miles). Westinghouse and Chrysler are also developing an advanced AC system for their next generation of EVs.

TEVan uses a "smart charging" system developed jointly between Chrysler and Norvik Technologies. Although not the first fast-charge system, it may be one

Fig. 5.30. Chrysler's TEVan is the first U.S. electric vehicle to receive Federal Safety Certification. Its fast charging system may be one of the most sophisticated developed to date. (Courtesy: Chrysler Corp.)

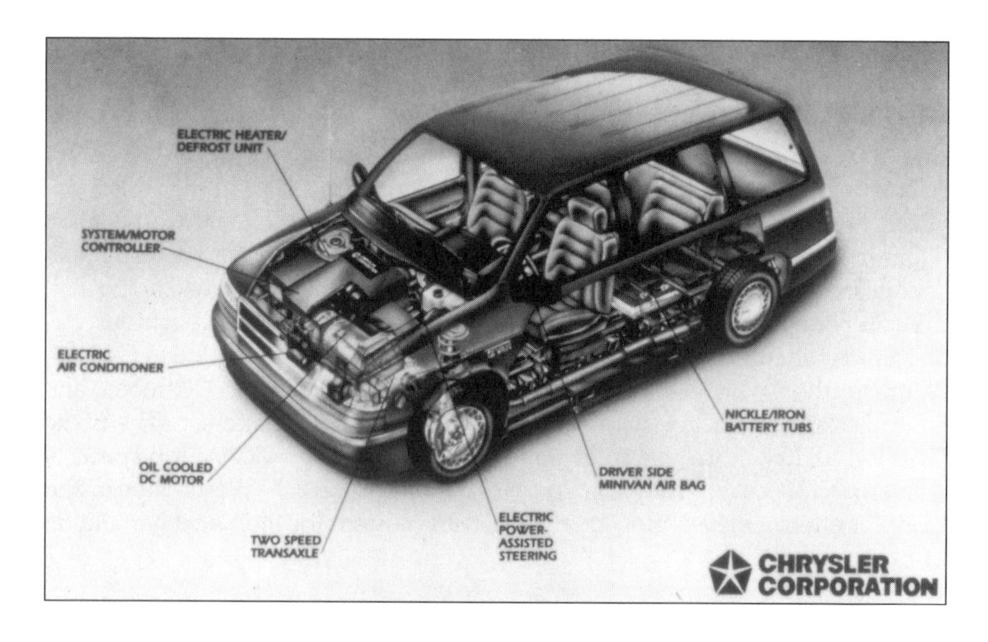

Fig: 5.31. TEVan Layout. (Courtesy: Chrysler Corp.)

of the most sophisticated. Previous systems were not designed for regular use because rapid charging can damage batteries. Chrysler's proprietary design senses the chemical makeup of a battery, as well as state-of-charge, then rapidly charges the battery according to tolerance. The key is in the diagnostics loop, which diagnoses the battery several times per second and adjusts the charge rate accordingly. It can be used with any battery system and has the ability to differentiate between battery couples and provide the appropriate charge. Battery life is extended by the system's ability to avoid overcharging.

GM's Impact

Impact is one of the most advanced EV designs to emerge from the automobile industry (Figure 5.32). This unique vehicle is a preproduction two-seater that utilizes specially designed lead/acid batteries developed by the company's Delco Remy Division to power a pair of advanced-design, 42-kW AC induction motors. The water-cooled motors, also developed by Delco Remy, independently drive the two front wheels. Since high-output batteries could not be built at an affordable price, GM approached the problem from the other end by concen-

Fig. 5.32. Impact high-performance electric car employs regenerative braking, reduced road load, and an advanced AC tractive system to keep energy consumption to a minimum. Range is 160 km (100 miles) on conventional lead/acid batteries. (Courtesy: General Motors Corp.)

trating on reducing road load. The vehicle's low-friction drivetrain, 448 kPa (65 psi), low-rolling-resistance tires, and an extremely low 0.185 coefficient of drag, are responsible for its diminutive appetite for energy. Impact has a highway range of 160 km (100 miles) and will accelerate from 0-96 km/h (0-60 mph) in eight seconds.

Ford's Ecostar

Ford's limited-production Ecostar is based on their European Escort van (Figure 5.33). Vehicles are converted to electric power in the U.S. Ecostar utilizes a 56-kW AC induction motor powered by advanced sodium/sulfur batteries. The 1400 kg (3100 lb) vehicle will accelerate from 0-80 km/h (0-50 mph) in 12 seconds. Range over the Federal Urban Driving Schedule is 160 km (100 miles), which is due primarily to the high specific energy of the batteries. Ford invented the sodium/sulfur battery in the '60s and development has since been carried on by Chloride Silent Power and Asea Brown Boveri in Europe.

Fig. 5.33. Ford's limited-production Ecostar uses advanced sodium/sulfur batteries invented by Ford in the 1960s. (Courtesy: Ford Motor Company)

References

1. Quanlu Wang and Daniel Sperling, "Energy Impacts of Using Electric Vehicles in Southern California," Institute of Transportation Studies, UCD-ITS-RR-92-13, May 1992.

2. Quanlu Wang and Mark A. DeLuchi, "Impacts of Electric Vehicles on Primary Energy Consumption and Petroleum Displacement," Institute of Transportation Studies, July 1991.

3. Stacy C. Davis and Sonja G. Strang, "Transportation Energy Data Book: Edition 13," Oak Ridge National Laboratory. (Because of reporting inconsistencies, fuel economy figures may not be directly comparable between countries.)

4. GM uses a figure of 170 W-h/mi to calculate Impact's city range. The author has used 180 W-h/mi in the preceding table, which is based on GM's published specifications of 26 propulsion modules supplying 16.2 kW-h energy, and 72 miles city range at 80% DOD. GM's internal figures would result in a slightly greater savings of primary energy. Conversion to SI is by the author.

5. "Electric Vehicles: The Beginning of a Bright Future," Electric Power Research Institute, Inc., Pleasant Hill, CA.

6. Jeffrey Skeer, "A Brief Survey of IEA-Sponsored Activity on Technology for Transport," published in OEDC Document, *The Urban Electric Vehicle: Policy Options, Technology Trends, and Market Prospects*, ISBN 92-64-13752-1.

7. Quanlu Wang and Danilo L. Santini, "Magnitude and Value of Electric Vehicle Emissions Reductions for Six Driving Cycles in Four U.S. Cities with Varying Air Quality Problems," paper presented at the 72nd Annual Meeting of Transportation Research Board, January 10-14, 1993, Washington, D.C.

8. Mark A. DeLuchi, "Greenhouse-Gas Emissions from the Use of New Fuels for Transportation and Electricity," *Transportation Research*, Vol. 27A, No. 3, pp 187-191, 1993.

9. P.J. McConachie, "Electric Vehicle Performance Requirements for Traffic Compatibility," Motor Vehicle Fuel Conservation Workshop, Australia Dept. of National Development and Energy, 1981.

10. Donald W. Kurtz, Theodore W. Price, and James A. Bryant, "Performance Testing and System Evaluation of the DOE ETV-1 Electric Vehicle," SAE Paper 810418, Society of Automotive Engineers, Warrendale, PA, 1981.

11. J.P. Zagrodnik, *et al.*, "Development of Advanced Battery Systems for Vehicle Applications," SAE Paper No 890783, Society of Automotive Engineers, Warrendale, PA, 1989.

12. "Battery and Electric Vehicle Update," *Automotive Engineering*, September 1992, p 17.

13. S.R. Ovshinsky, *et al.*, "Recent Progress in the Development of Ovonic Nickel Metal Hydride Batteries for Electric Vehicles," SAE Paper No. 921571, Society of Automotive Engineers, Warrendale, PA, 1992.

14. D. Reisner and M. Eisenberg, "A New High Energy Stabilized Nickel-Zinc Rechargeable Battery System for SLI and EV Applications," SAE Paper No. 890786, Society of Automotive Engineers, Warrendale, PA, 1989.

15. M. Altmejd and E. Spek, "Capabilities of Na/S Batteries for Vehicle Propulsion," SAE Paper No. 890784, Society of Automotive Engineers, Warrendale, PA, 1989.

16. Len G. Danczyk, Raymond S. Hobbs, and Robert L. Scheffler, "A High Performance Zinc-Air Powered Electric Vehicle," SAE Paper No. 911633, Society of Automotive Engineers, Warrendale, PA, 1991.

17. A.F. Burke, "Hybrid/Electric Vehicle Design Options and Evaluations," SAE Paper No. 920447, Society of Automotive Engineers, Warrendale, PA, 1992.

18. William Hamilton, "Basic Requirements for Urban Cars," SAE Paper No. 780219, Society of Automotive Engineers, Warrendale, PA, 1978.

Chapter Six

Three-Wheel Cars

Quincy-Lynn Tri-Magnum

What, sir, would you make a ship sail against the wind and currents by lighting a bonfire under her deck? I pray you excuse me. I have no time to listen to such nonsense.

Napoleon to Robert Fulton

The idea of urban cars and commuter cars seems to naturally introduce the three-wheel platform. Three-wheelers have been around since the inception of the automobile. They have enjoyed limited success in Europe and Japan, but have been largely ignored in North America. Opinions normally run either strongly against or strongly in favor of the layout. Advocates point to superior handling characteristics and opponents decry the three-wheelers' propensity to overturn. Both opinions have merit. A correctly designed three-wheel car can light new fires of enthusiasm under tired and routine driving experiences. But when subjected to poor design, improper application, or poor driver technique, a three-wheel platform is the less-forgiving layout on all counts.

Three-wheel overturn resistance is affected primarily by the same relationship of center of gravity height to effective half-tread that determines the resistance to overturn of four-wheel vehicles. A static margin of safety against rollover equal to that of four-wheelers can be attained by applying traditional engineering principles. However, three-wheelers are more sensitive to cg displacement, both longitudinally and vertically, and payload variations can have a significant effect on vehicle center of gravity. In addition, dynamic forces introduce a variety of resultants that affect the margin of safety against rollover. Under dynamic conditions, a reduction in the effective half-tread of a three-wheeler results whenever cornering or braking forces cause the cg to shift toward the end with the single wheel. Consequently, the higher forces that result from greater vehicle speeds are more likely to exceed the limitations of a three-wheel platform, given an equal cg-height-to-half-tread relationship. Designers of three-wheel cars must therefore provide a greater margin of safety, and vehicle loading must be more carefully controlled. A three-wheel platform is not a good layout for a large, heavy vehicle in which the payload is spread out along the length of the wheelbase. In addition, the design's application to high-speed vehicles, which are typically subjected to higher dynamic forces, should be more critically evaluated.

Mechanical advantages of the three-wheel layout result primarily from the elimination of one wheel. By eliminating the redundant fourth wheel the chassis becomes inherently less costly to manufacture. A three-wheel chassis is usually lighter and in many cases may be comparatively stronger. Torsional loads are reduced but still remain due to dynamic inputs and the location of the cg above the groundline. A chassis with only three wheels will have slightly less rolling resistance than a similar chassis with four wheels, due primarily to reduced bearing and brake drag. However, given equal coefficients of rolling resistance, drag eliminated by removing one wheel is essentially transferred through increased load to the remaining three wheels. Aerodynamic drag, however, can be significantly reduced by eliminating one wheel.

As a legal entity, three-wheel cars do not exist in the U.S. The Code of Federal Regulations currently classifies any vehicle with "not more than three wheels in contact with the ground" as a motorcycle. This, of course, releases any three-wheel vehicle from the need to comply with stringent passenger car safety regulations. An entrepreneurial company considering the market for a commuter/urban car in the U.S. could see this legal anomaly as a powerful argument in favor of the three-wheel configuration. However, the issue is neither settled nor is it limited to a simple release from safety regulations. It is true that crash testing can be successfully evaded. But to do so is not necessarily in the manufacturer's best interest. Ultimately, marketing and liability issues may backfire and create more problems than the current three-wheel classification resolves. Today's consumers are interested in vehicle safety and a manufacturer who appears to circumvent safety regulations may be judged guilty by default of building an unsafe product. Moreover, if a successful three-wheel design were to suddenly begin populating U.S. roadways, appropriate regulations would soon follow.

Mechanically Simple Design

The three-wheel configuration is blessed with a simplified chassis of fewer parts and reduced weight which translates into reduced manufacturing costs. The cost and weight savings are primarily derived from the simple elimination of one wheel. This feature results in a 25-percent reduction in the combined weight and cost of tires, wheels, brakes and suspension components when compared to a similar four-wheel design. A three-wheeler with the single wheel at the rear also makes it possible to eliminate the differential, both drive axles and some of the supportive structure. A single-front-wheel design simplifies front suspension and steering. Figure 6.1 compares the two three-wheel layouts with a conventional four-wheel layout.

Less obvious is the reduced need for torsional rigidity in the three-wheel chassis. With one wheel at each corner, the four-wheel vehicle is subjected to higher torsional loads due to uneven road surfaces and uneven bounce inputs at the four corners. Greater torsional loads require a more torsionally rigid chassis. This does not imply that a three-wheel chassis can dispense with torsional rigidity. Because mass is located some distance above the groundline, torsional loads are transferred to the chassis. Quincy-Lynn's Tri-Magnum required substantial cross-braces in order to adequately stiffen the frame.

Because a three-wheel chassis is mechanically simplified and lighter, less installed power will provide equal performance. Reduced power output places

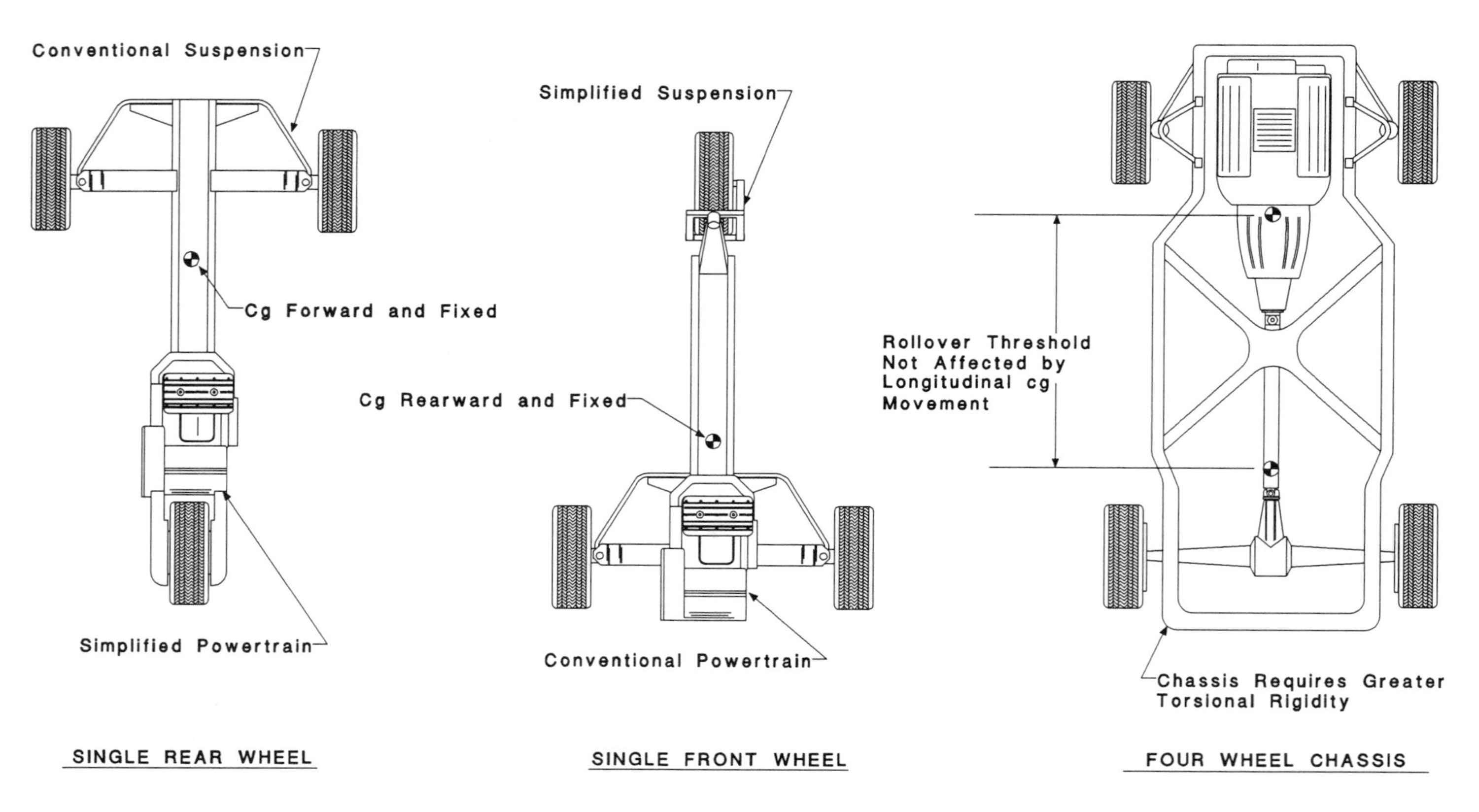

Fig. 6.1. Three-Wheeler and Four-Wheeler Layout Comparison.

less demand on the drivetrain which allows for lighter and less costly components. In a variety of ways, the simple act of eliminating one wheel has a cumulative effect throughout the design of the vehicle.

Sports Car Handling

Designing to the three-wheeler's inherent characteristics can produce a high-performance machine that out-corners many four-wheelers. A well-designed three-wheeler is likely to be one of the most responsive machines one will ever experience over a winding road. It is difficult to place a thrill value on the feeling of cornering response. In a study of three-wheeler stability for the U.S. Department of Energy (DOE), Dr. Paul Van Valkenburg documented very fast yaw response times that were far superior to four-wheel vehicles.

Yaw response time is the time it takes for a vehicle to reach steady-state cornering after a quick steering input. A softly sprung four-wheeler will have a yaw response time of about 0.30 sec. A four-wheel sports car will respond in about half that time. The most poorly designed three-wheelers tested by Van Valkenburg achieved steady-state cornering approximately 0.20 sec. after steering input. The best three-wheelers, however, reached steady-state cornering in as little as 0.10 sec., which is approximately 33 percent quicker than the best four-wheel car tested. Three-wheel cornering agility is a quality that must be experienced to fully appreciate.

The attributes responsible for the quick response to steering inputs have nothing to do with the number of wheels. Instead, they are by-products of the reduced mass and low polar moment of inertia that results when a three-wheel car is properly designed. The typical three-wheeler has approximately 30 percent less polar moment than a comparable four-wheel design. Most passenger cars are designed with a relatively high polar moment of inertia. Typically the engine is located over the front or rear axle and the fuel and luggage are located at the opposite end. The center of the vehicle is devoted primarily to the occupants.

A low polar moment of inertia results in a vehicle with more responsive handling, but it also tends to produce a more choppy ride. A vehicle with high polar mass is less nimble, but normally rides more smoothly. The relationship between polar mass and ride was discovered back in the '20s when it was found that cars rode better when the engine was placed far ahead over the front wheels. Sports cars tend to have a low polar moment of inertia for nimble handling, and they also tend to ride more roughly than passenger cars. Normally, a good balance between ride and handling can be achieved, regardless of the number of wheels

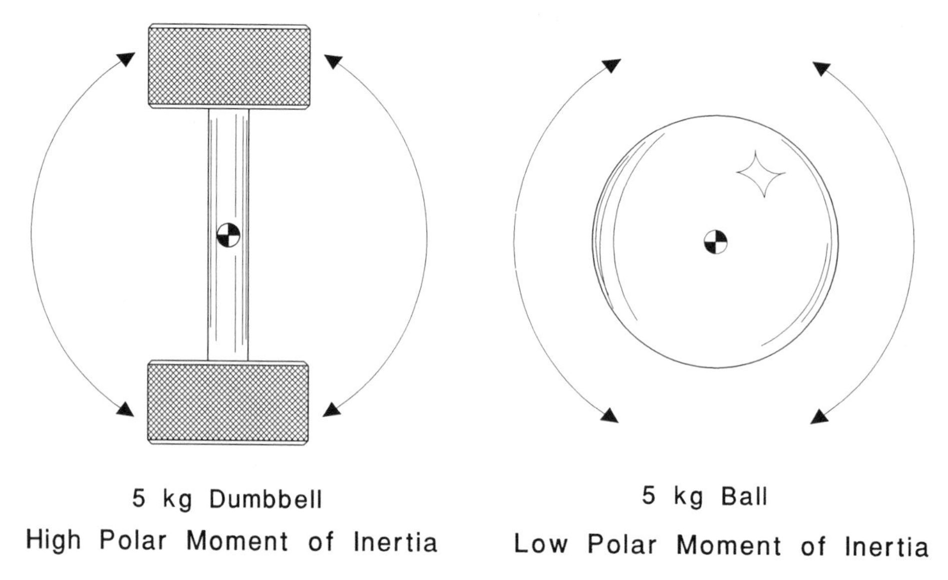

Fig. 6.2. Polar Moment of Inertia.

or the car's performance orientation. One does not have to choose either a rough ride or poor handling.

Rollover Threshold

A vehicle's rigid-body rollover threshold is established by the simple relationship between the height of the center of gravity and the maximum lateral forces capable of being transferred by the tires. Modern road tires have a friction coefficient on the order of 0.8, which means that the vehicle can negotiate turns that produce lateral forces equal to 80 percent of its own weight (0.8 g) before the tires lose adhesion.* The cg height in relation to the effective half-tread of the vehicle determines the length-to-height (L/H) ratio which establishes the lateral force required to overturn the vehicle. As long as the side-force capability of the tires is less than the side-force required for overturn, the vehicle will slide before it overturns. This analysis provides a useful figure for comparing the rollover threshold of various vehicles, as shown in Table 6.1.[1] Under dynamic conditions, a vehicle's rollover threshold is a more complicated issue.

* Suspension design has a significant effect on cornering forces. The discussion is necessarily simplified.

TABLE 6.1. RIGID-BODY ROLLOVER THRESHOLD COMPARISON

Vehicle Type	CG Height inch	(mm)[1]	Tread inch	(mm)[1]	Rollover Threshold (lateral g-load)[1]
Sports Car	18-20	(457-508)	50-60	(1270-1524)	1.2-1.7
Compact Car	20-23	(508-584)	50-60	(1270-1524)	1.1-1.5
Luxury Car	20-24	(508-610)	60-65	(1524-1651)	1.2-1.6
Pickup Truck	30-35	(762-889)	65-70	(1651-1778)	0.9-1.1
Passenger Van	30-40	(762-1016)	65-70	(1651-1778)	0.8-1.1
Medium Truck	45-55	(1143-1397)	65-75	(1651-1905)	0.6-0.8
Heavy Truck	60-85	(1524-2159)	70-72	(1778-1829)	0.4-0.6

[1] Metric equivalents inserted by the author.
Source: Ref. [1]

During dynamic conditions, overshooting the roll angle or tripping against an obstacle can cause the vehicle to overturn, even though it has a high rigid-body margin of safety against rollover. Rapid onset turns impart a roll acceleration to the body that can cause the body to overshoot the steady-state roll angle. This happens with sudden steering inputs, and it also occurs when a skidding vehicle suddenly regains traction and begins to turn again. A hard turn in one direction, followed by an equally hard turn in the opposite direction (slalom turns), also can cause the vehicle to overshoot the roll angle. Roll moment depends on the vertical displacement of the center of gravity above the vehicle's roll center. The degree of roll overshoot depends on the balance between the roll moment of inertia and the roll damping characteristics of the suspension. An automobile with 50-percent (of critical) damping has a rollover threshold that is nearly one-third greater than the same vehicle with zero damping.

Overshooting the steady-state roll angle can lift the inside wheel(s) off the ground. Once lift-off occurs, the vehicle's resistance to rollover rapidly diminishes, which results in a condition that quickly becomes irretrievable. The roll moment of inertia reaches much greater values during slalom turns wherein the forces of suspension rebound and the opposing turn combine to throw the body laterally through its roll limits from one extreme to the other. Inertia forces that result from overshooting the steady-state roll angle can easily exceed the forces produced by the turn rate itself.

Tripping is another cause of rollover in an otherwise rollover-resistant vehicle. Tripping occurs when a vehicle skids against an obstacle such as a curb. In this case, the lateral speed of the vehicle is suddenly arrested and high momentary

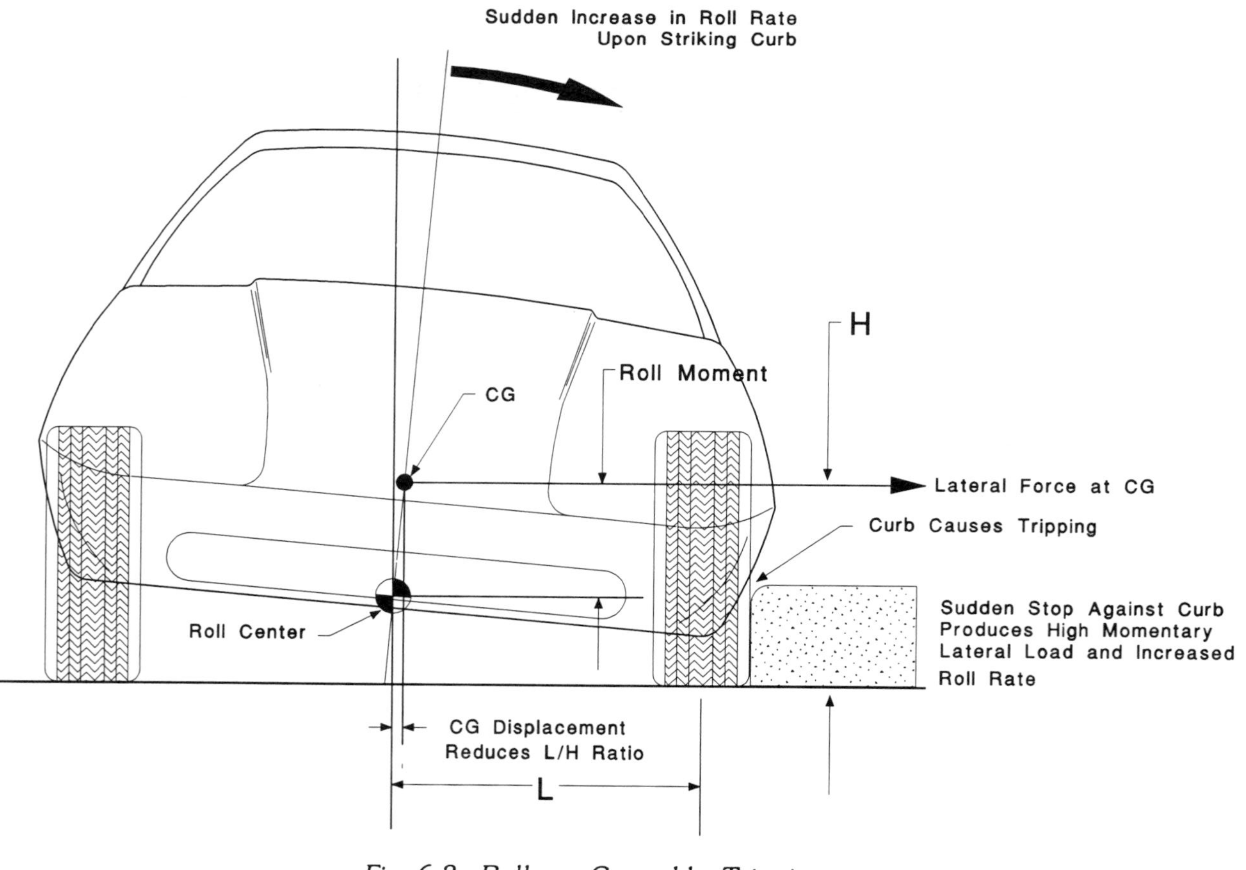

Fig. 6.3. Rollover Caused by Tripping.

loads are imposed across the vehicle's center of gravity (see Figure 6.3). If loads exceed the vehicle's rollover threshold, rollover will occur.

The conditions and their resulting forces are difficult to predict in random real-world events. Vehicles that appear to have an adequate margin of safety against rollover may nevertheless overturn under the right conditions. Consequently, the best design for rollover protection will include adequate roll damping and the greatest possible margin of safety against rollover. Roll damping in a three-wheeler can come only from the two side-by-side wheels at one end of the vehicle.

Single Front or Single Rear Wheel

The debate regarding the preferred choice between the single-front- or single-rear-wheel layout is primarily concerned with which configuration results in a more stable vehicle. Both configurations can be designed for a high rigid-body margin of safety against rollover. Under dynamic conditions, the two designs are no longer equal. With the single wheel in front, the vehicle's margin of safety against rollover is diminished during braking turns and increased during accelerating turns. If the single wheel is at the rear, its margin of safety against rollover is degraded during accelerating turns and enhanced during braking turns. Since braking forces can exceed acceleration forces, the single-rear-wheel configuration normally exhibits the best dynamic margin of safety against rollover.

The two configurations exhibit opposite oversteer/understeer characteristics as well. Vehicles with the single wheel in front inherently oversteer, and single-rear-wheel designs inherently understeer. Although chassis tuning can modify the layout's natural tendency, chassis tuning normally cannot make a single-front-wheel design understeer, or a single-rear-wheel design oversteer. Regardless of the inherent characteristics, neither configuration need result in a vehicle with handling characteristics that are different from those of a familiar understeering or oversteering four-wheel car. Oversteer with VW's Beetle was greater than with the single-front-wheel Trimuter. Tri-Magnum understeered less than the average, softly sprung American sedan. As for sheer performance, Tri-Magnum will undoubtedly out-corner most of the four-wheelers on the road today. (Photos are included in the Appendix.)

Single Front Wheel

The single-front-wheel layout is an inherently sleek design. The narrow front and wide rear of the body blends well with the natural shape of two occupants sitting side by side with legs extending forward. It is naturally wide across the occupants' shoulders where extra room is needed, and more narrow near the occupant's feet where width is not as important. This basic shape tends to be attractive and aerodynamically clean. Vehicle dynamics, however, are not as tidy.

Basic handling characteristics and the margin of safety against rollover result primarily from the relationship between the center of gravity location, the effective half-tread, and the wheelbase of the vehicle. The roll forces of a three-wheeler are taken exclusively by the two side-by-side wheels. Moving the cg toward the side-by-side wheels increases the effective half tread, which causes the vehicle to behave as if the side-by-side wheels had been moved further apart. The relationship of the effective half-tread to the cg height and the friction coefficient of the tires determine whether the vehicle will slide or overturn at high lateral g loads. With a three-wheel platform, a three-dimensional cone, calculated according to maximum lateral forces, models the effect of acceleration, braking and turning forces on the triangular wheel layout. As long as the base cone projects to the ground-plane inside the effective half-tread, the vehicle will slide before it overturns. If the base cone projects outside the effective half-tread, or moves outside because of dynamic forces, the vehicle will overturn before it slides. The margin of safety against rollover can be increased by increasing the effective half-tread, and by reducing the height of the cg. The effective half-tread increases when the two side-by-side wheels are moved outboard (increasing the actual tread), to a lesser degree when the wheelbase is increased, and when the cg is moved closer to the side-by-side wheels.

With the single-front-wheel design, the characteristics that produce a high resistance to rollover also create a rear-heavy vehicle, which causes oversteer. Oversteer can be moderated by placing larger tires at the rear and by appropriate suspension design, but the basic tendency cannot be eliminated. An oversteering vehicle is normally considered more difficult for the average driver to control. Additional disadvantages of the single-front-wheel configuration include an increased sensitivity to crosswinds, aerodynamic lift at the light (front) end of the vehicle, and an unfavorable shift in the cg forward and to the outside during braking turns.

The single-front-wheel design tends to yaw more easily when subjected to crosswinds. This is because the aerodynamic center of pressure is normally ahead of the center of gravity. Also, at high speeds, aerodynamic lift can cause

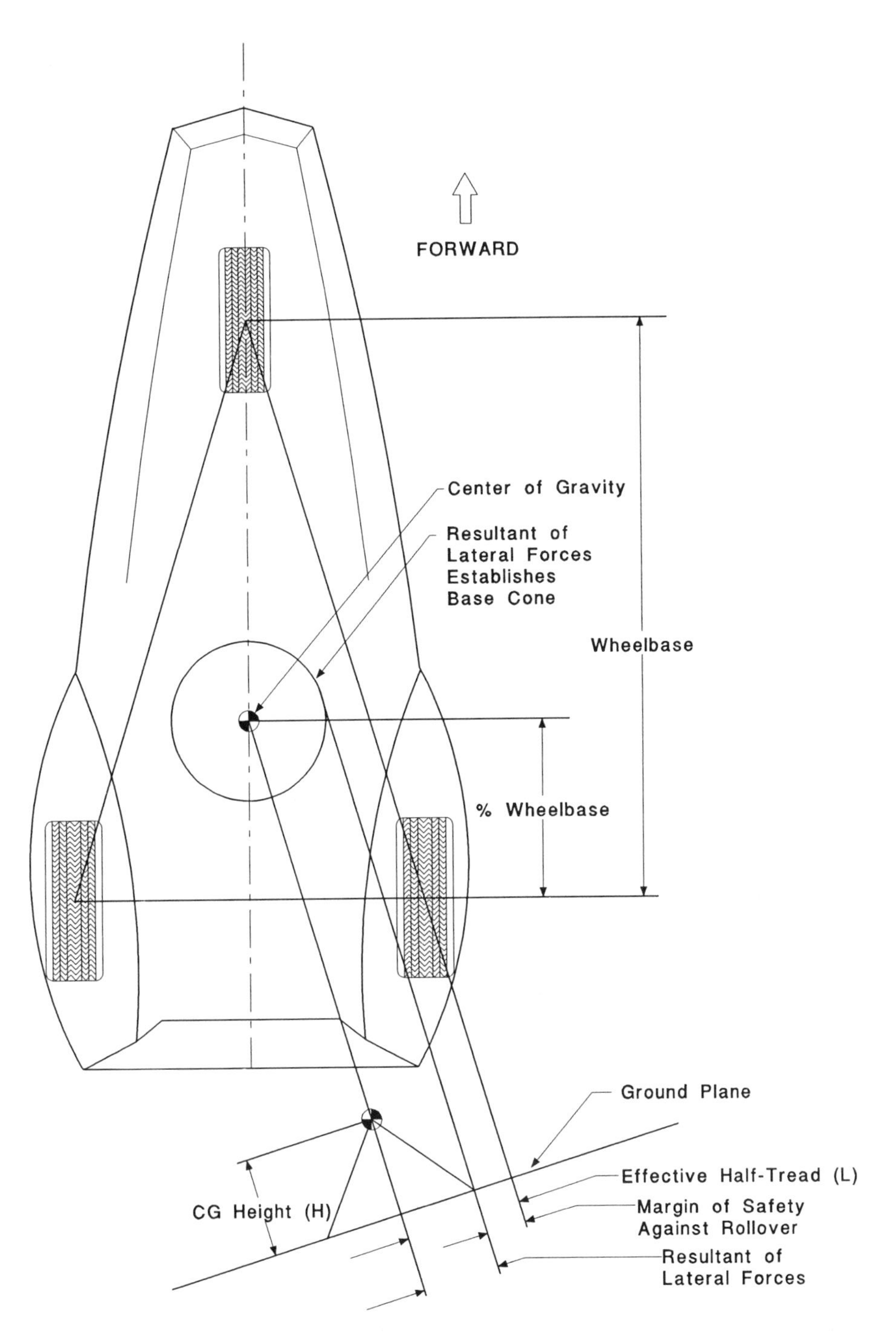

Fig. 6.4. Single-Front-Wheel Vehicle Rollover Margin of Safety.

the already light front of the vehicle to become dangerously light. Trimuter began to feel nose-light at approximately 130 km/h (80 mph). In spite of its limitations, however, the single-front-wheel configuration is still a viable layout for a relatively low-speed urban car. The design can be extremely stable and aerodynamic lift need not create a problem if maximum speed is limited and the body is correctly designed.

Single Rear Wheel

When the single wheel is located at the rear and the cg is moved forward toward the side-by-side front wheels, handling characteristics improve. The resulting front-heavy vehicle tends to understeer, just like the four-wheel cars with which consumers are familiar. Crosswind characteristics also improve when the center of gravity is located nearer the center of aerodynamic pressure (usually about 25-30 percent of body length). The relationships that determine the margin of safety against rollover are identical to those of the single-front-wheel layout, except they are reversed.

Braking turns no longer present a problem to the three-wheeler with the side-by-side wheels at the front. The forward resultant caused by a braking turn is toward the region of greater tread which tends to offset the negative effect of forces acting toward the outside of the turn. The net effect is that resistance to rollover is not necessarily degraded. Conversely, an accelerating turn produces an unfavorable resultant toward the outside rear. However, acceleration produces significantly lower forces than braking. Consequently, the magnitude of the unfavorable resultant is less with the single-rear-wheel configuration (Figure 6.5).

The disadvantages of the single-rear-wheel configuration include the difficulty of entering or leaving the vehicle over or behind the outboard wheel, reduced interior space resulting from the steering angle of the front wheels, and the limited ability to transfer power through a single powered rear wheel. In an effort to appropriately locate the cg, occupants are normally placed between the two side-by-side wheels. This location has the advantage of offering side intrusion protection. However, convenience of ingress and egress is sacrificed, along with substantial interior room which must be dedicated to the wheel house. The engine is often located behind the occupants where it powers the single rear wheel. This configuration was successfully used in a number of designs, including Ron Will's TurboPhantom, Walter Korff's Duo-Delta, and Quincy-Lynn's Tri-Magnum. Both the TurboPhantom and the Tri-Magnum are high-performance sports cars capable of speeds in excess of 160 km/h (100 mph).

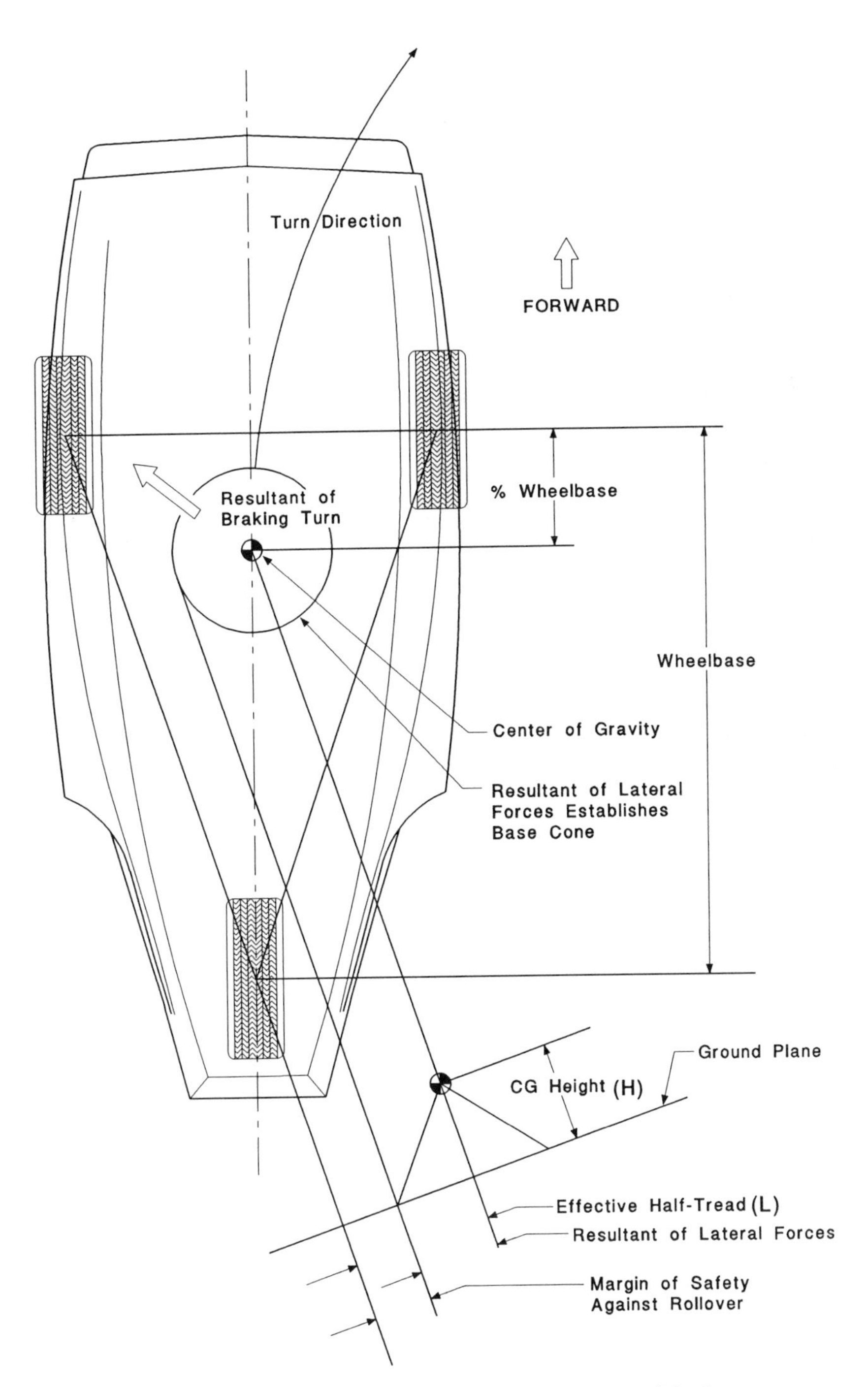

Fig. 6.5. Single-Rear-Wheel Rollover Margin of Safety.

A three-wheel design incorporating front-wheel drive and a transverse-mounted engine ahead of the wheels offers the ability to transfer more power to the road than is possible with a single, powered rear wheel. The drivetrain also creates a natural weight bias toward the two side-by-side wheels and thereby allows the passengers to move aft for easier ingress and egress behind the front wheels. The tradeoff with this layout is a drivetrain of increased complexity in comparison to the powered single-rear-wheel design. The VW Scooter and the Trihawk are two excellent examples of front-wheel-drive three-wheel cars.

Tandem or Side-by-Side Seating

With side-by-side seating, frontal area is not appreciably affected by reducing vehicle capacity from four to two occupants. However, placing one occupant behind the other in a tandem arrangement reminiscent of the German Messerschmitt can reduce frontal area by as much as 40 percent. The effect on aerodynamic drag can be significant. At 80 km/h (50 mph), aerodynamic drag of a four-wheeler typically accounts for approximately half of the total road load. In the case of a low-rolling-resistance three-wheel configuration, aerodynamic drag could represent an even greater portion of the vehicle's total road load. Using Trihawk's relationship between aerodynamic drag and rolling resistance, a 40-percent reduction in frontal area would theoretically translate into a 28-percent reduction in road load at 80 km/h (50 mph).

The tandem seating arrangement does have drawbacks. As a general rule, tandem seating tends to extend the vehicle's wheelbase and generally results in a greater cg displacement with different loads (one or two occupants). A tandem seater also tends to reduce the social feeling, which is enhanced by side-by-side seating.

Limitations of the Three-Wheel Configuration

The three-wheel configuration has a number of unique limitations. In a dynamic environment, a side impact or tripping against a curb can more easily result in a rollover with a three-wheel vehicle. The three-wheel platform does not work well with high profile vehicles, or vehicles in which cg location might be displaced by placing cargo at one end of the platform. There are conditions in which it is not desirable to rely on a single wheel for all the traction at one end of the vehicle. On snow-covered roads, the wheel in the center will normally have to traverse relatively undisturbed snow. Additionally, adapting a three-wheeler to an infrastructure built for four-wheelers presents challenges of its own.

Center-of-gravity location has a significant effect on the stability of a three-wheeler. The designer must therefore be able to rely on a stable and predictable cg location in order to design the vehicle for an adequate margin of safety against rollover. A four-wheel vehicle is not nearly as sensitive to longitudinal cg displacement. Consequently, if large displacements in payload are expected, such as with the four-passenger sedan or a vehicle that must carry a heavy cargo at one end of the platform, the four-wheel layout is the preferred design. In addition, the option of moving the cg location closer to the side-by-side wheels is inherently limited. Hardware and occupants cannot all occupy the same small space at one end of the vehicle. Load must be distributed along the wheelbase, if for no other reason than to provide room for mechanical components and occupants. Also, the single-wheel end of the vehicle cannot be too lightly loaded or the vehicle will become unstable due to inadequate adhesion.

Maneuvering at high speed over rough surfaces tends to destabilize a three-wheeler to a greater degree than a four-wheeler. A four-wheeler relies on the adhesion of two wheels at each end of the vehicle to maintain directional stability. Irregularities in the road can cause momentary losses of adhesion at the wheel encountering the disturbance. A critical loss of adhesion is likely to occur more often at the single-wheel end of a three-wheel vehicle.

Limitations on suspension tuning are another important consideration. One of the most effective tools for tuning a finished chassis is the ability to adjust the ratio of roll stiffness between the front and rear suspension. But roll stiffness in a three-wheel layout can come only from the end with the side-by-side wheels. This limits the options for tuning the suspension primarily to that of varying the size and inflation pressure between front and rear tires. A large difference in tire size would introduce the problem of selecting a spare tire that will work at either end of the vehicle.

Finally, blending with the existing infrastructure can be one of the most perplexing problems of any new product. If new solutions do not blend well, better technology often must give way to more marginal designs simply because the lesser technology is more amenable to the infrastructure. Automobile manufacturing, transportation, and service facilities are designed for four-wheel vehicles. As a result, trailers, hoists and pits, alignment racks, and inspection stations normally will not accommodate a vehicle with a wheel located in the center. Although fixtures for adapting to a single, centrally located wheel might easily be designed, a vehicle with a wheel at each corner would need no such adaptation.

Experimental Three-Wheelers

In Europe and Japan, three-wheelers have been relatively commonplace for most of the twentieth century. Today, however, their numbers are declining because of changes in taxation that eliminated their advantage of lower fees and taxes. Unfortunately, many three-wheelers have not been especially well designed. Even among production vehicles, an inadequate margin of safety against rollover has been a common flaw. The Bon Bug, which was manufactured for years in the UK and enjoyed success in the marketplace, would lift an inside wheel during hard turns. When Paul Van Valkenburg performed stability tests on three-wheelers in 1980, some of the vehicles had to be equipped with outriggers to keep them from overturning during cornering tests. A number of well-designed but less-known experimental three-wheelers have actually performed better.

GM's 511 Commuter

The 511 commuter is an experimental three-wheel car built by General Motors in the late '60s to test ideas for a special-purpose commuter. The design was a radical departure from the machines that were typically emerging from the advanced vehicle labs of Detroit. Little of the design was borrowed from the past.

Fig. 6.6. Early experimental commuter cars at GM. The three-wheel 511 commuter is shown in the foreground. (Courtesy: General Motors Corp.)

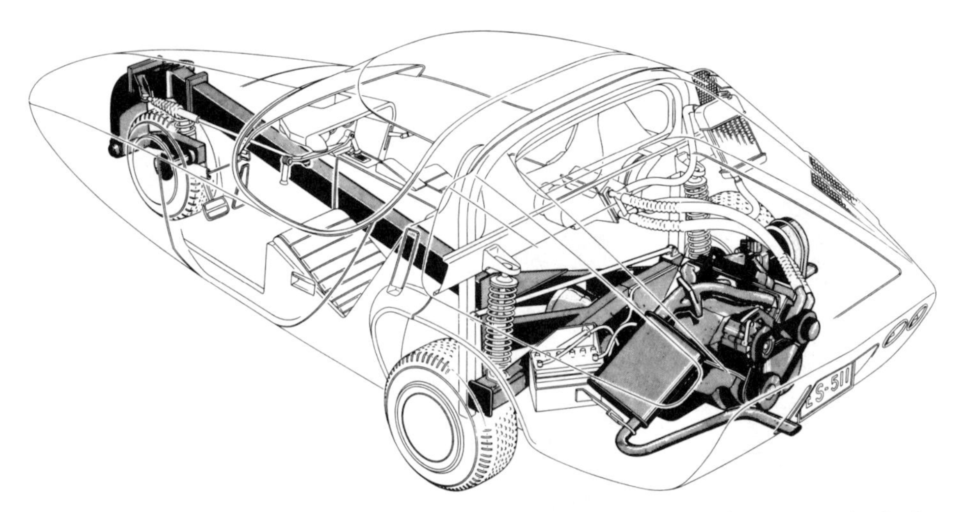

Fig. 6.7. The 511 Commuter featured a unique backbone frame in which the frame and drivetrain were unsprung weight. A large clamshell canopy and a direct-link handlebar steering system are other unusual design features. (Courtesy: General Motors Corp.)

The car's backbone frame was primarily unsprung and resulted in a relatively rough ride (Figure 6.7).

Direct-link handlebar steering was another innovative feature that actually worked quite well. A center-pivoting steering yoke was connected directly to the front suspension with a single link. A differential in the arms resulted in a 2.1:1 reduction between the yoke and the front wheel. The car was tested by engineers and race drivers who all adapted quickly to the steering system. Drivers reported no tendency to over-control.

The Lean Machine

General Motor's Lean Machine is a good example of what happens when a group of engineers are given free rein to develop an idea. The version shown in Figure 6.8 was the last in a series of 15 prototypes developed over a seven-year period. Lean Machine borrows much from the world of the motorcycle. A bubble-shaped driver's pod leans into turns while the narrow-tracked power pod follows along in standard, flat-cornering fashion. Lean angle is controlled by floor-mounted pedals. Direct-link handlebars steer the single front wheel. With practice the driver can balance the control pressures for effortless, gliding turns that emulate the leaning, zero-g turns of a motorcycle.

Fig. 6.8. GM's Lean Machine leans into turns for high-speed cornering. (Courtesy: Road and Track *magazine)*

Fig. 6.9. A narrow 711-mm (28-in) rear track would normally lead to poor cornering capabilities. With its leaning passenger pod, Lean Machine could negotiate 1.2-g turns. (Courtesy: Road and Track *magazine)*

A drag coefficient of 0.35 and a curb weight of 159 kg (350 lb) contributed to the Lean Machine's ability to deliver high performance on a reduced power budget. The 11 kW (15 hp) Honda ATV engine and 5-speed transmission located in the power pod pushed Lean Machine to 129 km/h (80 mph) and delivered fuel economy of 50 km/L (120 mpg) at a steady 64 km/h (40 mph). A 50-degree lean produced a stable 1.2-g turn on the a rear tread of only 711 mm (28 in).

The goal was to combine the economy of a motorcycle with the lean-into-the-corner thrill in a vehicle that resembled a full-bodied automobile. Officially, GM never had plans to produce the Lean Machine. However, facilitators who were present during Lean Machine product clinics conducted in the '80s report that consumers were exceptionally enthusiastic about the design. Market potential appeared to be about twice the level that had been expected.

Fig. 6.10. Project Engineer Jerry Williams lowers the canopy for a test drive. (Courtesy: Road and Track *magazine)*

TurboPhantom

Fig. 6.11. The dry lake bed at Lake El Mirage, California, forms a perfect backdrop for one of the most exotic three-wheelers ever built—the TurboPhantom. (Courtesy: Ronald J. Will)

Fig. 6.12. The steering wheel lifts up with the canopy to provide wide open access to the Phantom's interior. (Courtesy: Ronald J. Will)

TurboPhantom is one of the vehicles originally tested by Van Valkenburg when he studied three-wheeler stability. It is not one of the vehicles that needed an outrigger to stay upright in turns. TurboPhantom is a high-performance sports car that will maintain a 0.82-g corner with only two degrees of body lean. The vehicle's 60-percent margin of safety against rollover gives it greater resistance to rollover than most four-wheel vehicles.

TurboPhantom was designed around a Honda GL-1000 motorcycle. An 82 kW (110 hp) turbocharged version would reach a top speed of 209 km/h (130 mph). Curb weight was 680 kg (1500 lb), which put the TurboPhantom in the weight category of the early VW Beetle. Ronald J. Will, the car's designer, actually put the vehicle into production and marketed it as a "three-wheel luxury sports car." In 1981 the price started at $20,000. Today, Ron works for Subaru in New Jersey.

Fig. 6.13. TurboPhantom cornering at 0.82 g during handling tests at Edwards Air Force Base for Van Valkenburg's study on three-wheeler stability. (Courtesy: Ronald J. Will)

VW Scooter

Fig. 6.14. The VW Scooter, a state-of-the-art three-wheel prototype built by Volkswagen. Scooter features front-wheel drive, removable gull-wing doors, and safety features that meet European and North American safety standards. (Courtesy: Volkswagen AG)

VW's Scooter is a mid-'80s prototype built by VW Research and Development Division in Germany. One of the most impressive features of the Scooter is that the vehicle is said to meet European and North American safety standards. The vehicle is designed to pass the 50 km/h frontal collision test, which is the European equivalent of the U.S. 30 mph barrier test.

With the engine located ahead of the front drive axles, Scooter has exceptionally good handling characteristics. The 545 kg (1200 lb) three-wheeler can reach a top speed of 120 km/h (75 mph) and achieve fuel economy of 25 km/L (60 mpg) at 88 km/h (55 mph). Urban cycle fuel economy is on the order of 15 km/L (35 mpg). It is maneuverable in traffic, easy to park and has excellent cornering characteristics. Interior room across the shoulders is about the same as a Japanese Kei car. Gull-wing doors and the rear window are detachable to give Scooter a sporty, open-air feeling. Aerodynamic drag with the car closed up is 0.25. The concept behind the vehicle was to mix the flavor of the motorcycle with the safety and convenience of the automobile.

Fig. 6.15. VW Scooter undergoes skid tests in the rain. (Courtesy: Volkswagen AG)

Trihawk

Fig. 6.16. Trihawk three-wheel sports car is a modern-day, front-wheel-drive sports car that will reach 160 km/h (100 mph). (Courtesy: Harley-Davidson)

Fig. 6.17. Wide 1676 mm (66 in) stance and low center of gravity give Trihawk 0.87-g cornering capability before predictably understeering to the outside. (Courtesy: Harley-Davidson)

Fig. 6.18. Trihawk's business-like interior is reminiscent of early European sports cars. (Courtesy: Harley Davidson)

Trihawk is an excellent example of the potential of a front-wheel-drive three-wheel vehicle. The vehicle was designed in the early '80s by Bob McKee of CanAm and Indy fame. Styling was done by David Stollery, a former GM stylist who also established Calty, Toyota's U.S. design studio. The open-bodied roadster has a curb weight of 613 kg (1350 lb) and can accelerate from 0-96 km/h (0-60 mph) in slightly over 10 seconds. Maximum speed is nearly 160 km/h (100 mph). At 96 km/h (60 mph) Trihawk can be brought to a standstill in 45 m (148 ft), and in about 76 m (250 ft) at 129 km/h (80 mph).

Trihawk is noted for its exceptionally high turn rate. The center of gravity is just 305 mm (12 in) above the ground. *Road and Track* magazine recorded 0.83-g turns, *Car and Driver* recorded 0.87-g turns, and in independent tests Trihawk is said to have developed turn rates as high as 0.91 g before predictably understeering to the outside. For a while in the early '80s, Trihawk was manufactured under the direction of David Stollery at facilities in Dana Point, California. Price was in the $14,000 range. Ultimately, the project was purchased by Harley-Davidson and then subsequently shelved. According to officials at Harley, the potential for liability problems was the deciding factor.

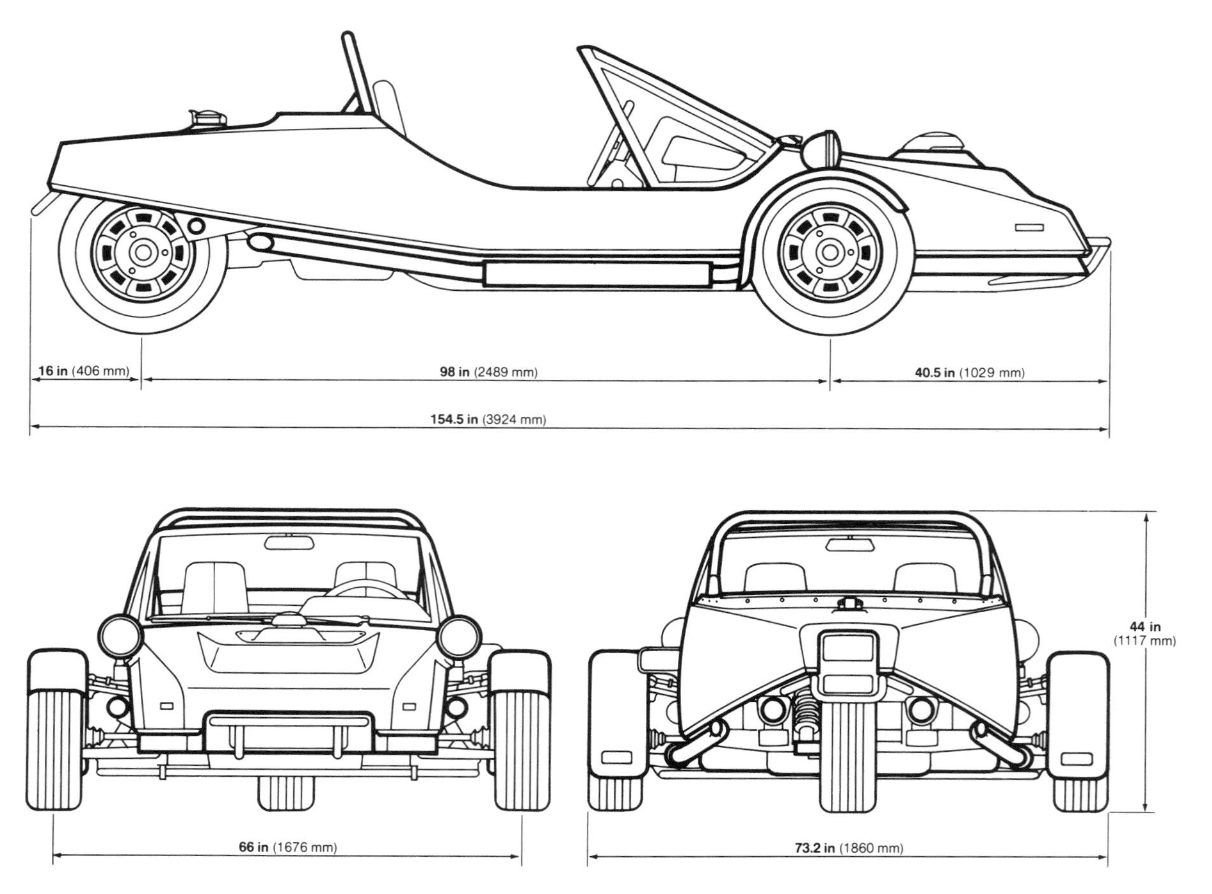

Fig. 6.19. Trihawk Specifications. (Courtesy: McKee Engineering)

Trihawk Design

Front-positioned driving wheels with 66" wide track

Low center of gravity only 12 inches off the ground

70% of total weight directly over front anti-roll system

Horizontally opposed Citröen four-cylinder single overhead cam, air-cooled 1300cc engine

Five-speed transaxle

Unequal length double A-arm independent front suspension with stabilizer bar and concentric coil springs over shocks

Rack and pinion steering

Front disc brakes

Trailing arm rear suspension with drum brake

5-1/2" x 13" cast alloy wheels with high-performance radial tires

Tubular steel perimeter frame with chrome moly roll bar

Full instrumentation with gauges

Racing-type fuel cell and filler cap

Fully adjustable contoured seats with lateral support

5.5 cubic foot enclosed luggage area and locking glove box

Superior body aerodynamics with a 0.41 coefficient of drag

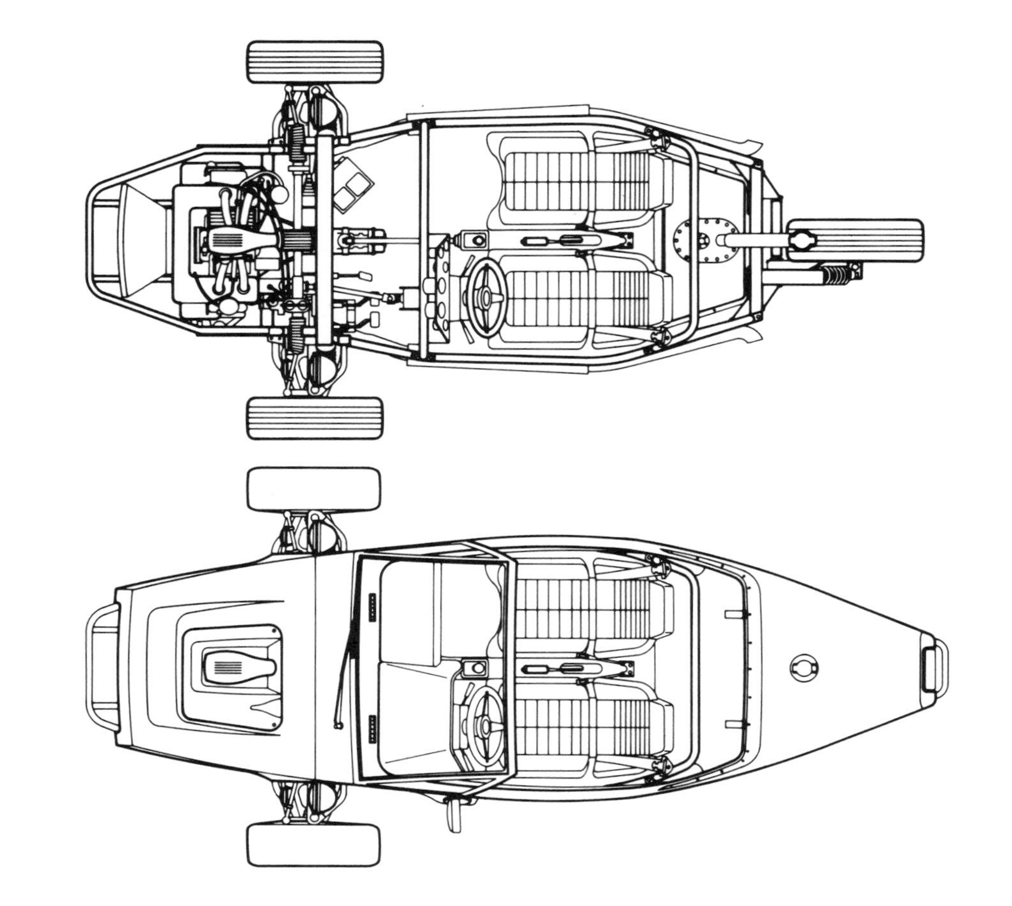

Fig. 6.19. Trihawk Specifications (continued). (Courtesy: McKee Engineering)

References

1. Thomas D. Gillespie, Fundamentals of Vehicle Dynamics, SAE R-114, Society of Automotive Engineers, Inc., Warrendale, PA, 1992.

Chapter Seven

Safety and Low-Mass Vehicles

Courtesy: TRW Safety Systems, Inc.

Fooling around with alternating current is just a waste of time. Nobody will use it, ever. It's too dangerous—it could kill a man as quick as a bolt of lightning. Direct current is safe.

Thomas Edison

The prospect of introducing unconventional low-mass cars into traffic with conventional cars brings up significant concerns about small-car safety. In a two-car crash the vehicle with the lowest mass receives the greatest blow, and matters get worse as mass differential increases. Smaller cars also overturn more often as a result of an accident. In addition, the implications of low-mass vehicles go beyond the issues of vehicle crashworthiness alone. Concerns about product liability exist whenever unconventional vehicle designs are considered. Engineering and legal precedents have been established with conventional cars. No such precedent exists with unconventional vehicle types. As a result, personal injury involving a lightweight three-wheel car, for example, could offer rich new opportunities for exploitation both in and out of the courtroom. Motor Vehicle Safety Standards that were developed for high-mass cars may not be appropriate for low-mass vehicles, especially with vehicle types that operate primarily in the urban environment. New vehicle types may require different, and in some regards, more stringent standards. New standards specifically designed for low-mass vehicles could provide engineering guidelines and help define accepted practices. And finally, designs for increasing low-mass vehicle safety must ultimately rely more on crash avoidance countermeasures. Given equal technology, a low-mass car will likely remain the less crashworthy design. Crash avoidance technology will tend to put all cars on more equal footing.

Vehicle Safety in Perspective

Automobile safety is an issue that is highly charged with emotion and complicated by the random and variable nature of the crash event. Crash tests measure the physical capabilities of the automobile structure to withstand a predefined event and protect the occupants inside. Accident statistics introduce the effects of the operating environment in combination with human behavior patterns. A number of factors associated with operating environment and human behavior determine the probability of an accident taking place, as well as the most likely magnitude and direction of forces, should an event occur. Once an event does occur, vehicle crashworthiness and crash survival systems determine the likelihood and severity of injury. Change at any level can affect the overall safety of the transportation system.

Accident statistics are often used to extrapolate the effects of operating environment and human behavior, as well as to evaluate the effects on safety of particular vehicle design attributes. However, evaluating statistical relationships for specific meanings is a difficult and demanding task. A simple review of data often leads to inconclusive results. The true nature of the risk of driving in traffic with the varied mix of drivers, potential encounters, and vehicle designs is clouded by many interwoven factors that are difficult to separate from the whole. Concurrent events are often causally unrelated, or related in ways that are not obvious. Accurate evaluations of risk and the factors that influence risk can easily become obscured by voluminous and often contradictory data.

Automobile safety is normally measured according to fatalities and injuries per distance traveled. In the U.S., transportation safety statistics are compiled as injuries and fatalities per 100 million "vehicle miles traveled" (vmt). Figures based on vmt (1 mi = 1.609 km) provide a measure of how safely the vehicle moves. It can be converted into "passenger miles traveled" (pmt) by multiplying the average vehicle occupancy rate (1.6 persons in the U.S.) by vmt. Regardless of how it is measured, all forms of transportation are becoming safer. Statistically, driving today is approximately 7.5 times safer than in 1930 and 2.25 times safer than in 1970. Cars built today are four times safer than vehicles built in 1969, and they are approximately 10-percent smaller and 20-percent lighter.*

* The apparent discrepancy between the safety level cited here and the one in the previous sentence has to do with the mix of older vehicles reflected in current accident statistics, as opposed to the independent safety level of new vehicles.

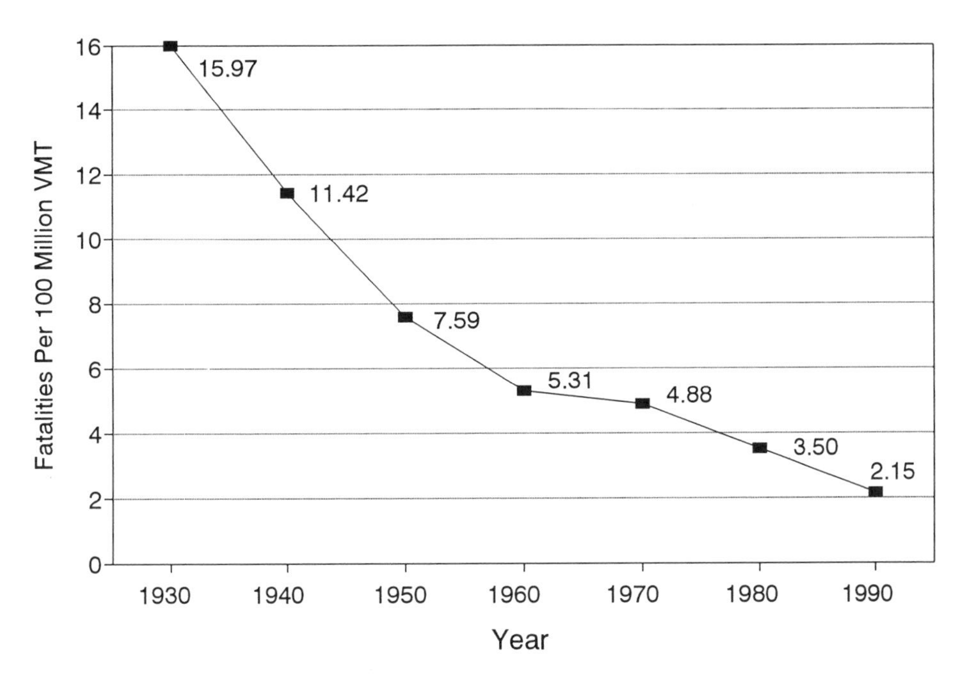

Fig. 7.1. U.S. Traffic Fatalities 1930-1990.

Figure 7.2 shows the risk relationship of a few everyday activities in the U.S., measured on a fatality-per-unit basis. Accident statistics reveal that bicycling is three times safer than walking. What is not obvious is that most pedestrian and bicycle fatalities are in fact automobile accident fatalities and have little to do with the intrinsic safety of walking or riding a bicycle. If that fact were not taken into account, any attempt to improve pedestrian and pedalcycle safety would be handicapped. Most causal relationships in accident statistics are significantly more obscure. The relationship between cause and effect may lie deep within statistical and vehicle crash performance data. Once accident data is deciphered, the next step would be to define an appropriate and realistic relationship between system costs, consumer convenience, and absolute system safety.

Several years ago I met a fellow who had purchased a home-built gyrocopter that wouldn't fly. The whole idea of such devices is to be able to soar with the eagles on a flea market budget. For that benefit, one is obliged to accept a certain level of risk, which this prospective pilot was all too eager to do. He had tried several times, but the machine simply would not leave the ground. Over the period of perhaps an hour, he recounted a string of stories about ill-fated attempts to fly that all ended with the gyrocopter on its side in the nearest ditch. For me, it

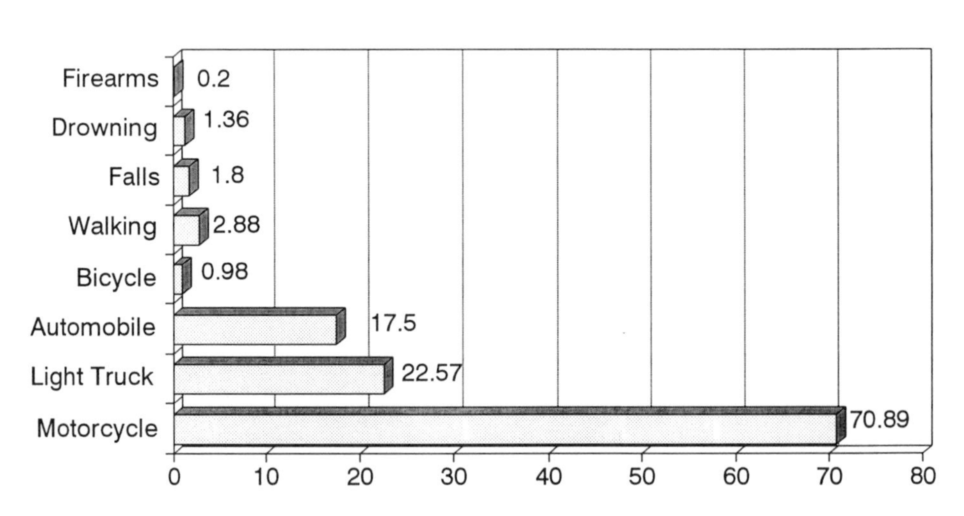

Fig. 7.2. Fatalities of Select Activities.

quickly became clear that here was a man who surely leaped before he looked. As it turned out, he had installed the rotor upside-down and overtightened the teeter-pin so the rotor could not pivot (the blade assembly could not flap). This guaranteed that the gyrocopter would remain forever earthbound. I remember thinking that his was probably the safest gyrocopter around and perhaps he ought to leave the rotor as it was and keep his safety record intact.

The problems experienced by the gyrocopter pilot resulted first from a desire to fly, and second from a decision to circumvent the system and fly on a shoestring. Appropriate training and reliable equipment would have been expensive and he was unwilling to pay the price or to give up the idea of flying. In reality he is not too different from the rest of us. An appropriate balance between cost and convenience, and the freedom to pursue activities, are relevant to all our decisions, including those involving personal safety. There is always a level of risk associated with activities that may otherwise add to the quality of life. The fellow with the gyrocopter simply functioned nearer the outer boundaries of the envelope on all counts.

Improved technology has already significantly reduced automobile accident injury and fatality rates. With advanced survival systems and crash avoidance technologies, it may soon be possible to eliminate the majority of occupant fatalities resulting from automobile accidents, regardless of the size of the vehicle. Because of the differences in operating environments, occupant protection in low-mass urban cars may require different priorities and techniques than might be appropriate for larger highway cars. Undoubtedly, the per-passenger cost of safety equipment is likely to grow as vehicle size shrinks. Low-mass cars will require a greater emphasis on safety engineering, as well as on technologies that can prevent vehicle crashes in the first place. As a result, a greater portion of first-costs will be allocated to safety features in significantly smaller vehicles.

Given equal technology, the smaller car will probably continue to be the less crashworthy vehicle. However, it is not necessary that low-mass cars actually equal the crashworthiness of larger cars in order to provide equally safe transportation. Operating environment has a large effect on the magnitude of forces that are typically experienced in an involvement. Vehicles that operate primarily in the urban environment are typically subjected to far less force when a collision occurs.

Accident Statistics and Automobile Size

In the U.S., as well as in most of the world (except for Japan), automobile fatalities are inversely related to car size. That is, the smaller the vehicle, the greater the occupant fatality rate. Figure 7.3 shows a comparison of size versus occupant fatalities, provided by The Insurance Institute for Highway Safety.[1]

In the early '80s, A.C. Malliaris reviewed the statistical association of fatalities and car size and developed Table 7.1 to correlate risk increase in relation to reduced wheelbase and reduced vehicle mass.[2]

TABLE 7.1. OCCUPANT FATALITY RISK VARIATION BY IMPACT TYPE AS A FUNCTION OF VEHICLE WEIGHT AND WHEELBASE

Impact Type	Percent Risk Increase Per 500 lb (227 kg) Reduction in Curb Weight	Per 5 in (127 mm) Reduction in Wheelbase
Frontal	16.6 ± 1.1	14.8 ± 0.9
Side	17.4 ± 1.1	14.9 ± 0.9
Rear	24.6 ± 2.0	21.8 ± 1.7
Other	32.3 ± 2.1	30.1 ± 1.6
All Types	18.6 ± 1.0	16.4 ± 0.8
	Per 500 lb (227 kg) Reduction in Curb Weight at Fixed Wheelbase	**Per 5 in (127 mm) Reduction in Wheelbase at Fixed Weight**
Frontal	-0.4 ± 3.2	15.1 ± 2.7
Side	5.3 ± 3.4	10.7 ± 2.8
Rear	1.6 ± 6.1	20.5 ± 5.2
Other	-14.6 ± 5.9	41.8 ± 5.0
All Types	0.9 ± 3.0	15.7 ± 2.5

Source: Ref. [2]

Fig. 7.3 and Table 7.1 are based on simple associations between vehicle size and weight, and fatality risk. Focusing on different statistical associations reveals a more comprehensive, but perhaps more inconclusive picture.

Accident Statistics and Other Variables

Automobile safety is statistically associated with a variety of conditions that are independent of vehicle type and design. For example, if data is controlled for driver age, large cars are often over-represented in accidents.[3] It turns out that drivers under 20 years of age are "at fault" in fatal crashes at twice the rate per

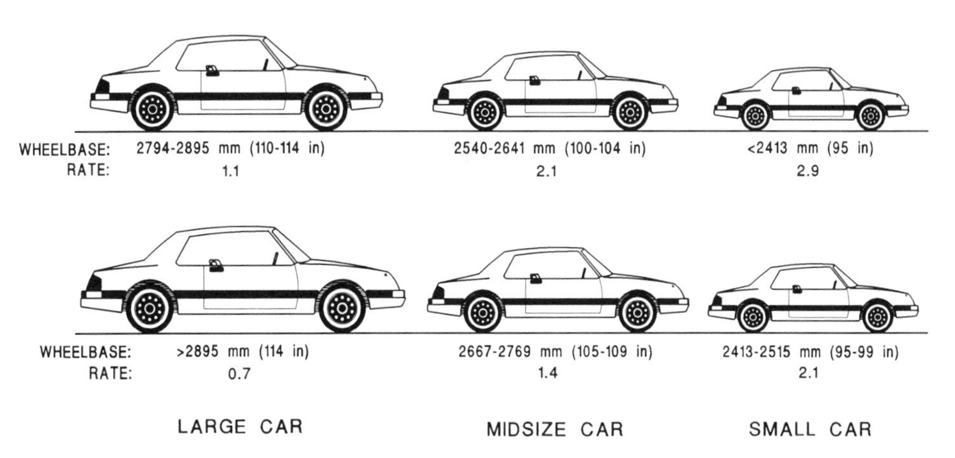

Fig. 7.3. Car Size Effect on Traffic Fatalities.

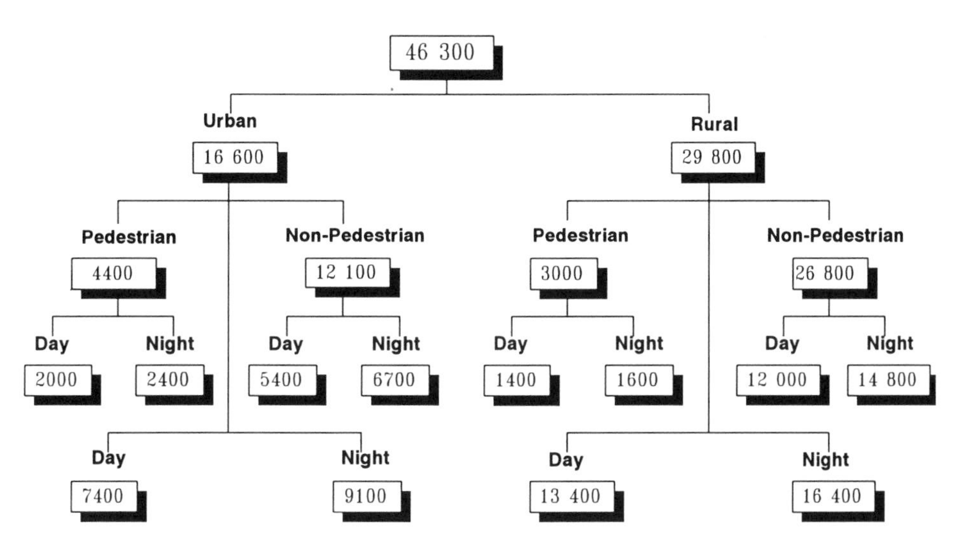

Fig. 7.4. Total U.S. 1990 Traffic Fatalities.

license holder of drivers aged 20-34, and five times the rate of drivers aged 35-64. These more youthful drivers are also over-represented as the drivers of the smallest cars. Moreover, the percentage of teenage drivers increases in proportion to the decrease in car size.[4]

Drinking drivers are another major contributing factor in automobile accidents and fatalities. Alcohol is involved in half of all traffic fatalities and younger drivers are over-represented in this group as well (Figure 7.5). Drivers under age 25 account for only 20 percent of total drivers but are involved in 34 percent of alcohol-related traffic fatalities. Many alcohol-related and driver-age-related fatalities may be incorrectly attributed to vehicle size.

Evidence suggests that drivers may adjust driving habits according to vehicle size. Wasielewski researched driver aggressiveness in 1983 and discovered that drivers of small cars take less risks in everyday driving.[5] Driver risk-taking was measured by close following in heavy traffic, greater vehicle speeds on a two-lane road, and the number of drivers who did not wear seat belts. Wasielewski found that driver risk-taking increased in proportion to vehicle mass, with the drivers of the largest cars driving in the most aggressive manner. This appears to be

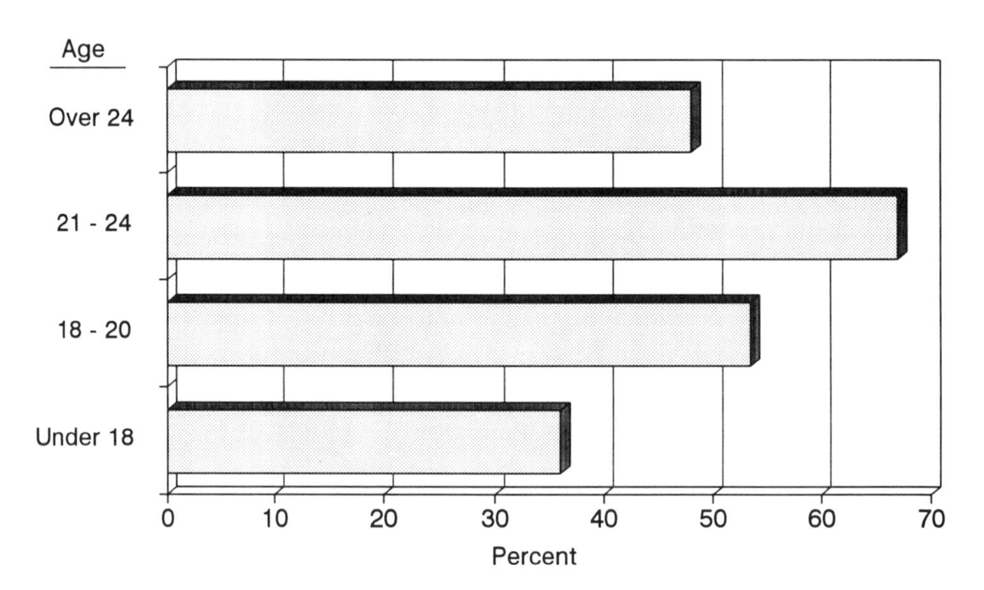

Fig. 7.5. Fatalities with Alcohol Involvement, Drivers and Non-occupants.

supported by statistics on pedestrian traffic involvements wherein large cars are responsible for the greatest number of fatalities. In the U.S. the smallest cars are involved in pedestrian deaths at about half the rate of the largest cars.[6]

Another consequence of increased large-car driver aggressiveness was pointed out by Sparrow in 1984. He noted that small cars are over-represented as the struck vehicle in tow-away rear end collisions and offered Table 7.2.[6]

TABLE 7.2. NCSS DATA ON REAR END COLLISIONS

# Rear Collisions	Struck Car Size				
(5, 6, 7 o'clock)	Med/Large	Compact	Subcompact	Super Sport	Total
Rear Collisions	225	50	69	1	345
Total Collisions	11,745	2484	2941	108	17,278
% Rear Total	1.91%	2.01%	2.35%	-	1.99%

Source: Ref. [6]

Rural driving is far more hazardous to travelers. Given that an accident occurs, statistics indicate that occupants are approximately five times more likely to be killed on a rural road than on an urban road, even though most accidents occur in the city. In 1989, urban areas accounted for 73.6 percent of all accidents but only 35 percent of traffic-related fatalities. Rural driving was much more lethal at 26.4 percent of the accidents and 65 percent of the total fatalities.[7] Another interesting observation is the difference between daytime and nighttime driving. On the basis of vehicle miles traveled in the U.S., urban commuting during daylight hours is ten times safer than nighttime driving on a rural road (Figure 7.6). Urban commuting during the day is nearly four times safer than the same trip at night. This variable alone has a greater impact on occupant safety than the statistical association between vehicle size and fatality rates.

Where, when, and how a vehicle is operated has an enormous effect on the safety and well-being of the occupants. All cars today are equipped with seat belts but many drivers do not wear them. Fatalities are approximately double for unrestrained occupants in a fatal automobile crash (Figure 7.7).

Many factors other than vehicle size and mass influence the level of driving hazard. Using the table developed by Malliaris, a 450-kg (1000-lb) commuter car with a 2032-mm (80-in) wheelbase would be assigned nearly three times the risk of a car of 2794-mm (110-in) wheelbase and 1589-kg (3500-lb) curb weight.

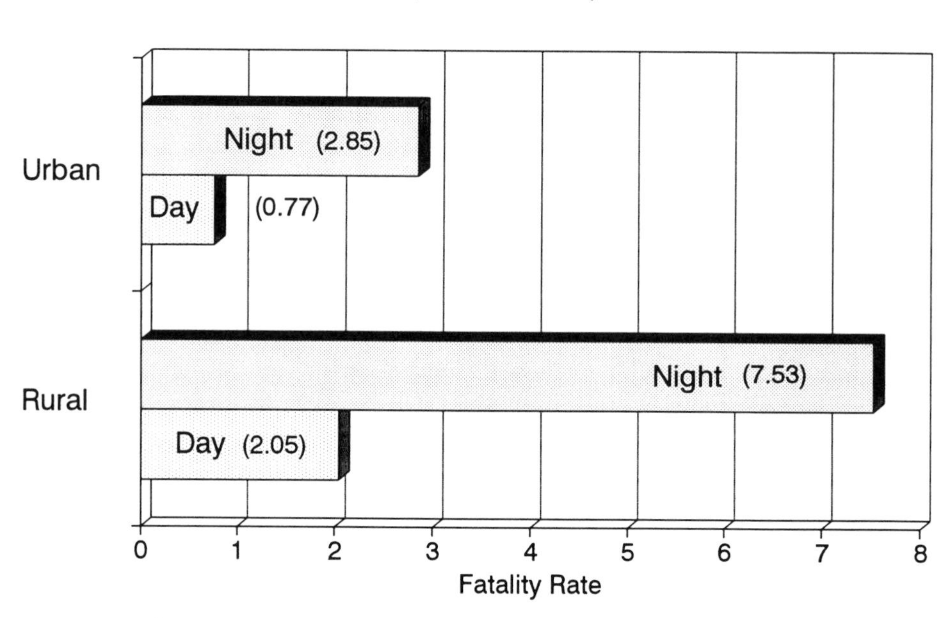

Fig. 7.6. U.S. Urban and Rural Fatality Rate (per 100 million vehicle miles traveled).

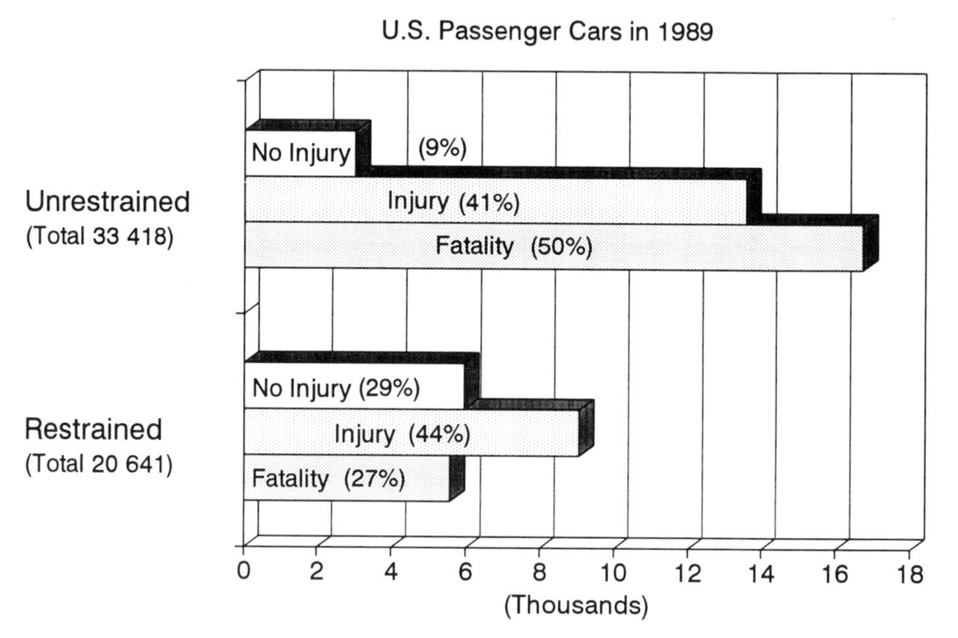

Fig. 7.7. Effect of Restraints on Injuries and Fatalities.

Based on the foregoing statistics, daytime commuting in the 450-kg Malliaris-rated vehicle would therefore be safer than nighttime urban driving in a full-size car. By controlling driver age, for example, daytime driving in the hypothetical commuter car with a driver aged 34 or more would be nearly twice as safe as the same trip in a full-size car with a teenage driver.

The Japanese Experience

In Japan the mix of vehicles by size category is much different than in the U.S. The average car in Japan is about the size of a U.S. mid-size car. The smallest Japanese cars are also much smaller than the smallest U.S. cars. The Japanese have established a special "Kei" (small) classification that includes vehicles having an engine of up to 660 cm^3 displacement and a curb weight of 590 kg (1300 lb) or less. These ultralight microcars are restricted to a top speed of 80 km/h (50 mph). Fuel economy is on the order of 21-25 km/L (50-60 mpg).

The Kei-car has a safety record in Japan that runs contrary to the U.S. experience with small automobiles. When Sparrow researched Kei-car safety in the early '80s, he found that they caused slightly fewer accidents and about 40 percent fewer fatalities per vehicle km traveled. On the basis of vehicle population, the Kei-car accident rate today is nearly 15 percent lower than the accident rate for regular-size cars in Japan. However, the accident rate is approximately 25 percent higher than it was when Sparrow researched the issue. The accident rate of regular-size cars in Japan is virtually identical to the 1982 rate.

Another characteristic that Sparrow noted had to do with driver aggressiveness. In general, the drivers of large cars were more aggressive, both with other cars and with pedestrians, and the drivers of small cars are less aggressive. Large-car aggressiveness and risk taking has been observed both in the U.S. and in Japan, but it is not well understood. As mentioned, Wasielewski documented that small-car drivers in the U.S. take less risks than drivers of larger cars.[5] He found that small-car drivers drive more slowly, wear their seat belts more often and do not follow as closely as do the drivers of large cars.

The relative aggressiveness of drivers can be measured by a number of characteristics, but most significantly the more aggressive drivers are responsible for the greatest number of accidents (Table 7.3). The magnitude of this characteristic becomes more significant when one considers the fact that younger Japanese drivers are the predominant users of the Kei-car and, as in the U.S., Japanese teenagers are considerably more accident-prone with their automobiles.

TABLE 7.3. JAPAN ACCIDENT RATE AND DRIVER AGGRESSIVENESS KEI-CAR COMPARED TO REGULAR-SIZE CAR, 1990

		Population		Accidents		Caused by	
		Count	% Total	Events	Rate	Events	%
Kei Car:	1990	2,060,591	6.3	29,958	0.015	16,951	56.6
	*1982	2,065,886	8.4	24,582	0.012	14,678	59.7
Std. Car:	1990	30,877,222	93.7	556,447	0.018	343,356	61.7
	*1982	22,512,638	91.6	392,890	0.017	258,381	65.8

*1982 figures taken from Ref. [6]
Source: Japan Ministry of Transport

When an accident does occur, the occupant of the Kei-car in Japan has a slightly better chance of surviving than the occupant of a regular-size car. This strongly contradicts the U.S. experience where the occupant of the smallest U.S. vehicle is approximately four times less likely to survive than the occupant of the largest vehicle. Injury rate, however, is higher in Kei-cars (Table 7.4). The increase in Kei-car injury rate is likely due to the vehicle's reduced crashworthiness while decreased fatality rate may be due to speed restrictions. Interestingly, Traffic Bureau statistics indicate that in 1982 approximately the same percentage of small- and large-car drivers that were injured in an accident were not wearing their seat belt (96 percent in both cases).

TABLE 7.4. JAPAN ACCIDENT AND INJURY RATE BY SIZE OF CAR, 1990

	Vehicles				Percent Accidents Resulting in Injury by Level of Seriousness					
	Population	%	Accidents	Rate	Fatal	%	Serious	%	Slight	%
Kei-Car	2,060,591	6.3	29,958	0.015	147	0.49	1.127	3.76	17,784	59.4
Regular Car	30,877,222	93.7	556,447	0.018	2831	0.50	15,612	2.80	277,060	49.8

Source: Japan Ministry of Transport

The pedestrian is always the loser in any encounter with an automobile. Although the Kei-car operates primarily in more congested urban areas where it is more exposed to pedestrian/automobile encounters, such encounters occur nearly 25 percent less often than with larger vehicles. When an accident does occur, the pedestrian has 1-1/2 times greater chance of surviving an encounter with a Kei-car. Therefore, given an equal number of exposures, the larger car is approximately twice as life-threatening to pedestrians (Table 7.5).

TABLE 7.5. PEDESTRIAN/AUTOMOBILE ACCIDENTS VERSUS FATALITIES, 1990 JAPAN

	Accidents	Rate	Fatalities	% of Accidents
Kei-Car	2202	0.0010	53	2.4
Regular Car	41,130	0.0013	1533	3.7

Source: Japan Ministry of Transport

The Japanese have achieved a remarkable safety record with very light-weight vehicles and much can be learned from their experience. Legally restricting the speed for all low-mass vehicles as they do in Japan may not be feasible in North America and Europe. For residents of many of the world's metropolitan areas, freeway/expressway driving is a necessary part of commuting. A special-purpose commuter car must therefore be capable of driving the freeways at the legal speed limit. Urban cars by definition operate at lower speeds. A separation between vehicle types, along with distinctions between the most effective strategies for safety engineering, may therefore be appropriate.

The Japanese experience with Kei-cars indicates that very small cars can be safely blended with larger vehicles, even though the smaller vehicles may be less crashworthy. If small urban cars were designed to protect the occupant during the most likely urban crash events, occupant safety could be optimized. Commuter cars designed for freeway driving obviously require more sophisticated crash survival technology.

Smaller Cars Require Better Safety Engineering

Accident statistics may indicate strategies in addition to improved vehicle crashworthiness that can reduce the casualty rate of smaller cars. Significantly smaller cars will benefit from lower urban speeds, and drivers appear to adopt defensive driving strategies with smaller vehicles. The difference between daytime and nighttime casualties may point to better lighting and lighter colors for smaller vehicles. Alarms or vehicle deactivation in the presence of alcohol might be mandated for significantly smaller vehicles. Special training and licensing for more youthful small-car drivers is another option. In addition, vehicle crashworthiness must be significantly improved as well.

Even with the reduced speeds in the urban environment, smaller cars will still experience greater deceleration forces in car-to-car collisions, especially during involvements with significantly larger vehicles. Moreover, freeway-traveling commuter cars will have the same exposure as larger cars, and the compounded risk caused by the smaller car's higher delta-V resultants cannot be ignored. A greater emphasis on vehicle safety engineering with strategies oriented toward the operating environment is therefore essential for smaller vehicles. Recent data indicate that improved vehicle design is already closing the gap between large- and small-car accident casualties. For example, statistics involving passenger cars of the 1976-1978 vintage indicate that cars weighing 900 kg (2000 lb) or less had a fatality rate that was 3.5 times higher than cars weighing 1800-2000 kg (4000-4500 lb). For cars of the 1986-1988 model year, the ratio is approximately 2.5 to 1.[8] In a recent report outlining options for improving automobile fuel economy, the U.S. Office of Technology Assessment (OTA) made the following observation about small-car safety:

> "It seems clear that, were significant downsizing of the fleet to occur, a good portion, and perhaps all, of any resulting loss in safety could be balanced by improvements in safety design."[9]

Given the different driver behavior patterns and the operating environments that may be characteristic of large, multi-purpose cars in comparison with smaller special-purpose vehicles, a less-crashworthy urban car could have the better overall safety record. The entire system, of which the vehicle is one component, establishes the ultimate level of safety. Nevertheless, smaller cars are destined to crash, and when they do the consequences are likely to be worse for the occupants unless countermeasures are designed into the vehicles. When an accident occurs, meaningful statistics are reduced to those associated with speed, impact vectors, vehicle crashworthiness, and crash survival technologies. Experience has shown that improved safety engineering produces a significantly greater reduction of harm in smaller cars than it does in larger cars.[10] With appropriate technology, it is possible to design lightweight vehicles that have greater occupant crash protection and greater overall safety than today's large cars. Safe small cars are technically feasible.

Automobile Crash Dynamics

Crash kinematics are primarily concerned with the forces of acceleration and the transfer of kinetic energy that take place during a collision. Designing a vehicle for occupant crash survival centers on technical measures for managing the kinetic energy in a way that does not cause injury to those inside.

When a vehicle stops or changes heading, the occupant continues on in the same direction and at the same velocity as before the maneuver. Under normal driving conditions this does not cause problems because the resulting forces are minimal. Steering and braking forces normally cannot exceed the friction coefficient of tires which might be on the order of 0.8. Consequently, the most abrupt maneuvers will rarely result in loads as high as one *g*. Unrestrained occupants can therefore change direction and speed along with the vehicle.

If the vehicle runs into an obstacle, loads imposed on occupants can rise dramatically. In this event, unrestrained occupants can no longer change speed and direction with the vehicle and instead, continue to move within the interior space at the pre-impact speed until colliding with one of the interior surfaces. In the case of a frontal impact, the most likely interior surfaces include the steering wheel, the windshield, and dash structures. Collision with the interior of the automobile is responsible for most of the occupant injuries and fatalities related

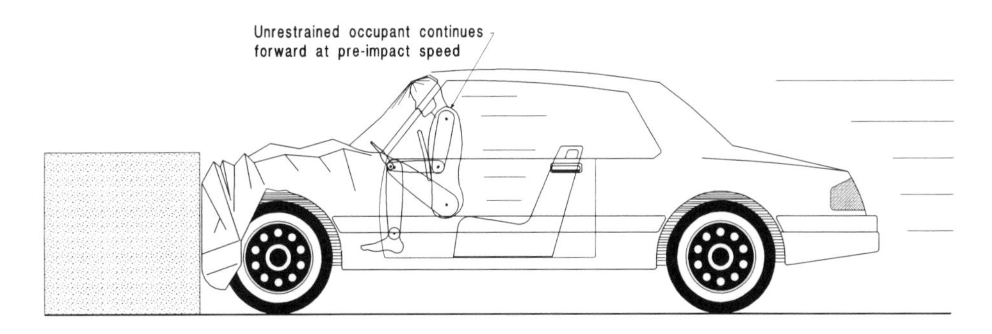

Fig. 7.8. Occupant Continues to Travel Inside the Vehicle.

to crashes. Methods for preventing casualties involve various techniques for restraining occupants from ballistic movement within the vehicle, controlling occupant deceleration to keep loads within tolerable limits, and cushioning potential contact areas inside the vehicle.

The magnitude of the forces depends on the difference between the pre-impact speed and the post-impact speed, and the time elapsed during the event. The more rapidly the vehicle changes speed, the greater the (acceleration) forces and the more dangerous it is for the occupants. Fortunately, the passenger compartment of an impacting or impacted vehicle stops over a period of time, rather than instantaneously.

Even when a fixed object is impacted, the vehicle does not instantly stop. Instead, it is brought to rest over the distance of the vehicle crush zone. The crush zone provides part of the time/distance required to decelerate the occupants. The remaining ride-down space comes from the free space inside the passenger compartment. If the vehicle did not crush, it would stop much more rapidly and greater forces would be transferred to the restraint system, requiring more ride-down space inside the vehicle in order to prevent injury. When an impact takes place with another vehicle, the kinetic energies of the striking and the struck vehicles combine to produce a resultant of forces which are affected by the mass and the combined crush characteristics of both vehicles. Most collisions occur at an angle and with vehicle-to-vehicle contact areas that react differently to impact. Consequently, the dynamics of a two-vehicle crash are complicated by the random contact characteristics and the mass differentials between impacting vehicles.

In the U.S., crash protection system performance is evaluated on the basis of the 30 mph (48 km/h) crash into a fixed barrier according to FMVSS 208. Most cars, however, collide with another car at an angle and therefore experience forces that are much less than those produced by a 30 mph (48 km/h) Barrier Crash Test. Ninety percent of accidents involving fatalities occur at a "barrier equivalent velocity" (BEV) less than that of 30 mph (48 km/h) Barrier Crash Test. BEV is determined by the velocity required to produce equal damage if the vehicle were crashed into a fix barrier, instead of the actual event.

The most lethal impact, and one that will probably never have a high survival rate, is the direct head-on collision. In 1990, head-on collisions in the U.S. accounted for three percent of all car-to-car involvements and 29 percent of all car-to-car events involving fatalities.[11] Fortunately, the largest percentage of these collisions took place at lower, and consequently more survivable speeds in urban traffic. Rural collisions, which occur at greater speeds, accounted for 72 percent

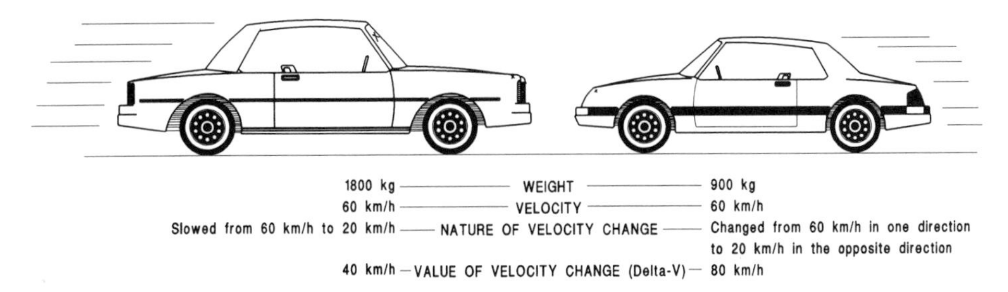

Fig. 7.9. Head-On Collision Between Large Car and Small Car.

of fatal head-on, two-car involvements. A head-on collision produces extremely high forces, and a large-car/small-car involvement transfers the largest share of energy to the smaller car. Delta-V expresses the magnitude of the velocity change and serves as a measure of the difference in energy transfer between the two impacting vehicles.

At lower urban speeds, markedly improved survival rates are possible with improved safety engineering, and fatalities need not increase even if vehicles on the order of 450-kg (990-lb) curb weight are introduced. According to Friedman of Minicars, Inc., with established techniques it is possible to achieve 80 km/h (50 mph) frontal and oblique barrier performance for all cars, and 160 km/h (100 mph) small-car-to-full-size-car oblique impacts without injury.[12] Friedman refers to conventional small cars, not to vehicles on the order of 450 kg (990 lb). Nevertheless, low-mass vehicles do not preclude safe vehicle designs.

Occupant Crash Protection

Anyone who has watched an Indy race has undoubtedly witnessed the results of state-of-the-art crash engineering. Essentially, the driver is tightly restrained inside a rigid enclosure designed to remain intact while the rest of the vehicle is sacrificed to the process of dissipating kinetic energy. This technique accounts for the high oblique-impact survival rate at speeds on the order of 320 km/h (200 mph). Passenger-car crash engineering is based on the same principle.

Items such as improved safety glass, recessed controls, frangible protrusions, and high-density padding contribute to the passenger car's safety. As a result, the interior of a modern automobile is a relatively safe environment, even for the unrestrained occupant, at speeds up to 25 km/h (15 mph). At greater speeds, other means are required in order to prevent injury.

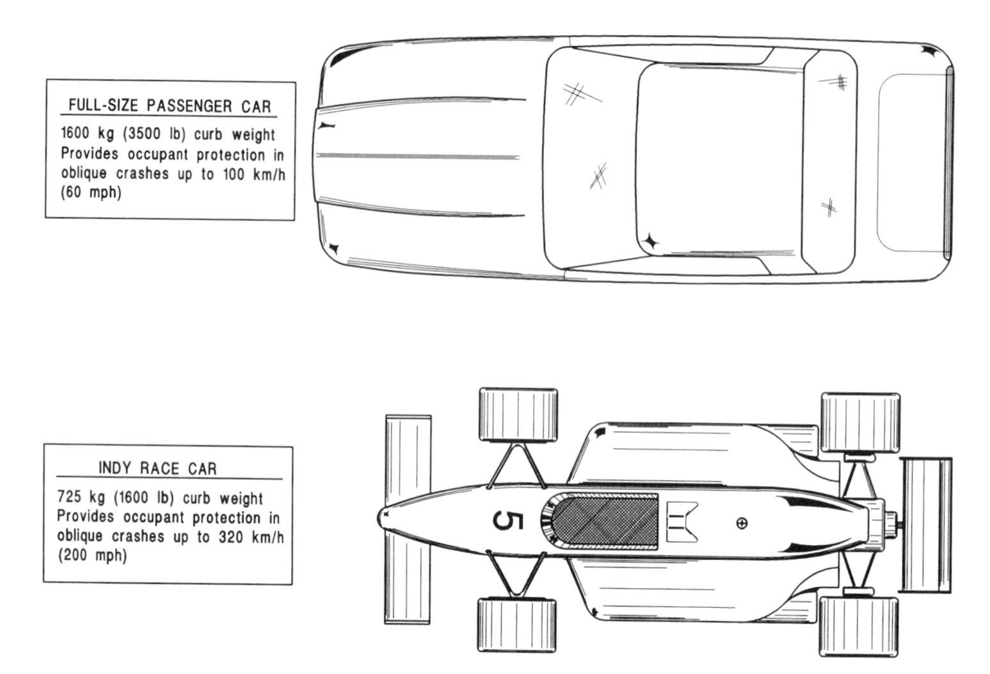

Fig. 7.10. Passenger Car and Race Car Share Crash Protection Strategy.

Traditionally, the restraint system and the vehicle structure work together as a complete system to protect occupants during a crash.[13] A relatively rigid passenger compartment serves a purpose similar to that of the race car enclosure. It is designed for minimal deformation and to prevent intrusion into the occupant zone. The parts of the vehicle that lie outside the passenger compartment are designed to crush at a controlled rate in response to impact forces. Restraints have an elongation rate (in the case of seat belts) which functions to decelerate the occupant at a controlled rate while evenly distributing loads over the body. Restraints also prevent ejection and collision with the interior of the vehicle. Saving occupants from injury therefore depends on having enough total stroke or ride-down space (crushing of cars and interior stroke space) to allow the occupants to decelerate over a time/distance interval that will keep loads from exceeding the maximum tolerable level.

Stated in abbreviated terms of whole-body deceleration limits, the U.S. National Highway Traffic and Safety Administration (NHTSA) has determined that the human body can tolerate acceleration loads up to a maximum of 60 g without injury, provided that loads are properly distributed.[14] (A complete description of the load limitations specified for barrier crash test procedures are contained in the Code of Federal Regulations, Title 49, Part 571-FMVSS and Part 572-ATD.) This 60-g limitation establishes the necessary vehicle crush rate and the performance criteria of the restraint system. If the occupant is decelerated over a greater time/distance, loads will be reduced and the likelihood of injury due to improper load distribution or load spikes will also be reduced. If deceleration time/distance is reduced, loads will climb above the 60-g level and injury will become increasingly more severe.

In actual practice, it is difficult to achieve a uniform deceleration rate during an automobile collision. When a vehicle strikes an obstacle it immediately begins to decelerate, but not at a uniform rate. Figure 7.11 shows the actual deceleration curve of a Ford sedan during a 60 km/h (38 mph) frontal barrier crash. When variations in vehicle deceleration result in occupant deceleration loads in excess of 60 g for longer than 3 ms, the design fails to meet NHTSA's performance requirements. Historically, vehicle restraint systems have been designed around a 30-g curve in order to account for deceleration variables and insure barrier test performance.[10] Experience has shown that modifications to the vehicle structure can eliminate the extremes in deceleration and produce a much more uniform crush rate.

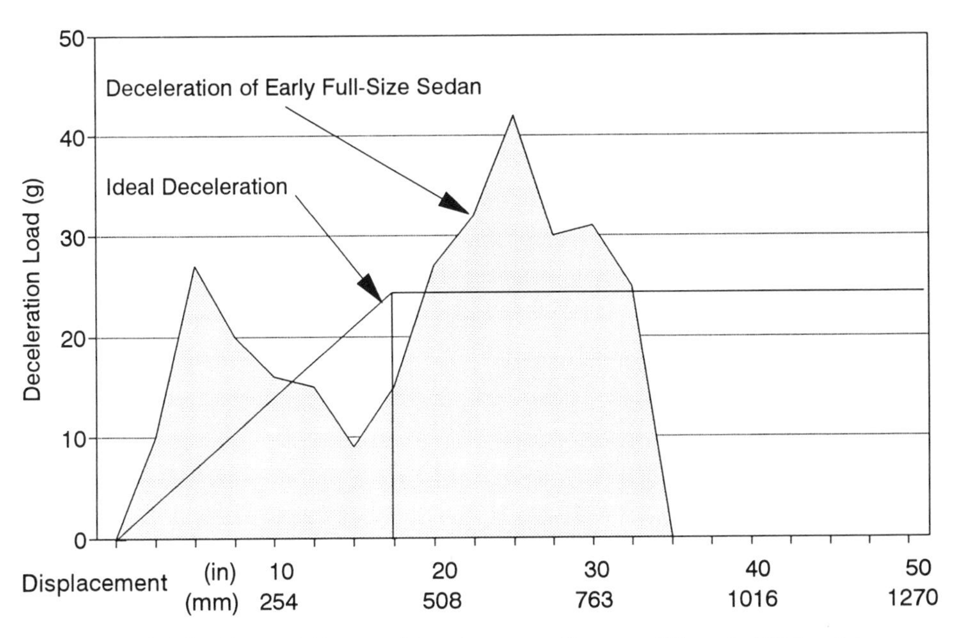

Fig. 7.11. Frontal Barrier Deceleration Curve—38 mph (61 km/h).

Upon impact, the occupant does not immediately begin to decelerate with the vehicle. Instead, the occupant continues on at the pre-impact speed until the free-flight distance is taken up in the restraint system and the loads can reach the maximum level throughout the system. As a result, the amount of time elapsed between obstacle impact and maximum occupant deceleration varies depending on the onset rate, free-flight distance, restraint system characteristics, and the position of the occupant at the time of impact. A smaller car has a smaller crush zone and is generally subjected to higher forces during a car-to-car collision. As a result, demands on the vehicle structure and the restraint system are greater. Smaller cars also have less interior space, which demands a more efficient use of the available time/distance. Optimal efficiency throughout the vehicle structure and restraint system is essential with smaller vehicles. Figure 7.12 shows a uniform 50-g deceleration schedule. Ride-down schedules on the order of 50 g are technically feasible and tolerable, provided that the vehicle crush zone and the restraint system operate together at optimal efficiency.[13] In general, however, building a uniform crush rate into the vehicle structure, which must fulfill a variety of other requirements, is much more challenging than building the same degree of uniformity into the restraint system.

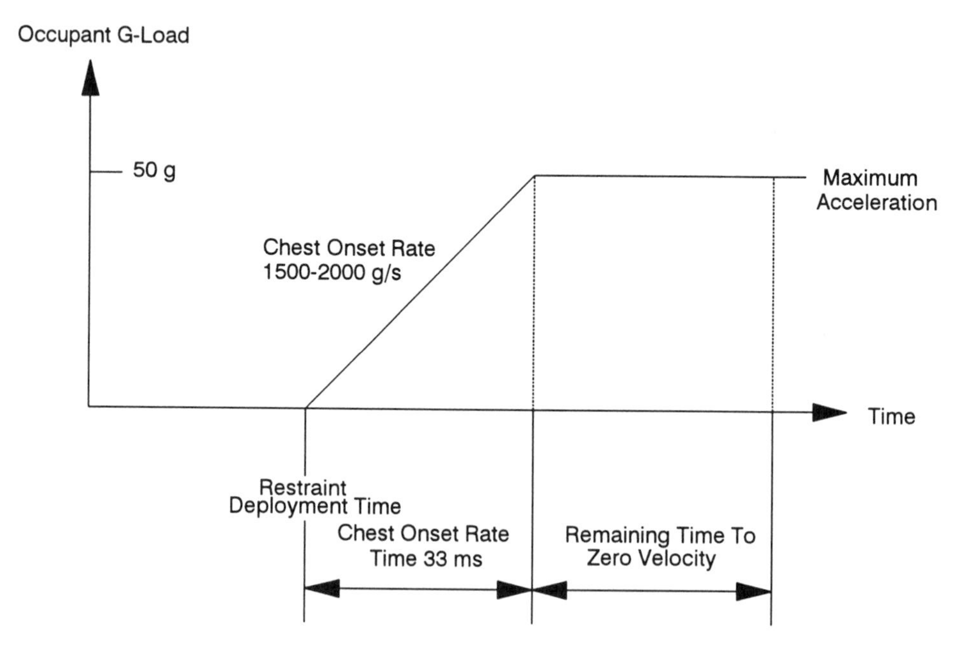

Fig. 7.12. Load/Time Curve.

Restraint Systems

State-of-the-art restraints include the mandatory three-point seat belt, airbags and the knee-bolster. After two decades of experimentation it is generally agreed that a combination of approaches is necessary to provide the broadest possible protection.

Restraint systems are designed primarily to control loads imposed from the front. Since the vehicle is presumed to be traveling forward, collision with an obstacle will produce loads on the occupants that are predominantly opposite to the direction of travel. However, if the vehicle is the struck vehicle rather than the striking vehicle, loads can be imposed from any direction. A review of fatal accidents in relation to the direction of the principal force reveals the degree to which frontal loads are the predominant cause of fatalities.

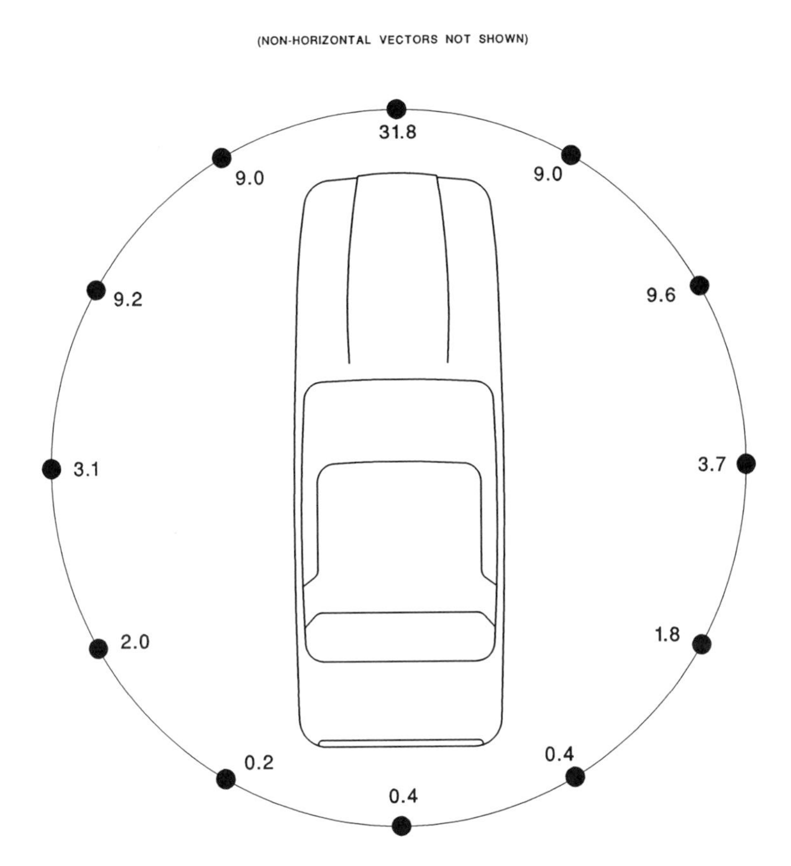

Fig. 7.13. Percent Fatalities Based on Principal Direction of Force.

Forces from the five, six and seven o'clock directions are managed primarily by the seats and headrests. These forces are usually the product of relatively low BEVs and seats should be designed to evenly transfer loads to the occupant and prevent head and neck injuries. Approximately 35 percent of the loads occur from one side or the other. Loads imposed from the 8, 9, 10 and 2, 3, 4 o'clock directions are also predominantly at low BEVs. They are also much more difficult to protect against because of the minimal stroke available in the sideways direction. According to the National Crash Severity Study, the highest injury rates result from impacts toward the side of the struck vehicle, even though loads are the result of sideswipes and angular impacts which may not transfer high energy values. Side intrusion members are designed to prevent intrusion into the occupant zone and seat belts prevent ejection and ballistic movement within the vehicle. However, at high BEVs, protection is largely inadequate in the sideways direction. Impacts that result in loads from the frontal 60 degrees include the highest BEVs and result in the greatest number of fatalities (approximately 60 percent). An adequate vehicle crush zone and a properly designed restraint system can eliminate the majority of fatalities resulting from these collisions, even with significantly smaller vehicles.

Seat Belts

The integrated three-point (lap/torso) safety belt with the emergency locking retractor (ELR) is standard equipment on today's cars. A self-retracting reel allows relatively free movement while maintaining minimal tension on the belt. This lets the belt spool out as the occupant leans forward then retract upon return to a normal seating position. Retractors are equipped with inertial locks to prevent spool-out during a collision. Even mild braking forces are adequate to actuate the locking mechanism.

The three-point seat belt is an exceptionally effective device. Statistics indicate that wearing seat belts reduces the number of fatalities by half. Seat belt use can also result in a 300-percent reduction in injury rates for those occupants who survive fatal vehicle crashes. After years of encouragement, and recent mandatory requirements on a state level, approximately 60 percent of vehicle occupants now wear restraints. That means that approximately 40 percent do not; and as a result, fatalities and injuries are much greater than if seat belts were faithfully used. Seat belts restrain occupants and prevent ejection or impact with the interior surfaces of the vehicle. A belt's elasticity also softens the transfer of energy to the occupant. A "soft" belt works somewhat like an airbag in that the system, either by belt elongation or by a force-limiting device, has a built-in elasticity that reduces the maximum load on the occupant.

Modern seat belts also have characteristic problems that sometimes prevent performance from matching theory. The lap portion of the belt sometimes slips above the top of the pelvis (submarining) during a collision and can thereby produce serious internal injuries. In order to perform correctly, the lap belt must engage the pelvis during the event. Belts that are appropriately angled toward the floor usually perform correctly. However, improper placement or an out-of-position or undersize occupant can result in submarining during a collision.

In general, slack in the belt assembly is the single major contributing factor in the failure of the seat belt to perform at maximum effectiveness. Slack is caused primarily by the portion of the belt that is loosely wound around the retractor reel. Consequently, the belt is essentially ineffective during the first few milliseconds after an impact as the build-up of forces locks the retractor and the forward movement of the occupant pulls the belt tightly around the spool. Forward movement unfortunately eliminates valuable interior stroke space that might otherwise provide room in which to decelerate the occupant. If seat belts were worn race-car-tight, belt slack would cease to be a problem and system efficiency would be much greater. Since a tight-fitting belt is not acceptable to consumers, an alternative approach is to leave the belt loose and tighten it immediately upon impact. Modern pre-tensioning devices utilize propellants or pre-tensioned springs that activate upon impact to tighten the take-up spool and eliminate belt slack. Spring-actuated pre-tensioners can remove 100 mm of slack in about 8 ms.

A final problem has to do with the belt itself. The single narrow strap across the chest is limited in its ability to safely transfer loads to the occupant. At higher loads the sternum and ribs have been broken by forces imposed across the chest by the torso restraint. Wider belts and even inflatable pads built into the belt (often referred to as the air belt) have all been tried with some success. However, one of the difficulties has to do with system bulkiness. Consumers are not especially fond of the torso belt in its present form. A wider, more bulky version is likely to be even less popular.

Airbags

Airbags began life some twenty years ago as the panacea of automobile accident injuries. The idea of a large pillow that would protect occupants in a collision was irresistibly simple and almost too obvious. Today, airbags are being installed on cars as rapidly as possible, and as a result, their limitations may begin to advertise themselves. Stories about someone's aunt Mildred who bought a car that was fully outfitted with airbags and got hit from the side already abound. With public exposure, airbags will be seen for what they are—excellent devices that work very well in certain types of collisions and not at all in others.

Airbags are designed to protect the occupant during single impact events that result in frontal loads across an arc of approximately 60 degrees (Figure 7.14). Forces from outside the arc may not activate the system, or the airbag may provide little or no protection if it is activated. Secondary impacts are also largely unprotected by inflatable restraints because of the system's rapid deflation.

Airbags are not stand-alone restraint systems. In conjunction with knee-bolsters, they provide excellent protection in a frontal collision. For the broadest protection, however, airbags must be part of a system that includes the three-point harness as the primary restraining device.

Correct airbag operation requires that a series of precisely timed events take place over the duration of the collision. The timing of these events will be different in different applications. In general, a smaller vehicle with reduced interior space will require a system that works within a more condensed schedule. Consequently, a generic inflatable restraint system cannot be designed to work in a multitude of automobiles. Each system is designed according to the crash characteristics of the vehicle.

The sequence of events begins when a sensor detects rapid deceleration and closes contacts to energize the propellant. A solid propellant then ignites and

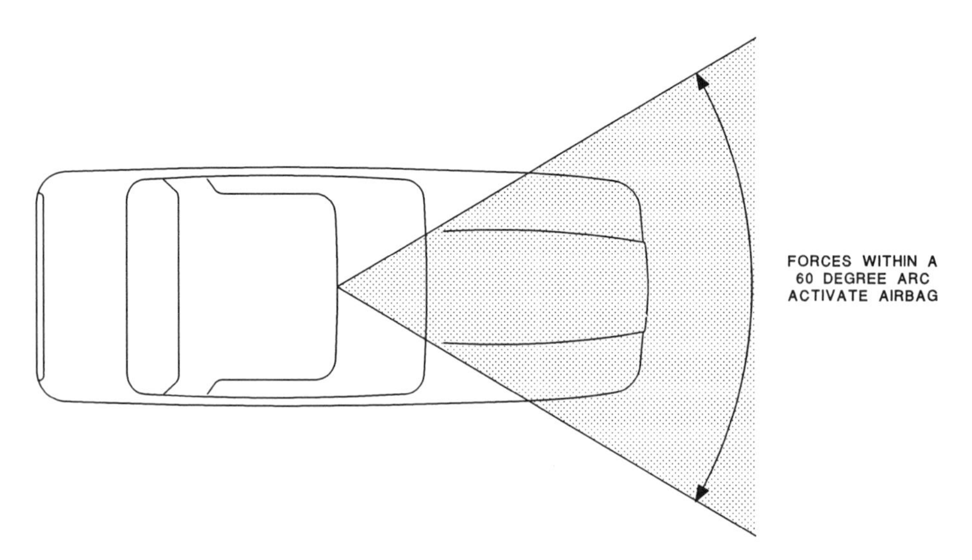

Fig. 7.14. Activation Zone for Airbags.

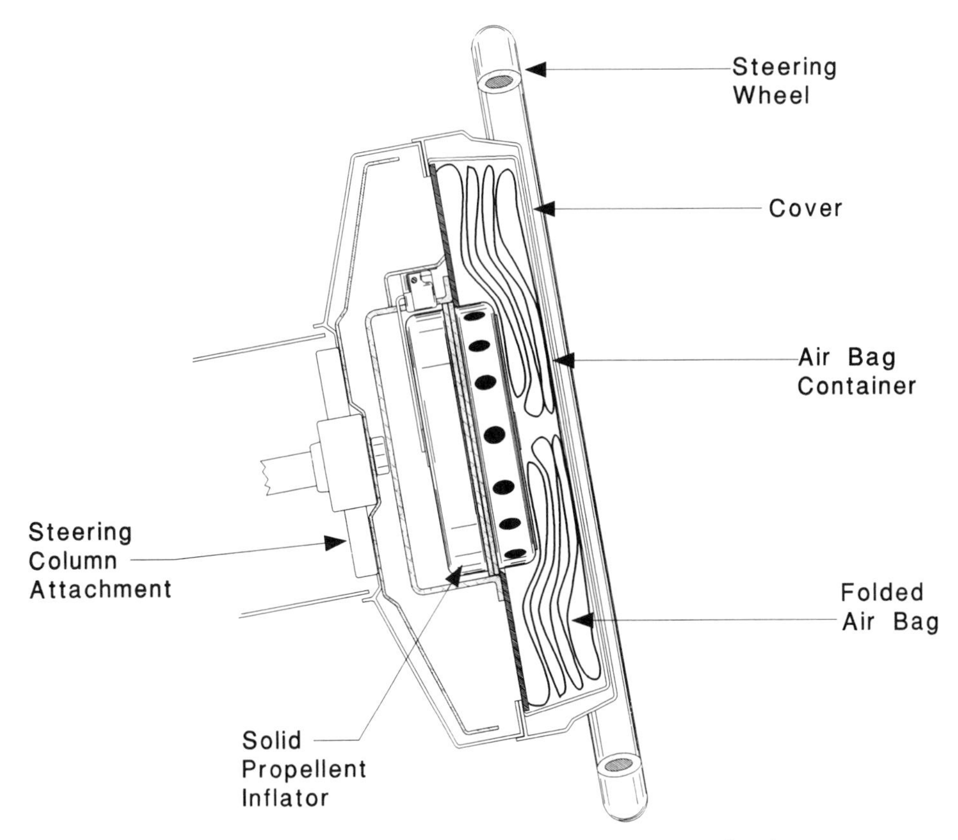

Fig. 7.15. Driver's Side Airbag in Steering Wheel.

inflates the airbag. Forward occupant excursion is arrested by the inflating bag which also prevents contact with the supporting structure. The airbag dissipates the energy to avoid undue rebound, then rapidly deflates. In a typical 30 mph (48 km/h) Barrier Crash Test, the entire sequence of events take place within a window of approximately 150 ms. Airbag deployment through the beginning of deflation occurs in as little as 30 ms.

The smaller the vehicle, the less time is available in which to complete the operations. This is due to the reduced crush distance and the more limited interior space of the smaller vehicle (Figure 7.17). During the crash event, time and distance are convertible, and the time/distance value is inversely related to the magnitude of the load. As the time/distance value decreases, the load experienced during the event becomes greater.

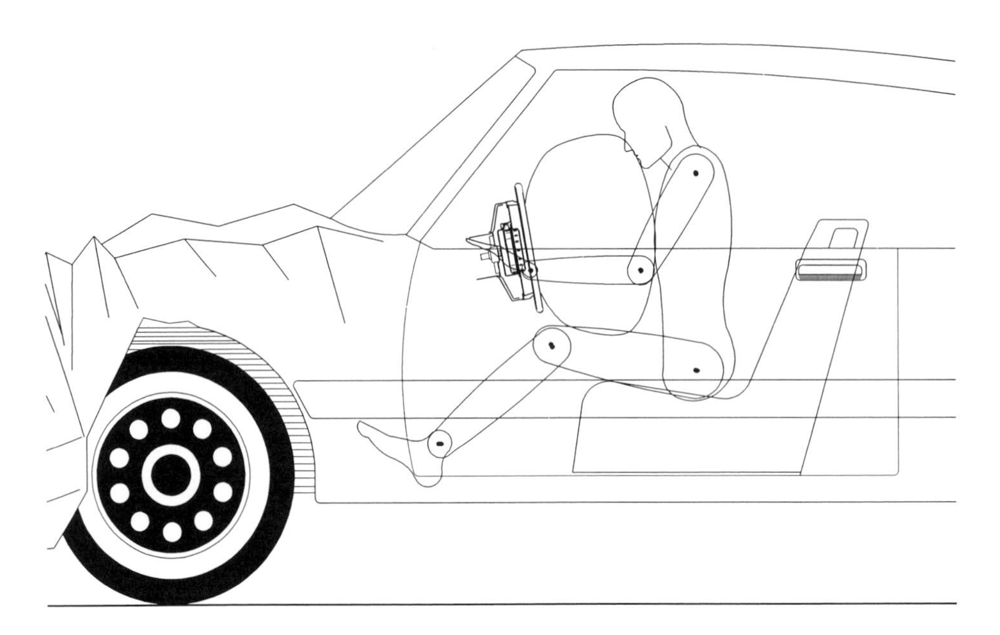

Fig. 7.16. Vehicle Crush Zone and Airbag Work as a System.

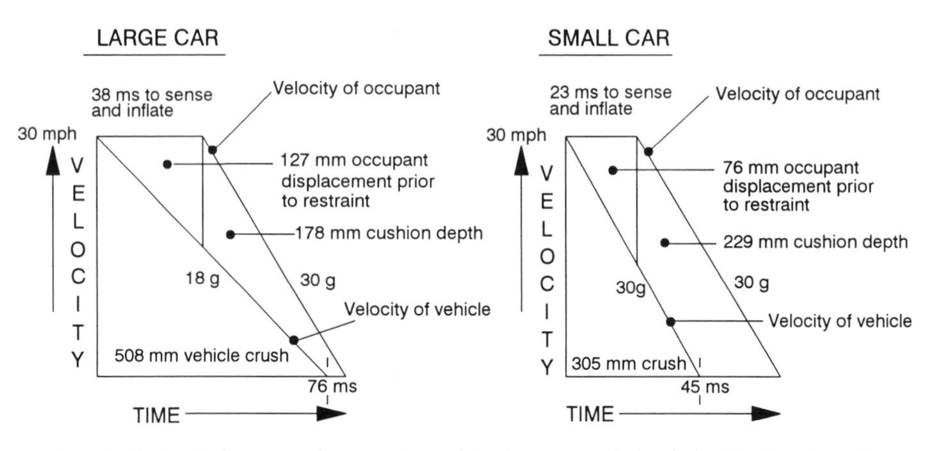

Fig. 7.17. Airbag Deployment and Ride-Down Schedule (U.S. 30 mph Barrier Test).

Stephen Goch of TRW Vehicle Safety Systems, Inc., described the following sequence of events for a large-car inflatable restraint system during a typical 30 mph (48 km/h) Barrier Crash Test:[15]

T_0: Front face of bumper makes contact with the barrier and begins to decelerate the vehicle. The occupants continue to travel at 30 mph (48 km/h).

T_{10}: An arming sensor located in the occupant compartment detects rapid deceleration and closes contacts. Electrical signal is then transmitted to the discriminating sensors wired in series and located in the engine compartment.

T_{15}: The discriminating sensors detect rapid deceleration and close contacts. Electrical signal is then transmitted to the diagnostic module where it is ultimately transmitted to the airbag modules.

T_{20}: Initiators in airbag modules receive electrical impulse and ignite, initiating bag deployment.

T_{25}: Airbag covers open and bags begin to fill.

T_{30}: Retractors lock, and occupants begin to load the lap and shoulder portions of the three-point systems.

T_{50}: The airbags fill and pressurize, and begin contributing to the restraining force already being provided by the belt systems to the occupants.

T_{60}: Knee-bolsters begin to contribute restraining force. On the driver's side, the steering column begins to absorb the additional energy of the occupant.

T_{90}: The occupants reach maximum forward excursion and begin rebound.

T_{120}: The vehicle is at maximum excursion into barrier and begins to rebound.

T_{150}: The vehicle and occupants come to rest.

Crash Management Strategy for Low-Mass Vehicles

Both commuter cars and urban cars can be designed to fulfill ECE and U.S. Barrier Crash Test requirements. Differences between low-mass and high-mass vehicle crash management strategies have to do with the condensed time budget and different methods of providing occupant ride-down space in the low-mass car. A typical barrier crash of a 450-kg (990-lb) commuter car will be a condensed event resulting in relatively high loads. Consequently, efficiency throughout the crash protection system must be optimized in order to prevent injury. Load spikes caused by a non-uniform deformation rate must be minimized so that occupant deceleration can be carried out at maximum loads without crossing into the injury zone.

Studies at the Swiss Federal Institute of Technology suggest that ride-down space in low-mass vehicles is best provided by increased free space inside the passenger compartment, rather than relying on an exterior deformation zone.[16] Conventional cars are designed with an exterior deformation zone in combination with a rigid passenger compartment. Stroke or ride-down space comes partially from vehicle deformation and partially from free space inside the passenger compartment. Studies suggest that low-mass vehicles might be designed with an exterior that is largely identical to the rigid passenger compartment of a conventional car. Occupants would be decelerated by airbags and elastic restraints within the vehicle. The more rigid exterior of the low-mass vehicle would cause the less rigid deformation zone of the larger vehicle to yield during a collision. Inside, the low-mass car would be relatively open, smoothly contoured and padded. Advanced vehicle control systems could eliminate the conventional steering column and pedal controls.[17]

Horlacher AG, Mohlin, Switzerland, has developed a number of promising low-mass vehicle designs that incorporate an impact belt to improve vehicle crashworthiness. The company has also provided vehicles for crash testing and evaluation by the Zurich Institute of Technology in cooperation with the Swiss Institute of Forensic Medicine and the Department of Accident Research at Winterthur Insurance Company in Switzerland. Tests indicate that Horlacher's "hard shell" concept, in which the vehicle is built for minimal deformation during a collision, provides maximum protection and best utilizes the crush zone of the high-mass vehicle. Barrier test results for Horlacher's design and two other low-mass vehicles, one unmodified and the other modified structurally and by the addition of a steering wheel airbag, are shown in Table 7.6.[18]

Lightweight construction and occupant safety are not necessarily mutually exclusive concepts. Low-mass vehicles utilizing improved materials and restraint systems, in conjunction with a more rigid exterior, can provide a comparatively safe environment for occupants during a crash. Crash tests also were conducted with a hard-shell Horlacher vehicle sent head-on against an Audi 100 of approximately twice the mass. Photographs of the crash dramatically illustrate that a small-car/large-car involvement need not result in a demolished small car (Figures 7.18 through 7.20). With the hard-shell concept, ride-down space

TABLE 7.6. ECE BARRIER CRASH TEST OF LOW-MASS VEHICLES WITH EUROSID1-DUMMY

	Solec 90 (unmodified)	Solec 90 (reinforced)	Horlacher "Impact Belt"
Impact vel.	11 m/s	11.2 m/s	9.3 m/s
Vehicle Mass	621 kg	684 kg	552 kg
Head Accel. (3 ms)	128 g	51 g	45 g
HIC (Head Injury Criterion)	2230	421	292
Max Floor Accel.	351 g	86 g	84 g
Max Impact Deformation	340 mm	298 mm	138 mm
Remaining Deformation	310 mm	230 mm	20 mm
Mean Force During High Energy Absorption	160 kN	260 kN	250 kN

Source: Horlacher AG, Mohlin, Switzerland

Fig. 7.18. A hard-shell Horlacher impacting an Audi 100, head-on. Horlacher speed is 52 km/h (32 mph); Audi speed is 26 km/h (16 mph). (Courtesy: Horlacher AG)

Fig. 7.19. Horlacher after the crash. Passenger compartment is intact. Windshield damage is due to a crash recorder that broke loose. (Courtesy: Horlacher AG)

Fig. 7.20. The Audi 100 after the crash. (Courtesy: Horlacher AG)

comes from the vehicle's interior. Occupant deceleration is controlled by the restraint system.

A basically hard-shell, low-mass vehicle can also be equipped with a relatively small exterior deformation zone, perhaps on the order of 300 mm (12 in). Interior ride-down space and/or deceleration loads might thereby be reduced. Passenger-side stroke is available from the typically large free space between the occupant and the dash. Driver-side stroke might include the controlled forward displacement of a more conventional steering assembly.** There are a variety of ways in which ride-down space can be provided. The percentage that comes from vehicle excursion into the barrier, in relation to the percentage that comes from the occupant's forward excursion inside the passenger compartment, is unimportant. Taken as a sum, the total stroke represents the total working space (time/distance) available in which to bring the occupant to rest uninjured. The primary consideration would be the degree to which the engineer deems it necessary to rely on the larger vehicle's crush zone to decelerate the smaller vehicle. A 30 mph Barrier Crash of a hypothetical low-mass vehicle is shown in Figures 7.21 and 7.22.

The hypothetical crash is essentially the same as a large-car barrier crash except events take place within a more condensed time budget. Such a condensed schedule is technically feasible. Rapid-deployment airbags have been demonstrated, and application techniques are outlined in available literature on airbag

** Side-by-side seat is assumed. A tandem arrangement introduces interesting possibilities for protecting the rear passenger with the back of the driver's seat.

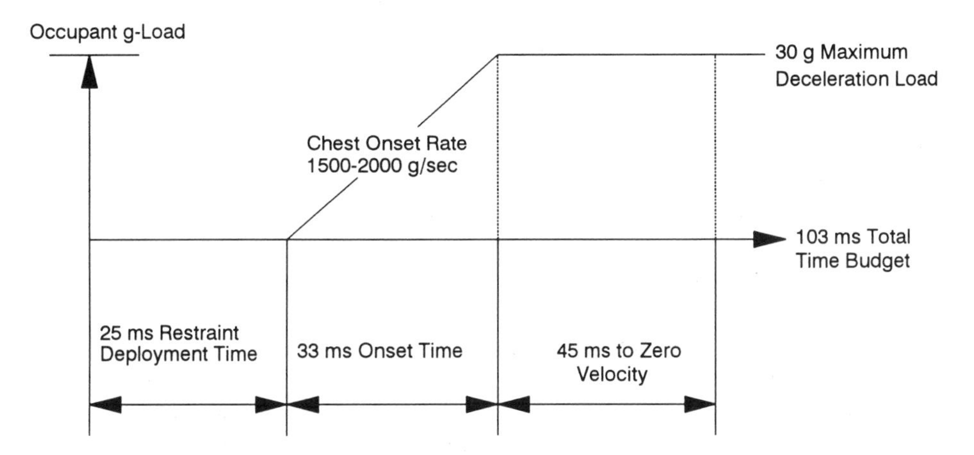

Fig. 7.21. Hypothetical Low-Mass Vehicle Deceleration Time Budget.

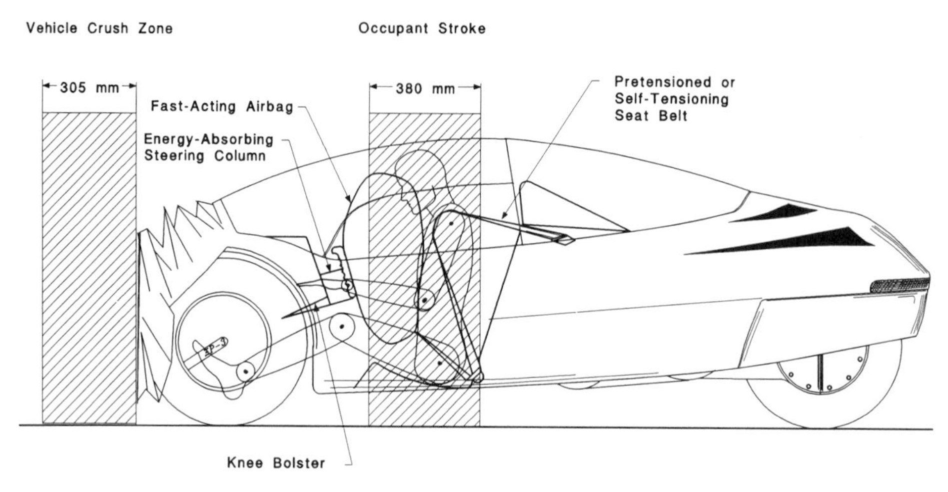

Fig. 7.22. Hypothetical Low-Mass Vehicle 30 mph Barrier Crash.

technologies. Hazards of rapid-deployment systems primarily involve the potential for injury to out-of-position occupants during rapid inflation.[10] Different vehicle standards for urban and commuter cars could resolve such difficulties by mandating designs that keep occupants in place.

Advanced Systems for Improving Automobile Safety

Several years ago at Quincy-Lynn an envelope arrived containing a drawing that illustrated a rather humorous idea for surviving automobile crashes. Occupants were strapped into a spherical compartment which rested in a receptacle in the vehicle on top of a large shoehorn-like device. The shoehorn handle extended forward so it slightly protruded at the front of the vehicle where it would be the first item to make contact in a frontal collision. Impact would cause the shoehorn to pivot and catapult the occupant sphere bodily away from the crash event. An elastic umbilical cord then arrested the sphere's ballistic flight and kept the occupants from being catapulted out of the neighborhood! Even now, the idea produces a chuckle whenever it comes to mind. However, the image of modern automobiles careening out of control, sacrificing pieces of their structure as occupants are slammed against restraints at bone-crushing g-loads seems all too primitive, and perhaps not much different from the bizarre idea proposed in the drawing. Crash survival technology is but an interim step toward the much more civilized goal of avoiding automobile crashes in the first place. The potential to avoid vehicle crashes exists within the technologies and subsystems that are being developed under the Intelligent Vehicle/Highway Systems (IVHS) program. Such technology is especially important to low-mass vehicle safety.

Several programs worldwide are underway to develop and deploy IVHS. IVHS is envisioned as a multi-layered assemblage of subsystems designed to provide a variety of traffic management and advisory functions, along with trip navigation and vehicle location, and ultimately vehicle guidance and control on automated highways. In the U.S., it is expected that a comprehensive development plan may be completed by 1997, and deployment could then begin as technology matures.[19] The Automated Highway System (AHS) is a central component. AHS will allow cars to travel along highways in platoons of 10 to 20 vehicles. Vehicle control will be assumed by on-board computers and cars will be guided within a special lane at a set speed in tightly packed groups with approximately three meters headway. Reduced headway and improved traffic management are expected to produce a 300-percent increase in highway capacity. Collision avoidance is a predominant aspect of IVHS. A number of stand-alone subsystems and integral technologies have the potential to significantly reduce automobile collisions. In combination with Advanced Vehicle Control Systems (AVCS), new technologies may ultimately make collisions a thing of the past. Neither IVHS nor AVCS is a particular technology but rather a number of technologies that can be applied in a variety of ways to improve road conditions and reduce driving hazards. Technologies that can improve future vehicle safety include the following:

Obstacle Detection
Blind Zone Detection
Forward Looking Radar
Driver Condition Monitoring
Headway Control
Collision Detection (warning and active)
Head-Up Display
Voice Activation
Automatic Lane Change
Automatic Braking

Electronic safety features are not necessarily devices of the distant future. Technologies such as anti-lock braking systems (ABS) are already prevalent and others are on the way. Early versions of obstacle detection and warning systems may be optional equipment as early as 1995. Such systems continuously monitor near-vehicle obstacles in side blind spots and to the rear. A simple version might provide head-up icon display and audible warning upon initiation of lane change or when shifting into reverse. Ultimately, systems will be integrated into AVCS to automatically brake or steer the vehicle away from an impending impact. Headway control may show up on production cars near the end of the '90s as part of adaptive cruise control. Adaptive cruise control will

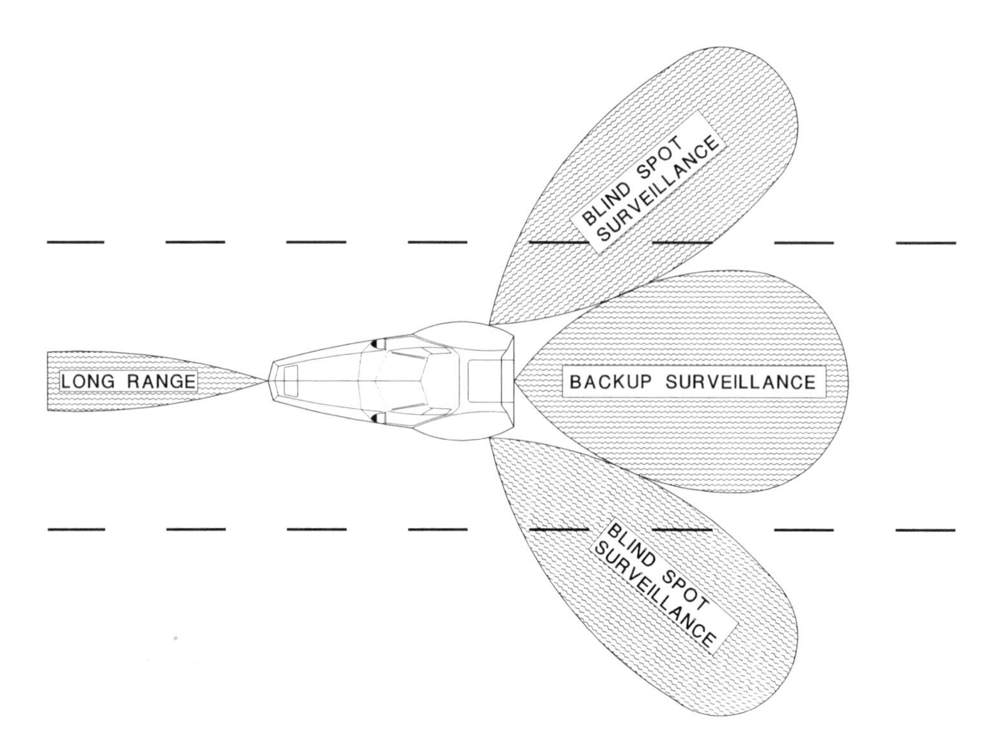

Fig. 7.23. Surveillance Warning and Avoidance Systems.

detect a vehicle ahead and adjust speed as needed to maintain the driver-selected headway. More advanced systems will detect an impending collision and steer around the obstacle or brake to avoid it.

Sensors for obstacle and blind zone detection are relatively inexpensive and compact. A 180-degree field of vision for intersection surveillance is also possible but requires more advanced technology. In order to see past traffic, the sensor must be located high on the vehicle, or even above it. Existing sensors are likely to be cosmetically unacceptable, but future designs are sure to be smaller. Again, early versions will likely be limited to head-up icon display and audible warning. Later versions will brake or steer the vehicle to avoid an imminent collision. The potential of collision avoidance to reduce roadway casualties becomes apparent when one considers that approximately 90 percent of automobile accidents are due to human error. Systems that warn or intervene to avoid collisions could significantly reduce driver-error accidents.

Driver monitoring is another technology that has far-reaching implications. A system developed by Nissan can detect driver fatigue by monitoring steering inputs.[20] Distinctive steering patterns are produced by the fatigued driver and

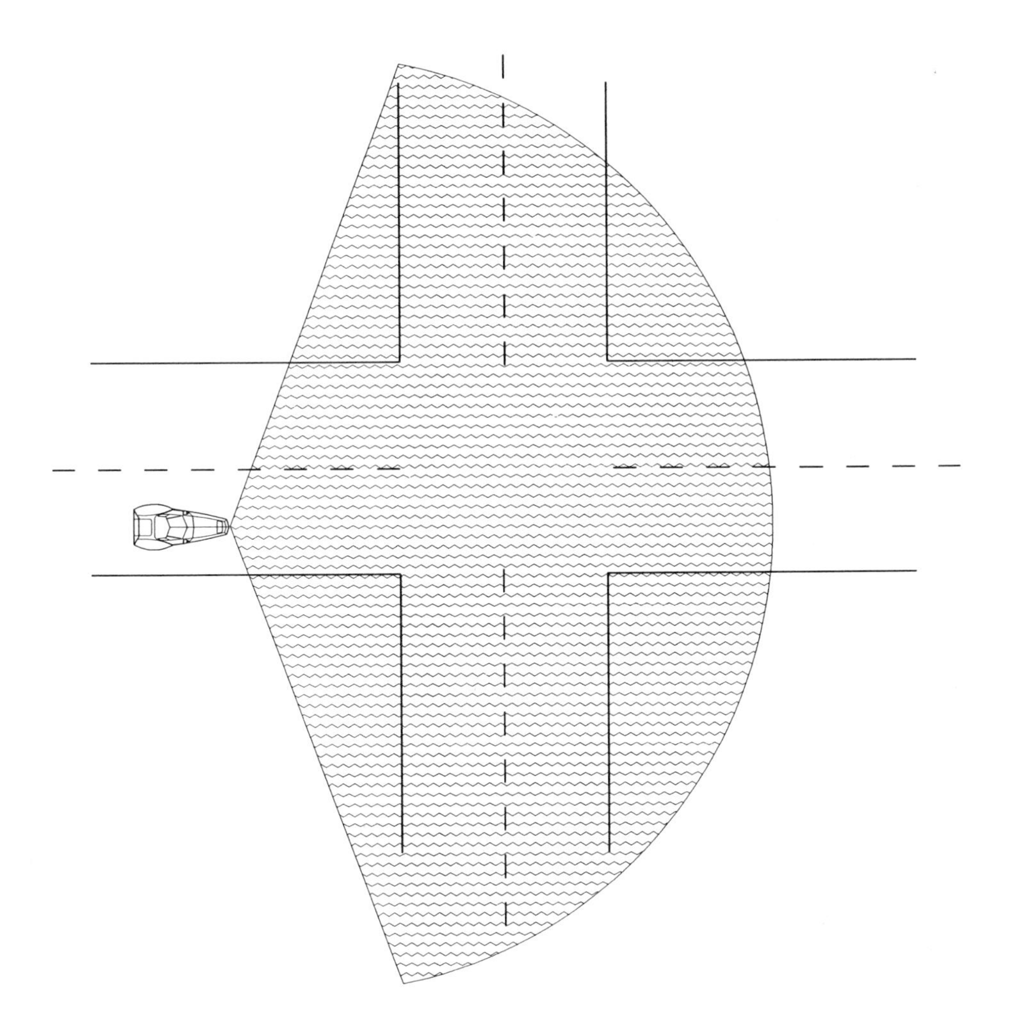

Fig. 7.24. Collision Detection and Avoidance at Intersections.

a continuous state of zero input is characteristic of drivers who have dozed off at the wheel. An audible warning could alert the fatigued driver and automatic braking could stop the vehicle if the driver dozed off. It is also possible to remotely monitor electroencephalograms, heart rate, and body temperature with sensors in the steering wheel. In the event of a medical emergency, a heart attack for example, the system could take control of the vehicle and bring it safely to a stop. It could also automatically send a signal to notify an emergency medical center. Drinking drivers might also be sensed and warned, or the vehicle could be deactivated if the operator is intoxicated. This feature alone has the potential to eliminate roughly half of the fatalities due to automobile accidents.

New Standards for New Vehicle Types

Proliferation of unconventional and inappropriately regulated new vehicle types would surely not contribute to safer highways. Existing standards for crash tests tend to create a barrier and make such an event unlikely. In the past, there have been discussions about the advisability of changing the Code to include a blanket exemption for all vehicles under a specified weight (450 kg or 1000 lb, for example). However, NHTSA has consistently resisted the idea of creating a categorical exemption based on vehicle weight, and instead, has issued limited exemptions to low-volume specialty vehicle manufacturers in certain instances. But new vehicle types are likely to introduce new problems that must be offset by new solutions. Neither a blanket exemption nor an individual exemption addresses the question of how unconventional, low-mass vehicles might be safely integrated into traffic. Ultimately, a new category may have to be created to account for the special considerations of unconventional vehicle types.

Crash management systems for sub-cars, urban cars and commuter cars require different approaches, both in relation to conventional passenger cars and in relation to each other. Open-bodied vehicles in the sub-car category may be so varied in design as to make consistent safety standards difficult to develop. Consequently, manufacturers might be required to demonstrate "reasonable precautions" in vehicle safety engineering. NHTSA might then individually evaluate each design and provide certification on a case-by-case basis. At the very least, two-wheel designs should require occupant/rider protective head-gear. Standards for three- and four-wheel open-bodied designs could specify occupant restraints, roll bars, protective side members and high-density padding at points of potential contact with the vehicle structure. However, a certain level of risk is probably unavoidable and might therefore be accepted under the "obvious and voluntary" rationale that NHTSA cites in safety issues involving motorcycles.

As for urban cars, the present barrier test may not be appropriate for a vehicle that operates primarily in the urban environment.*** Within all fatal or disabling accidents (including both urban and rural involvements), 97 percent involve an angular impact and a full 90 percent occur at speeds below a 48 km/h (30 mph) barrier equivalent velocity.[13] The (30-mph) barrier crash is a murderous event that equates to a real-world head-on crash between two cars that are each traveling at 48 km/h (30 mph), which translates into a 96 km/hr (60 mph) closing rate. Cars designed to be less dangerous in such a crash may actually be

*** Urban cars are envisioned as restricted to surface streets. A commuter car is assumed to be a vehicle of unrestricted use. Therefore it will likely operate in other than an urban environment.

more dangerous in the far more common low-speed, angular-impact urban event. Average speeds are substantially lower in the urban environment. Consequently, current standards should be re-evaluated as they apply to urban cars.

Electric cars, most likely urban cars, present unique problems. High-mass batteries in an otherwise low-mass vehicle require additional structural considerations. Batteries should be retained and protected within the vehicle during a crash. If batteries break away, vehicle mass will suddenly decrease and the kinematics will no longer be the same. Also, occupants must be protected from spilled electrolyte. I am aware of one case involving a conventional car in which the overturned vehicle caused electrolyte to spill from a battery located under the rear seat. As a result, the occupant sustained additional injuries after the event because of exposure to sulfuric acid. Runaway electrical fires, hydrogen explosions, and electrical shocks are additional hazards. Although the technology for containing electrolyte and preventing electrical shocks, fires and explosions is reasonably straightforward, new standards for electric urban cars may be necessary. In anticipation of increased numbers of electric cars in general, NHTSA issued an Advanced Notice of Proposed Rulemaking in which comments are solicited on potential safety-related issues of electric vehicles. The purpose is to determine if new standards should be developed specifically for electric cars.[21]

Urban cars might be restricted to surface streets and 80 km/h (50 mph) in a manner similar to the Kei-car in Japan. Commuter cars can have no such restrictions because of the freeway operating environment. Additionally, electronic crash warning and avoidance technology could be mandated according to existing technology, with systems emphasis given according to conditions of the most likely operating environment. Rollover accidents tend to increase in relation to the decrease in vehicle wheelbase, but are less affected by reductions in vehicle mass. Rollover accidents also are significantly more common in rural areas than in cities, primarily due to greater speeds on rural roads. NHTSA estimates that the vehicle is out of control in 50-80 percent of rollover accidents and that the risk of ejection is 3.5 times greater in fatal single-vehicle rollover accidents.

Countermeasures might therefore concentrate on preventing ejection and loss of control, as well as improving the dynamic margin of safety against rollover. Vehicles with ABS have a significantly lower incidence of loss of control. A mandated margin of safety against rollover would reduce the potential incidence of such accidents involving commuter cars and three-wheel cars. In 1992, NHTSA issued an Advanced Notice of Proposed Rulemaking which solicited comments on a variety of crash avoidance and crashworthiness options that have the potential to reduce rollover hazards for all cars. Tilt-table tests and side-pull tests with minimum stability requirements are options that are under consideration. Similar tests are under review in Europe. Regardless of the consensus on standard cars, such tests might be appropriate for low-mass cars of significantly reduced wheelbase.

Other options for commuter cars might include improved restraints, better headrests and mandated electronic crash avoidance systems. A five-point harness with a torso web would improve occupant survival. Headrests are typically too far from the head and do not provide support for the neck. Low-mass vehicles tend to be followed more closely, struck from the rear more often, and are subjected to higher deceleration forces when struck. Requirements for headrest designs should therefore be more stringent. In addition, close-following drivers might be warned by flashing taillights. Information from a rearward sensor could be processed by the vehicle's computer to determine when a closing-rate/speed/headway threshold is breached. Better engineering and a different orientation in standards can compensate for many of the disadvantages that result from reduced vehicle mass. Emphasizing improved technology in low-mass vehicles has the potential to largely offset the hazards associated with reduced vehicle mass.

Three-Wheel Cars

Three-wheelers are now classified as motorcycles in the U.S., which essentially places them outside the domain of meaningful safety standards. However, three-wheel cars may be more deserving, rather than less, of carefully developed engineering guidelines. A dynamic margin of safety against rollover is not addressed in existing standards, and it should be addressed if three-wheel cars are to be safely integrated into traffic. In addition, the potential for longitudinal center-of-gravity displacement in three-wheel cars must be considered in developing appropriate standards. The idea of placing three-wheel vehicles in the passenger car classification may not be appropriate either. Three-wheelers require a number of special engineering considerations. A brief review of the regulatory history and the prevailing attitudes toward motorcycle/three-wheeler safety will help bring three-wheeler classification into perspective, and perhaps clarify the more broad issue of special standards versus blanket exemptions for unconventional, low-mass vehicles.

In the late 1960s the U.S. Code of Federal Regulations contained an exemption for all four-wheel cars under 1000 lb (454 kg) curb weight. The exemption was removed in 1973 because on closer examination it appeared that smaller cars actually have a greater need for safety features than larger cars. At the same time, NHTSA also proposed to redefine the motorcycle classification to exclude three-wheel vehicles that have essentially the same end use as automobiles; that is, three-wheel vehicles that are built with some protective structure resembling a body. Through a series of legislative false starts, intense debate, and industry resistance, three-wheel vehicles have not been excluded from the motorcycle definition as was originally intended. Today a motorcycle is still any vehicle "...having a seat or saddle for the use of the rider, and designed to travel with not more than three wheels in contact with the ground....."

Over the years, the issue of three-wheel vehicle classification has been closely associated with arguments in favor of a blanket exemption for low-mass four-wheel cars. According to the argument, if motorcycles are allowed, and three-wheelers can operate under the motorcycle classification, it does not make sense to ban ultralight four-wheelers, just because they may not meet passenger car safety standards. Ultralight four-wheel cars would surely be safer than motorcycles, and perhaps even safer than three-wheel cars that are presently allowed under the motorcycle classification. NHTSA affirms that the idea of a blanket exemption is not valid because of the concept of "obvious and voluntary" risk.

The difference in risk types is the pivotal distinction behind the decision to allow motorcycles on the roadways, relatively unregulated, while regulating inherently safer automobiles. According to this line of reasoning, people are entitled to take voluntary risks such as climbing a mountain, skydiving or riding a motorcycle. These activities carry a certain level of "obvious" risk and the person who participates in them is presumed to "voluntarily" accept that risk. Involuntary risks have to do with using ordinary household appliances or driving automobiles. The risks involved in these activities may not be as obvious and therefore their safety must be more carefully regulated. Automobiles are equipped with a substantial body that might lead one to believe they are safer than they really are. Motorcycles, on the other hand, are obviously hazardous and therefore people are not likely to be misled into believing that they are protected from injury when, in fact, they are not.

Arguments based on obvious and voluntary risk do not support the idea of placing full-bodied three-wheel vehicles in the motorcycle category. Quite the contrary, such arguments seem to imply that three-wheel vehicles should be subject to appropriate safety standards. At the same time, it seems clear that three-wheel cars do not belong in the passenger car category. As discussed above, three-wheel vehicles require unique engineering and safety considerations that are not addressed by existing passenger car safety standards. Moreover, new standards that address the unique safety implications of unconventional vehicle types are probably necessary if extremely low-mass vehicles, both three- and four-wheel designs, are to match the safety record of larger cars.

In the past, debates have often focused on exemptions; however, exemptions do not improve highway safety. New classifications should not circumvent safety standards, but instead, they should define appropriate guidelines.

The final issue concerns product liability. Without appropriate regulations, the most qualified manufacturers may actually be discouraged from marketing three-wheelers or other unconventional vehicle designs in order to avoid potential liability difficulties. This is a sensitive issue. On the one hand, manufacturers are reluctant to imply that marketing decisions are subject to legal intimidation. On the other hand, the realities of product liability are certainly relevant. Concerns regarding product liability were an important consideration in the decision by Harley-Davidson to shelve the three-wheel Trihawk project, even though the vehicle had excellent handling characteristics and the market appeared strong. Regulations that adequately address the safety implications of low-mass and unconventional vehicle types would open the way for manufacturers to develop new, environmentally friendly and energy-efficient vehicle types. In this regard, appropriate safety regulations for low-mass cars deserve a thoughtful and thorough review.

References

1. "Shopping for a Safer Car," Insurance Institute for Highway Safety, 1992.

2. A.C. Malliaris, *et al.*, "Discerning the State of Crashworthiness in the Accident Experience," Office of Vehicle Research, NHTSA, paper presented at the 10th International Conference on Experimental Safety Vehicles, DOT, 1985.

3. Evans L., "Accident Involvement Rate and Car Size," General Motors Research Lab., Publication GMR-4453, August 1983.

4. Allan F. Williams, *et al.*, "Cars Owned and Driven by Teenagers," *Transportation Quarterly*, April 1987, pp. 177-188.

5. Paul Wasielewski, "Do Drivers of Small Cars Take Less Risk in Everyday Driving?" General Motors Research Laboratories, GMR-4425, July 13, 1983.

6. F.T. Sparrow, The Coming Mini/Micro Car Crisis: Do We Need a New Definition?, Pergamon Press Ltd., 1984.

7. National Safety Council, "Accident Facts," 1990 edition.

8. E. Chelimisky, "Automobile Weight and Safety," GAO/T-PEMD-91-2, U.S. General Accounting Office, Washington, D.C. (Weight categories are reported in pounds and have been converted by the author.)

9. U.S. Congress, Office of Technology Assessment, "Improving Automobile Fuel Economy: New Standards, New Approaches," OTA-E-504, U.S. Government Printing Office, Washington, D.C., October 1991.

10. Harold J. Mertz and James F. Marquardt, "Small Car Air Cushion Performance Considerations," SAE Paper No. 851199, Society of Automotive Engineers, Warrendale, PA, 1985.

11. National Safety Council, "Accident Facts," 1991 Edition.

12. This was predicted as early as 1973 based on technology known at that time. See final statements in Ref. [13].

13. Donald Friedman, "Development of Advanced Deployable Restraints and Interiors," paper presented at the Vehicle Safety Research Integration Symposium, Washington, D.C., May 30-31, 1973.

14. Test criteria of FMVSS 208 and NCAP procedures include a Head Injury Criterion formula which includes load/time limitations, a maximum chest acceleration of 60 g, a maximum chest compression of 3 inches, and a maximum femur load of 2250 pounds (10 kN). These limitations are approximated by assuming a maximum full-body limitation of 60 g. Refer to Code of Federal Regulations, Title 49, Part 571-FMVSS and Part 572-ATD for complete description of tests and criterion.

15. Stephen Goch, Thomas R. Krause, and Allen Gillespie, "Inflatable Restraint System Design Considerations," SAE Paper No. 901122, Society of Automotive Engineers, Warrendale, PA, 1990.

16. Robert Kaeser and Felix H. Walz, MD, "New Safety Concepts for Low Mass Electric/Hybrid Cars," paper presented at the 25th ISATA meeting, Florence, June 1992; and Robert Kaeser, "Safety Potential of Urban Electric Vehicles in Collisions," paper presented at The Urban Electric Vehicle Conference, Stockholm, May 25-27, 1992.

17. Felix H. Walz, MD, *et al.*, "Occupant and Exterior Safety of Low-mass cars (LMC)," paper presented at the IRCOBI-Conference, Berlin, September 1991.

18. Robert Kaeser, *et al.*, "Collision Safety of a Hard Shell Low Mass Vehicle," paper presented at the IRCOBI Conference, Verona, Italy, September 9-11, 1992.

19. "Department of Transportation's IVHS Strategic Plan Report to Congress," Department of Transportation Publication No. FHWA-SA-93-009.

20. Shigeo Aono, "Electronic Applications for Enhancing Automotive Safety," SAE Paper No. 90137, Society of Automotive Engineers, Warrendale, PA, 1990.

21. *Federal Register*, December 27, 1991, p. 67038, NHTSA Docket 91-49, Notice 1.

Chapter Eight

Alternative Cars in Europe

Renault's "Zoom." (Courtesy Renault)

The mass transit system of choice world-wide is the automobile. To the dismay of idealists and urban saviours, it has no serious competitors, even though its continued use depends on a precarious supply of elaborately refined petroleum products.

Syd Mead

Ultralight minicars, vehicles that might be considered alternative cars under the definitions presented in Chapter Two, are relatively commonplace outside North America. Over the decades, designs have run the gamut from open-bodied three-wheel motorscooter cars at the lower extreme to the more recent Austin Morris Mini and Japan's Kei-car at the upper extreme. Following World War II, minicar designs in Europe proliferated with vehicles like the so-called "bubble cars" produced by Heinkel, Isetta, and BMW, and Messerschmitt's "scooter car" representing the greatest departure from the conventional automobile while still remaining an automobile of sorts. These economical vehicles emerged in response to high motor fuel prices, high automobile taxation, and the poor economy that characterized the region after the war. Curb weight on the order of 225-375 kg (500-825 lb), and fuel economy of 20-25 km/L (50-60 mpg) were typical. First costs as well as operating and maintenance costs were substantially below the costs of owning and operating a conventional car. But regardless of their cost-saving benefits, bubble cars and scooter cars ultimately became extinct, leaving larger multi-passenger cars like the Austin Morris Mini and its derivatives to satisfy the low-end market.

The fate of these prototype alternative cars in Europe seems to ask a fundamental question about the idea of limited-capacity, low-mass cars today. Might their demise in an improving economy indicate that consumers are likely to reject urban cars and commuter cars in a modern, affluent society? Is environmental degradation and resource depletion only a crisis of a different name, and might attempts to market significantly smaller cars in response to it produce similar disappointing results?

The European experience with minicars can provide important insights for potential manufacturers of urban cars and commuter cars today. Undoubtedly the issue is more complex than simply counting their successes and failures and projecting the results into modern transportation products of similar curb weight. A variety of attributes having to do with regional economies, taxation policies, consumer attitudes, and vehicle theme played significant and complex roles. Small size and limited utility were part of the equation. Perhaps more important is the unmistakable vehicle theme of sacrifice and scarcity, which these cars offered to consumers who had their sights set on better economic times. In an improving economy, consumers first wanted improved performance and greater utility from their single vehicle, and ultimately larger, more luxurious automobiles. Minicars occupied a position at the bottom of the economic ladder. Moreover, multi-passenger cars like the Mini offered a better overall value in terms of transporting maximum payload with minimum vehicle expenditure, which is a primary requisite for vehicles in the low-end market. In addition, minicars existed long before terms such as "ecological breakdown," "greenhouse effect," and "ozone holes" were invented. Today, new consumer attitudes, new societal concerns, and new economic conditions point to a different product and a different marketing approach for a successful alternative car.

The potential consumer market for limited-capacity urban- and commuting-specific cars in Europe will grow in relation to the growth in multi-car households, which is primarily a by-product of increased consumer purchasing power. Lower purchasing power is more favorable for multi-passenger economy cars like the Mini, and greater purchasing power will tend to increase the potential market for smaller, mission-specific vehicles. Once purchasing power exceeds basic survival needs, complicated personal factors having to do with lifestyle and self-image become much more significant factors in purchasing decisions. A limited-capacity alternative car must therefore appeal to more affluent consumers. Neither Europeans nor Americans are likely to be attracted by a vehicle theme that implies a compromise in lifestyles. A high-end consumer product must offer significant intangible values in combination with real societal benefits (the so-called "green" appeals), as well as reduce operating costs. A low-end equivalent may have to rely more on lower first costs and a market created artificially by higher fuel prices and/or preferential taxation.

While consumers are universally influenced by the intangible values of style, prestige, and creature comforts, and sensitive to taxation, first costs, and operating costs, the effect of these attributes on purchasing decisions varies according to regional values and economic conditions. In general, consumers are more sensitive to first costs than to operating costs and will often purchase a lower-priced vehicle even though long-term operating costs may be greater. Consequently, reduced operating cost is a relatively poor motivator in consumer purchasing decisions. Vehicle styling and ambiance can be a powerful motivator but is highly subjective and varies according to cultural differences. In Japan, for example, consumers do not want to stand out as different, and as a result, the somewhat overstated individuality typical of many American products is often seen as garish. Product styling and packaging generally reflect regional values and preferences. The vehicle theme is the broad medium by which the product ceases to be a purely mechanical item and takes on a set of values that have meaning to consumers. Bubble cars and scooter cars embodied disproportionate negative economic values and insufficient positive benefits, and can therefore serve more as an antithetical model for themes and implied values that might be appropriate for modern consumers.

Early European Minicars as a Metaphor of Their Environment

If there is one characterizing attribute of Europe in the twentieth century it might be that of volatility. In America, both the political history and the course of development have been comparatively stable and consistent. Although Americans may still remember Vietnam as a shock to their national self-confidence, Europeans have had a series of Vietnams in their own backyards throughout much of the twentieth century. And while the second World War may have been a boon to the U.S. economy, it was a disaster for economies in Europe. After the war, countries that were not physically ruined were in economic shambles. Despite the poor economy, new cars were nevertheless taxed at 40-60 percent of the purchase price and already punitive license fees typically increased in proportion to engine displacement and vehicle weight. Gasoline was rationed, and what little was available (6 gallons per month in England) was extremely high priced. Rationing also occurred in the '50s, long before Americans received a similar dose of reality with the OPEC oil embargo in the early '70s.

In addition, Europe has never enjoyed the enormous consumer base that exists in the U.S. Although the population of Western Europe as a whole was (and still is) slightly larger than that of the U.S., this population of consumers was divided into 14 main national units, each with its own frontiers and stringent trade barriers. Within this varied and difficult-to-reach market, too many manufacturers were competing for too few sales. Ninety percent of Europe's cars were built by 14 principal factories, and the remaining 10 percent of cars emerged from an additional 36 separate manufacturers. Yet the combined sales of automobiles in Western Europe in 1953 barely equalled the sales of Chevrolets in the U.S. during the same year.[1] Benefits that flow from the economy of scale were therefore not available to European manufacturers, at least in regard to the domestic market. Naturally, the automobile market was ripe for entrepreneurial ventures that may not have resulted in the best products, nor been good for established manufacturers that were in for the long haul.

Against this backdrop, a number of differences have traditionally existed between North American and European populations and transportation needs. The densely packed cities of Europe are typically more adaptable to public transportation. As a result, transit systems have been better financed and more readily used by travelers. The proliferation of the automobile after WWII can be at least partially blamed on the U.S. As soon as the economy began to recover, Europeans launched their own version of America's highway mania, partly because of the American idea of unbridled mobility, but also because it is much less expensive to build roads than public transportation systems. Today, Europe has an extensive network of modern roads and the cars to occupy them, as well as public transportation systems that rate among the world's finest.

As for automobile designs, European designers have traditionally been more innovative than their American counterparts. In 1945, one in four European cars already incorporated the front-wheel-drive powertrain, approximately a quarter-century before Americans discovered the inherent benefits of the layout. The live rear axle had been virtually abandoned in Europe even before the war; after the war, new ideas in vehicle configuration were rampant. Small cars developed by German aircraft manufacturers were marvels of mechanical simplicity. The Heinkel and the Messerschmitt were vehicles that carried the name of their airborne predecessors. Many came from existing automobile manufacturers, and a number of designs came from small companies with no history in automobile manufacturing. Far too many were prime examples of what can happen when an inexperienced and under-financed company rushes to market with an unrefined product. And companies with little or no service or sales organizations often left products on their own once they were in the field.

Isetta and Heinkel

Isetta and Heinkel are two examples of well-engineered minicars, popularly referred to as "bubble cars." Designed in Italy and ultimately manufactured by BMW in Germany, the 343-kg (755-lb) diminutive Isetta was destined to become an inexpensive transportation workhorse for thousands of Europeans during the '50s. The less popular Heinkel was designed and built in Germany by the manufacturers of the famous Heinkel aircraft. Isetta and Heinkel were strikingly similar vehicles and both utilized a single front door. However, the more powerful and deluxe four-wheel Isetta featured a pivoting steering column that attached to the door and would thereby swing out of the way when the door was opened. Heinkel's steering column was fixed and as a result partially blocked entrance into the cabin.

Isetta began life with a 236-cc two-stroke engine. Later, a more powerful 300-cc engine was offered. The engine was as unorthodox as the rest of the car. Its two parallel cylinders shared a common combustion chamber. The inlet port was in one cylinder and the exhaust port was in the other: a design that reportedly gave the engine excellent scavenging characteristics. Power from the rear-mounted engine was delivered through a four-speed, syncromesh gearbox, then transferred by a duplex chain drive to the two rear wheels. The drive wheels were set on a common axle and a single drum brake provided stopping power to both rear wheels.

Fig. 8.1. Isetta, one of the most popular of the bubble cars, weighed only 343 kg (755 lb) and would achieve 21.25 km/L (50 mpg). (Photo K557A(4) reprinted with permission of Quadrant Picture Library)

Some people mistakenly believed that Isetta was a three-wheel vehicle. Close inspection actually revealed four wheels. However, the rear wheels were placed only 216 mm (8-1/2 in) apart in order to eliminate the need for a differential. As a result, the four-wheel Isetta ended up with the worst of both worlds, three-wheel rollover characteristics at the price of four wheels. Heinkel, on the other hand, was a true three-wheeler.

In a road test of the Iso Isetta manufactured in Milan, Italy, *The Motor* magazine, in September, 1954, reported a top speed of nearly 50 mph (80.5 km/h) and fuel economy slightly greater than 50 mpg (21.25 km/L). In those days there was no official driving schedule, so fuel economy tests were done by simply driving the vehicle over a random course deemed representative of regional motoring habits. "Driven hard" for 219 miles (352.37 km), *The Motor* reported 50.4 mpg (21.43 km/L). In a later test published in the magazine's April 11, 1956 issue, top speed had been increased to 60 mph (96.5 km/h) and fuel economy had dropped to 44.2 mpg (18.8 km/L) over a similar varied course. Total drag at 10 mph measured 29 lb. At 60 mph drag was extrapolated at 83

Fig. 8.2. Heinkel was manufactured by the German aircraft manufacturer of the same name. Both Heinkel and Isetta were designed with a single front door. (Photo 13124154 reprinted with permission of Quadrant Picture Library)

lb, which equates to a road load of 13.28 hp (9.9 kW). Of the era's so-called "bubble cars," Isetta had one of the best reputations for economical and trouble-free operation.

Messerschmitt

Having been put out of the aircraft business, Messerschmitt began manufacturing three-wheel cars under the same name. An unconfirmed rumor claimed that early designs used the same canopy installed on the famous German fighters of World War II. The car's close-fitting canopy was noted for quickly fogging over in cold weather. Pivoting the canopy to one side exposed a gaping interior that was blocked only by the relatively low sides of the vehicle. Power came from a 174-cc single-cylinder, two-stroke Fitchel & Sachs engine. Later, displacement was increased to 191 cc for more power. Power was delivered to the single rear wheel through a four-speed transmission which had no reverse gear. In order to propel the car in reverse, the engine was stopped, then restarted in the reverse direction. Top speed was 50 mph (80.45 km/h) with the original 174-cc engine. *The Motor* magazine, in an article entitled "Air Cooled Outings," April 1956, reported fuel consumption over a varied course of 72.5 mpg (30.8 km/L) with the smaller engine.

Fig. 8.3. Three-wheel Messerschmitt was one of the most unorthodox designs of the period. Note the blown canopy and curved windshield. The engine was located in the rear. (Photo 13124151 reprinted with permission of Quadrant Picture Library)

Fig. 8.4. The canopy swings to one side to expose the entire interior. Direct link steering was via downturned handlebars. (Photo 19083(7) reprinted with permission of Quadrant Picture Library))

Messerschmitt carried two passengers in tandem. However, the rear seat was wide enough for one adult and a small child. Curb weight was 462 lb (210 kg). Messerschmitt was equipped with a fully independent suspension, and all three wheels were sprung with rubber elements mounted in torsion, rather than conventional springs. Advantages of rubber springs include low cost and inherent self-damping properties.

The Market was Heading in the Opposite Direction

Designs of the period were numerous and diverse. Although many were well-engineered, minicars were positioned exactly 180 degrees from where the market was heading. In addition, consumers never really accepted the minicar as a legitimate alternative to a "real car." In 1956, miniature car sales in Germany and France had risen to nearly 15 percent of total new car sales. In Europe as a whole, 386,742 of the 1,888,000 total cars sold in 1956, or one in five cars, were miniatures.[2] But the majority of these so-called miniatures were cars like the Citroen 2CV and the Fiat 500 and 600 series, which were actually smaller, underpowered versions of more conventional cars. The possible exception would be the Isetta which sold about 30,000 units in 1956, despite its unorthodox design.

Reliant was another successful manufacturer of three-wheel miniature cars. However, Reliant's success came primarily from the fact that its product line grew up with the market, while still offering consumers the benefits of reduced taxation. The tax advantages of three-wheelers were substantial. For example, the purchase tax in the UK on a three-wheel car was 30 percent of the nominal wholesale value, compared to a 60 percent purchase tax on a four-wheel car. Road fund tax on a three-wheeler, which is similar to the yearly state-level license fee in the U.S., was about half the rate of a four-wheel car. Early on, Reliant discontinued its miniature models and concentrated on larger, more conventional vehicles. The 1978 Reliant in Figure 8.5 is essentially a conventional car, which happens to have only three wheels.

Fig. 8.5. Reliant was one of the early contenders in the post-war miniature car boom. The company's products evolved along with the market while still retaining three wheels. This 1978 three-wheel Reliant Robin Saloon is very much a conventional car, except for the number of wheels. (Photo 13124160 reprinted with permission of Quadrant Picture Library)

If a marketing appeal can be attributed to the minicars that emerged in post-war Europe, it is undoubtedly that of sacrifice and basic utility as a way to cope with the prevailing economic conditions. But the economy was improving and consumers were looking forward to better times. As early as 1952 an article in *The Motor* magazine stated:

> ".....There seems to be sufficient evidence to show that the general public prefers a roomy car of presentable appearance to a 'baby car,' regardless of the higher initial costs and running expense..... Questions of prestige and 'keeping up appearances' also play an important part, and the customer's aim of displaying (or assuming) by his car, especially when used mainly for business purposes, a certain degree of prosperity, frequently seems to override economic considerations."

Although consumer rationales may be based on utility and costs, appeals to bare utility tend to be weak motivators, even in a poor economy. Moreover, modern consumers are far more affluent and products must satisfy complex psychological needs and reflect modern lifestyles. Today's alternative car must therefore attract consumers primarily on its own merits while suggesting through its design the social goal of conserving energy and preserving the environment. It must satisfy personal needs and provide utility in a way that supports personal values. An alternative transportation product that is positioned as a low-cost substitute for a "real car" is likely to produce disappointing sales. When personal values are packaged in an upscale and intrinsically appealing product, market potential will be significantly improved. Correct positioning and vehicle theme are crucial to the product's success.

Transportation in a Revitalized Europe

Europe's high population densities, shorter distances, growing pollution, and mounting traffic congestion all suggest a good environment for small, energy- and space-efficient runabouts. Automobile tax structures also generally favor smaller cars. Conversely, fewer multi-vehicle households, a consumer preference for larger, more luxurious cars, and a greater reliance on public transportation for primary and secondary transportation needs point to just the opposite result. The prevalence of leasing and the widespread use of business cars are other important factors that could either encourage or discourage mission-specific cars.

Transportation systems in Europe are universally oriented toward the private automobile. Car ownership in Europe has grown rapidly through the last half of the twentieth century. But Europeans still own fewer cars per capita (or per household) and drive only 60-80 percent as much as Americans on a trip-average basis. The shorter distances typical of most European cities, along with higher automobile operating costs and greater access to public transportation, all work together to reduce total vehicle kilometers traveled by car. When reduced automobile ownership is combined with shorter distances, Europeans actually drive only 40 percent of the distance driven by Americans, on a kilometer per capita basis.[3] Still, automobiles in Europe account for approximately 80 percent of the total passenger kilometers traveled. Table 8.1 provides a comparison of roadway, automobile, and population densities in Europe, Japan, and the U.S.

TABLE 8.1. AUTOMOBILE AND POPULATION DENSITY OF SELECTED COUNTRIES

	Area (km^2)	Population	Population Density (pop/km^2)	Cars	People/ Car	Cars/ km^2
UK	244,046	75,701,000	236	20,069,437	2.9	82
Germany	365,755	79,753,000	218	34,051,299	2.3	93
France	547,026	56,614,000	103	23,010,000	2.4	42
Italy	301,225	57,663,000	191	24,300,000	2.4	81
Japan	377,708	123,537,000	327	32,621,046	3.8	86
U.S.	9,155,579	249,633,000	27	143,081,443	1.7	16

The most significant constraint affecting the potential market for mission-specific cars is undoubtedly that of reduced multi-car households. In Europe, far fewer households have two vehicles. In the UK in 1990, 33 percent of households had no car, 44 percent had one car, and 23 percent had two or more cars. In France approximately 28 percent of households have two or more cars. Statistics are similar throughout much of Western Europe. This compares with approximately 58 percent of U.S. households that have two or more cars, and only 9 percent of households that have no car.

When only one vehicle is available, the single vehicle must serve the broadest possible utility. Vehicle ownership and travel patterns in Europe have therefore tended to discourage limited-utility vehicles. In the UK, even the two-passenger sports car has essentially become extinct. Because prestige and self-image are important motivators, and these needs must often be satisfied by a single vehicle, vehicles have tended to become larger, more powerful, and more luxurious. The low-end market, where consumers are necessarily concerned more with vehicle utility, is satisfied by multi-passenger economy cars such as the front-wheel-drive Austin Mini.

Another important attribute of European car-purchasing habits has to do with the prevalence of company cars. In West Germany and Sweden, one-third of new cars are purchased as company cars. In the UK, more than 50 percent of new cars purchased in recent years have been company cars. Holland also has significant numbers of company-purchased cars. Business cars accounted for 27.7 percent of the 13 million plus new cars sold throughout Europe in 1990, and are expected to comprise 35 percent of the market by the year 2000. Most companies allow employees to use their company cars for private purposes, and only 17 percent charge the employee for private use. Employees typically pay a flat fee in their income tax for the benefit of these cars. However, many employees who drive company cars do not pay for gasoline and are not taxed according to how much they use. Different motivations undoubtedly apply when a car is part of a company's payroll expenses and part of an employee's compensation for services.

A survey of company car policies conducted by the British Institute of Management revealed that 90 percent of the organizations surveyed allocated cars to managing directors, 86 percent to senior managers, and 60 percent to middle managers.[4] For junior managers, the need for an automobile in business was the primary criterion for car allocation in 32 percent of the companies. The type of car provided is usually determined by job status and internal hierarchy, salary, and the need to project a company image. During a business downturn, companies seek ways to reduce their fleet costs, which generally include withdrawing company cars or switching to vehicles that are less expensive to purchase and operate. While the economics of mission-specific cars may be appealing to businesses, job requirements and hierarchy disputes may tend to work in the opposite direction, unless the alternative product also has prestige. By providing the option to expand car allocations or forego withdrawal during tough times, urban cars and commuter cars might thereby penetrate the business market. The second car market that remains when the primary car is provided by the employer also offers a niche for low-mass cars.

The leasing market is another important factor. Leased cars are much more prevalent in Europe. In the Netherlands, more than 35 percent of new cars are leased. In the UK, leased cars account for 25 percent of new car sales, and in Belgium 15 percent are leased. The lease market is rapidly growing in Germany, France and elsewhere. The most popular lease arrangement is "contract for hire" in which total vehicle operating expenses, minus fuel, are included in the monthly payments. This trend has significant implications for manufacturers of electric cars in general, as well as for unconventional IC vehicles. Independent lessors may inflate the projected maintenance costs of EVs in contract for hire agreements, or companies may be reluctant to extend contract for hire leasing on non-traditional vehicle types. Residual values are another important aspect. Residual values may not be as high with an unconventional vehicle type until it becomes widely accepted and established. In the case of EVs, battery replacement costs are likely to have a negative impact on electric car residual values.

New automobile purchase tax policies can encourage a move to smaller, less-powerful cars. Ireland, Germany, and Italy already tax cars according to engine size and/or power. Existing tax policies could have a mixed effect on electric cars because of the higher taxes that might result from potentially higher initial costs, and the lower taxes due to having circumvented IC-engine-based taxation. In Europe, cars in general are taxed much more heavily than in the U.S., and the rate and type of tax varies between countries. Value added tax (VAT), which is based on purchase price, favors less-expensive cars. Table 8.2 shows automobile taxation in selected European countries.

TABLE 8.2. CAR TAX RATES IN SELECTED COUNTRIES

Country	VAT	Car Tax	Vehicle Excise Duty
Belgium	25%/33%	None	£26.53 - £675.44
France	22%	None	Regional tax
Germany	15%	None	Engine size tax
Ireland	21%	21.7% < 2 litre 24.5% > 2 litre	Engine size tax
Italy	19% < 2 litre 38% > 2 litre	None	Horsepower tax
Netherlands	18.5%	16%-24% based on list price	Av. £171.47
UK	17.5%	5% on 5/6 of price	£110

Source: European Vehicle Leasing 1992/93

Available parking space and living arrangements are also relevant. In the UK, most of the population live in single-family homes; but most homes are older and many do not have garages. In many areas streets are narrow and street parking is limited or prohibited. People are therefore more restricted in the number of cars due to restricted or inadequate parking space. In Germany and Italy, 72 percent of households live in apartments. As a result, the recharging infrastructure for electric cars cannot rely as heavily on home recharging. Shorter distances make electric cars more viable in Europe; reduced availability of home recharging in regions of greater apartment dwellers and fewer multi-car households tend to make them less viable.

European Transit Systems

Although transportation in Europe is strongly oriented toward the private car, the region's more densely packed populations are more adaptable to transit systems, as well as to cycling and walking. Consequently, Europeans travel by bus and rail far more than do commuters in the U.S. In Europe, transit systems (all modes) typically carry 15 percent of the total passenger-kilometers traveled, compared to about 2.5 percent in the U.S. Also, more Europeans walk or cycle to work, services and leisure activities. In comparison to Americans, Europeans generally do not view cars as the same necessity to living and working.

In addition, the proliferation of private automobiles may soon begin to peak in many areas of Western Europe. Experiments by a number of European communities with prohibiting or restricting automobiles in inner cities could change attitudes and ultimately reverse the trend toward cars over the coming decades. In the U.S., wide open spaces and low land values have encouraged the ever-increasing spread of the suburbs, which in turn has made American transportation systems more committed to the private automobile.

Rail systems are much more developed and universally accepted in Europe than in the U.S. In the U.S., railways have been much maligned since the ascendancy of the automobile. After WWII, rail services were generally neglected in Europe as well. By concentrating on new road networks, policies favored trucks for short-haul freight transport and the private automobile for passenger transport. However, the oil embargo in the early '70s sent transportation planners back to the railroads nearly everywhere except in the U.S. Most developed countries have a greater density of railways than exists in the U.S., both in terms of kilometers of rail per square kilometer of land area and in relation to the density of roads.

Table 8.3 provides a comparison of roadway and railway densities in a number of European countries, as well as those in the U.S. and Japan. Table 8.4 compares passenger transport by roadway with transport by railway. Although Europeans travel far less distance than their counterparts in the U.S., they travel 7-15 times more distance by rail.

TABLE 8.3. 1990 ROAD AND RAIL INFRASTRUCTURE OF SELECTED COUNTRIES

Country	Roads Kilometers (× 1000)	Roads Density (km/1000 km²)	Rail Network Kilometers (× 1000)	Rail Network Density (km/1000 km²)
UK	382	1561	16.9	69
Germany [1]	549 [2,3]	1538 [2]	44.0 [3]	123
France	806	1481	34.1	63
Italy	304 [4]	1009 [4]	19.6	65
Japan	1115	3015	-	-
U.S.	6244	667	192.7	21

[1] Includes former Federal Republic of Germany (East Germany).
[2] Excludes urban roads in former Federal Democratic Republic (West Germany).
[3] Estimated from previous year.
[4] Excludes urban roads.
Source: Transportation Statistics Great Britain 1992, The Department of Transport

TABLE 8.4. 1990 PASSENGER TRANSPORT PER HEAD OF POPULATION IN KILOMETERS

Country	Travel by Road: Passenger Cars and Taxis	Travel by Road: Buses and Coaches	Rail (Excluding Metro)	All Modes
Great Britain	10,586	817	595	11,997
West Germany	9388	885	709	10,982
France	10,413	732	1130	12,274
Italy	8555	1472	788	10,815
Japan	4491	867	2915	8272
U.S.	17,002	156	85	17,243

Source: Transportation Statistics Great Britain 1992, The Department of Transport

Although railways are much more costly to build than roadways, they are much more efficient carriers, both in terms of space and in terms of energy per passenger kilometer. A passenger traveling by rail uses only a fifth of the fuel used to traverse the same distance by air, and less than half the fuel of an equivalent trip by car. Trains are also one of the safest modes of travel. In the UK in 1991, there were 1.4 fatalities per billion passenger kilometers due to railway accidents. Motor vehicle accidents during the same year resulted in 57 fatalities per 100 million vehicle kilometers, or over 300 fatalities per billion passenger kilometers, which makes travel by car about 225 times more hazardous than the same trip by rail.

The Future of Urban Cars and Commuter Cars in Europe

A purely utilitarian rationale would point to transit systems as the most space- and energy-efficient transportation alternative. But as Renault's spokesman, George Cagnard said, "Now that people have the freedom to pick up and go where and when they please, do you really think they will give it up?"[5] Regardless of the practical and economic reasons to use transit systems, even in transit-adapted Europe people are unwilling to abandon the freedom of private cars. True, the measure of freedom available in gridlocked traffic, or while confined to a vehicle in search of a parking place, may be more illusory than real. But if one's city were walled off and inhabitants were prohibited from leaving, the entire population would suddenly be overtaken by a compulsion to escape. Choice and a sense of freedom are basic to human nature and, in general, automobiles provide both a sense of freedom, as well as real and discretionary mobility.

Today, when one speaks of alternative cars in Europe, the subject is generally that of electric cars. With that having been predefined, discussions can then progress to the sub-categories of commercial fleets, conventional-size passenger cars, and significantly smaller urban-specific vehicles. For many years, electric vehicles have been more extensively used in Europe than in the U.S., but more on a commercial and fleet level rather than for private transportation. It is only within the last decade that the idea of consumer EVs has become more seriously discussed as the ultimate answer to national energy sufficiency and inner-city air pollution. The European Electric Road Vehicle Association (AVERE), Europe's counterpart to the Electric Vehicle Association of the Americas (EVAA) in North America, estimates that European roadways could be populated with half a million EVs by 1999.

AVERE is active in a number of areas to insure that the EV infrastructure develops along sound lines and is coherent throughout all of Europe. Items on the agenda encompass a broad range of technical, legal, and political factors that are both necessary for market development and best considered before there are large numbers of EVs in service. On a technical level, the goal is to promote safe product design and standardization among suppliers, manufacturers, and public recharging stations. Ideas for special credit cards for recharging stations are also in the early stages. On a legal and political level, AVERE works to encourage tax incentives, special roadway and parking accommodations, and government participation in EV development programs. Efforts have been largely successful.

Utilities companies are already firmly behind the impetus for a switch to electric cars. Enthusiasm on a government level is also high, perhaps even higher than in the U.S. The combination of shorter distances and higher population densities makes EVs even better suited for European cities than for those in the U.S. Acceptance of electric cars on a consumer level is the final hurdle to a widespread move toward EVs. However, a consumer market for electric cars does not naturally exist. Just as with urban- and commuting-specific IC vehicles, there is little or no pent-up consumer desire waiting to be filled. Consumers are not hungry for transportation products of less capacity and performance and, in the case of EVs, higher first costs. European attitudes, economic conditions, and transportation preferences and infrastructures differ between regions, and differ from those in the U.S. On a region-by-region basis, transportation patterns and local infrastructures provide niches and marketing opportunities for manufacturers of alternative cars. But the fundamental challenge in Europe is essentially identical to the challenge in other developed regions. If there is to be a large consumer market for special-purpose cars and electric vehicles, it must be artificially created, either by taxation policies and preferential roadway accommodations, and/or by innovative product design and new marketing appeals.

The idea of marketing special-purpose vehicles to a universe of consumers who are oriented toward existing product goes against the traditional approach in which product design is driven by the market. And in the abstract, there is no reason to persuade consumers to move toward unconventional and significantly smaller transportation products. But when the growing problems of resource depletion and environmental degradation are considered, reasons begin to emerge. The cost of energy itself as well as the costs associated with preserving and repairing the environment are likely have a negative economic impact on transportation industries. Framed strictly in terms of a business rationale, a non-traditional approach may be necessary in order to preserve the health of the market. Business may have to lead the market with new appeals that address global limitations. Business must invent products that contribute to the long-term health of the market and the society, then frame those products in ways that have emotional appeal and make sense to consumers.

The best environment for creating a market for urban- and commuting-specific cars exists where the choice of individual vehicle utility is highly discretionary and trip distances are relatively short. These conditions are most characteristic of affluent consumers residing in areas of high population density. The economies of Western European populations have been improving and multi-car ownership is growing. A successful European Economic Community may ultimately produce the world's strongest economic block and improve living conditions throughout the region. Conditions necessary for cultivating a market for mission-specific cars in Europe already exist to a significant degree and, over the long term, favorable conditions will likely grow.

The practical advantages of significantly smaller vehicles have already been explored. The hypothetical commuter car mentioned in Chapter Two (Fig. 2.2) could achieve triple the fuel economy of the average new car in Europe and thereby equal or surpass the energy efficiency of transit systems. Transportation expenses would be lower or held in check, and the transportation system would become more efficient and less polluting. Arguments in favor of electric cars are perhaps even more powerful. But in terms of first costs to consumers, EVs will undoubtedly be more costly alternatives, at least until the technology is more mature and volumes are much greater. And in terms of vehicle performance, electric cars may also entail a greater compromise in vehicle capabilities. A shift in emphasis away from multi-purpose conventional vehicles and toward urban-specific vehicles may be the most essential factor in persuading consumers to purchase electric cars. Consumers must see electric cars, not in comparison to the performance capabilities of conventional cars, but in relation to actual trip requirements. Conversely, overselling consumers on expected improvements in EV technology may prove counterproductive.

Today, the so-called "green market" in combination with a real and growing crisis in inner-city traffic congestion and air pollution provide a unique opportunity to pioneer cleaner and more space- and energy-efficient personal transportation products. Although urban cars and commuter cars powered by clean-running internal combustion engines can provide similar benefits with greater vehicle capabilities, popular attitudes toward electric cars tend to give EVs a psychological advantage. Electric cars are the benefactors of a new personal transportation paradigm that is already forming in the minds of a growing number of transportation planners and consumers. The electric car's technical distinction creates a natural shift in consumer perceptions, which should be encouraged through marketing appeals that center on product differentiation and global benefits. Vehicle themes must suggest an upscale orientation that is compatible with modern lifestyles.

A product's emotional value becomes increasingly more important as consumers become more affluent. Ultimately psychological values tend to direct the choice of facts on which purchasing rationales are based. Alternatives to high-mass private cars become marketable first to the degree that multi-car households provide a universe of consumers with discretionary vehicle needs, and second to the degree that consumers' psychological needs are satisfied by the new product. Multi-car households as well as appealing product attributes were both in short supply when minicars were marketed after World War II. Minicars appealed to consumers' utilitarian needs and relied on a poor economy to create a universe of potential consumers.

The battle between utilitarian and psychological values is exemplified by the historic switch that took place between Henry Ford and GM's Alfred Sloan in the early days of the automobile industry. The automobile began essentially as a plaything for the wealthy, rather than the consumer product that it is today. Ford envisioned a much larger market if the average consumer could purchase a car. He therefore produced a single, low-priced car and appealed to consumers on the basis of price and utility. By the early '20s, Ford Motor Company dominated the market with 60-percent market share compared to GM's 12 percent. Sloan was an admirer of Ford's pioneering efforts, but he also understood that Ford's preoccupation with the idea of a basic, low-cost car was his greatest weakness in the emerging consumer market. Sloan began his onslaught on Ford by revamping GM's line to offer consumers a choice of cars, starting with a slightly higher-priced but more stylish version of Ford's low-priced car and progressing through a line of products, each offering increased luxury and style. The yearly model change was introduced and cars could be purchased in a choice of colors. Although Ford's product still carried the lowest price, within five years GM was the dominant and profitable car maker and Ford was losing money.

Ultimately, Ford had to adopt Sloan's methods or perish. In the face of improving economic conditions and greater consumer expectations, appeals to consumers' psychological needs had created a combination that proved fatal to the idea of the purely utilitarian automobile. A modern design based on a similar appeal to basic utility is likely to meet with a similar fate.

As futurist Syd Mead poetically states in his collection of renderings, Oblagon: "Designers in all disciplines have the enviable task of creating possibilities beyond the limits of the mundane and the predictable, to invent not just the idea but the rationale for futures not yet within the grasp of available technique." In order to successfully imbue an ultralight automobile with high-end values, bold new ideas in product design are essential. Product differentiation is a fundamental ingredient in the task of unlinking consumer expectations from those associated with the traditional automobile and creating a new category of transportation products where none existed before. But an unlinking from traditional concepts must first occur in the visions of product planners and designers. Creating a new personal transportation paradigm through innovative vehicle configuration is a challenge that lies closer to the grasp of available technique than the technical solutions required to double or triple the fuel economy of conventional vehicles.

Fig. 8.6. Rendering by Barbara Munger.

References

1. Laurence Pomeroy, "The Size, Structure and Shape of European Automobiles," *The Motor*, March 17, 1954.

2. Laurence Pomeroy, "Miniature Motorcars," *The Motor*, February 5, 1958.

3. Lee Schipper, *et al.*, "Fuel Prices, Automobile Fuel Economy, and Fuel Use for Land Travel: Preliminary Findings from an International Comparison," Lawrence Berkeley Laboratory, 1993 (draft).

4. Michael Woodmansey, "Business Cars Survey 1985," British Institute of Management.

5. Quote cited by Stephen Kindel, "What Price Freedom," *Financial World*, August 22, 1989.

Appendix

The Quincy-Lynn Cars

Quincy-Lynn Enterprises, Inc., was a product development firm co-founded by the author in the mid-'70s to develop a number of internally financed projects. Over its 12-year history, the firm developed a variety of recreational land vehicles and watercraft, as well as health and fitness consumer products. Pet projects included a number of special-purpose vehicles that might be classified as urban cars and commuter cars. Beginning in 1974 and ending in 1982, the firm's alternative cars evolved from the original concept of basic personal transportation represented by Urbacar, into the high-performance sports/commuter vehicle, Tri-Magnum. Early on, vehicle packaging began to center on the idea that special-purpose personal transportation products would become marketable only through strong product differentiation and appealing new vehicle themes that captured the imagination of consumers. This approach is most evident in the design of Trimuter (Figures A.10 through A.17).

Urbacar

Fig. A.1. Designed in 1974, the 295-kg (650-lb) Urbacar was powered by an unmodified 12-kW (16-hp) single-cylinder industrial engine and would run 96 km/h (60 mph) and achieve fuel economy of 23 km/L (55 mpg).

Fig. A.2. Urbacar next to the lake at Lakewood Village, California.

Fig. A.3. The engine is located ahead of the oil bath chain drive housing. A Salsbury continuously variable transmission transfers power. The starter mounted at the rear of the final drive housing engages a ring gear opposite the CVT-driven pulley to power the car in reverse.

Fig. A.4. Swing-axle rear suspension. Reverse motor is visible at top of the final drive housing.

Fig. A.5. A view from underneath shows the layout of four plateform vibration dampers that carry the drive package. The drive package could be removed from the bottom, and vibrations were effectively isolated.

Urba Electric

Fig. A.6. Urba Electric was designed in 1976 and pioneered the continuously variable transmission as a speed control system for the electric car. Shift input was from a floor-mounted pedal that served as a conventional accelerator. Transmission shift position tracked the pedal position and thereby controlled vehicle speed from zero to 96 km/h (60 mph) while the motor ran at a steady-state speed. GM's "Drive I," built in the late '70s by Delco Remy, was based on the Urba Electric design.

Fig. A.7. Urba Electric's simple, clean lines and foam-filled bumper. Urba Electric's 48 V lead/acid battery pack was typical of the low-voltage systems of the period.

Fig. A.8. The air-cooled, compound-wound DC motor connects to the Electromatic Drive Transmission. The electronically controlled transmission utilized a hysteresis brake on a lead screw to force the movable face against the fixed face of the driver pulley. A standard speed-sensitive Salsbury unit served as the driven pulley.

Fig. A.9. Batteries were arranged in the reverse "T." The Urba Electric chassis was essentially an upgraded version of the Urbacar chassis. The anti-roll bar extends forward to account for increased longitudinal loads due to greater curb weight.

Trimuter

Fig. A.10. Trimuter represented Quincy-Lynn's idea of the personal transportation product of the 21st century. Mechanically, the design was unsophisticated, but stylistically was ahead of its time. Consider that the vehicle was designed near the end of 1978 when the Bertone styling influence was at its peak. Maximum speed was 100 km/h (63 mph) with a 12-kW (16-hp) industrial engine. Approximately 70 percent of the weight was carried by the two side-by-side rear wheels. Curb weight with its foam/FRP body was 386 kg (850 lb).

Fig. A.11. Simple lines and clamshell canopy create a striking combination. Turn signals are located in the side mirrors. An audible motorcycle turn-signal beeper helps convey a sense of differentiation from the automobile.

Fig. A.12. Clean lines are apparent with the canopy closed. The polycarbonate windshield made it impractical to install wipers. Dual headlights are mounted inside the door at the front.

Fig. A.13. The interior is snug and form-fitting. Seats are formed into the body. Floor controls include brake and accelerator pedals. There is no shift lever. The transmission automatically engages on throttle-up. Reverse is electric. A 65-liter (17-gallon) fuel tank fills the tunnel, which was designed to house batteries in an electric version.

Fig. A.14. The simple powertrain layout places the engine above the final drive. Notice that the differential and the drive and driven CVT pulleys are installed backwards. The unmodified Briggs & Stratton industrial engine achieved 21 km/L (50 mpg). A modern automobile drivetrain could easily deliver double the fuel economy.

Fig. A.15. Leading arm front suspension provided anti-dive braking. The steering damper was installed as a precautionary measure, but was probably unnecessary. Leading arm is carried by Moog bushings. Hole in end of axle is for VW-Beetle-style speedometer cable connection to the dust cover on the opposite end.

Fig. A.16. The chassis provided room for batteries in anticipation of an electric version that was never built. The battery layout would have maintained the rear biased cg. Two parallel frame rails formed a central cradle that would accept either a fuel tank or a row of batteries.

Fig. A.17. The finished gasoline-fueled chassis, minus the fuel tank. The fuel tank occupied the space between the parallel frame members and ran the full length. Its 65-liter (17-gallon) capacity provided a cruising range of 1368 km (850 miles).

Town Car

Fig. A.18. The unconventionally styled hybrid/electric Town Car is perfectly framed by the alien landscape of California's Lake El Mirage. The vehicle was designed in 1980 and based on a VW Beetle chassis. Maximum range at a steady 56 km/h (35 mph) was approximately 160 km (100 miles). On batteries alone, range was approximately 105 km (65 miles) at a steady 56 km/h (35 mph).

Fig. A.19. Town Car chassis is equipped with a 72-volt 6-kW Baldor series motor, which was mounted directly to the VW transaxle. All gears including reverse were usable. The motor was controlled by an off-the-shelf 600-amp chopper controller. To leave room for passengers at the center of the platform, batteries were located at both ends. Two, however, ended up on top of the tunnel between the rear passengers. The generator was powered by a 3-kW (4-hp) IC engine. It delivered 1.6 kW and consumed gasoline at the rate of 1.55 liters per hour (0.41 gal/h).

Fig. A.20. Town Car was designed on a safety-vehicle theme with substantial rollover and side intrusion beams as predominant styling features. Styling was to have the look of brushed stainless-steel perimeter and roll members to which painted body segments were attached in the built-up effect reminiscent of Star Wars creations. In the prototype, these features were primarily cosmetic and not supported by the necessary sub-structure.

Fig. A.21. The rear was dominated by the large, foam-filled bumper. Taillights almost didn't make it onto the design, but Town Car was rescued by lights from an Oldsmobile station wagon.

Centurion

Fig. A.22. Centurion was based on a Triumph Spitfire chassis which was fitted with a 23-kW (17-hp) naturally aspirated diesel engine. It was designed in 1981 and was the most conventionally styled car of Quincy-Lynn. Random city traffic resulted in 28 km/L (65 mpg), freeway cruising at 88 km/h (55 mph) produced about 36 km/L (85 mpg), and 54 km/L (128 mpg) was obtained with a single acceleration to steady 56 km/h (35 mph). The engine could be loaded for economy at almost any speed due to its four-speed transmission with overdrive in each gear.

Fig. A.23. Clean lines continue across the rear. The aft panel is foam-filled. The lightweight chassis and foam/FRP body combined to produce a curb weight of 545 kg (1200 lb), which gave it a power-to-weight ratio of about 70 pounds per horsepower. In layman's terms, that means that almost everything else on the road could beat it away from a stoplight.

Fig. A.24. Retractable headlights, flush glazing and a small air intake below the foam-filled bumper helped keep aerodynamic drag to a minimum.

Fig. A.25. The chassis was modified to carry the fuel tank just behind the seats. Radius rods normally attached to the body of the Spitfire. Simple outriggers welded to the frame provide new anchor points. The three-cylinder Kubota engine mates to the Spitfire with a simple adaptor plate.

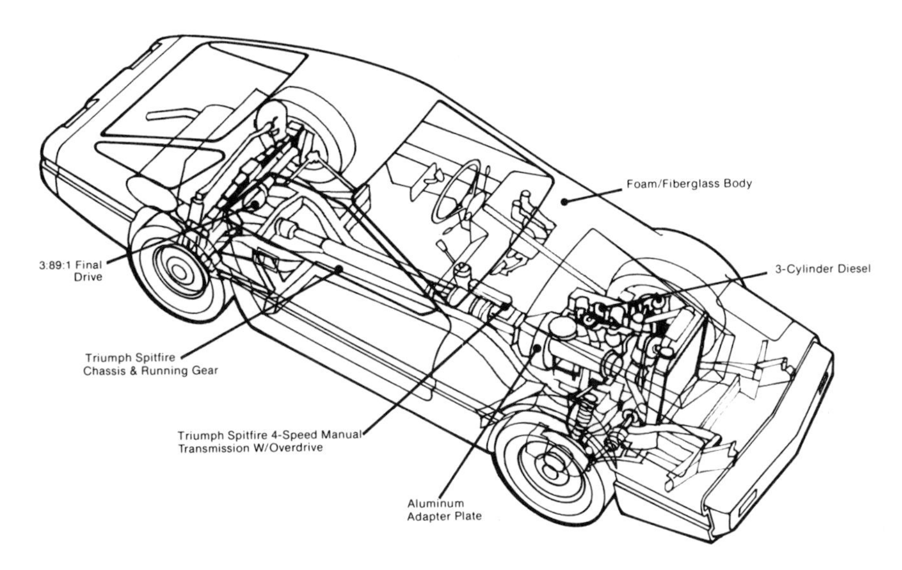

Fig. A.26. Ghost view of Centurion.

Tri-Magnum

Fig. A.27. Designed in 1982, Tri-Magnum would out-corner most production cars on the road today. Maximum speed was 160 km/h (100 mph). Tri-Magnum was the final car in the Quincy-Lynn series. Fuel economy was roughly equivalent to the original motorcycle even though curb weight was much greater at 510 kg (1125 lb). The forward cg produced a predictable understeer.

Fig. A.28. This high-angle view shows the car's sleek lines. The rear hatch is removable to expose a Kawasaki KZ900 motorcycle inside. The clamshell canopy with the high beltline resulted in a long first step, which added a flavor of adventure to the act of entering and leaving the vehicle.

Fig. A.29. Retractable headlights were partially nested in the back of the bumper. The foam-filled bumper was a familiar feature.

Fig. A.30. A shifter located between the occupants carried the clutch lever and most of the switches. The steering wheel tilts forward for entering and leaving the vehicle. "T" handle under the cowl operates the retractable headlights.

Fig. A.31. Taillight bays provided a way to rapidly eliminate body width in order to create an appropriate look with the single rear wheel. Rearward-facing scoop on top and shark-gill louvers on body sides extract hot air from the engine room. A scoop underneath forces air inside.

Fig. A.32. The assembled Tri-Magnum chassis. Fixtures at the center of the VW axle beams are designed to adjust ride-height. A simple framework attaches to a Kawasaki KZ900 motorcycle to create the three-wheel chassis. The VW Beetle front suspension is welded in place. Occupants' knees are positioned over the transverse suspension beams.

Index